The Men Who
United the States

ALSO BY SIMON WINCHESTER

Atlantic

Skulls

West Coast: Bering to Baja

The Man Who Loved China

A Crack in the Edge of the World

The Meaning of Everything

Krakatoa

The Map That Changed the World

The Fracture Zone

The Professor and the Madman

In Holy Terror

American Heartbeat

Their Noble Lordships

Stones of Empire

Outposts

Prison Diary, Argentina

Hong Kong: Here Be Dragons

Korea: A Walk Through the Land of Miracles

Pacific Rising

Small World

Pacific Nightmare

The River at the Center of the World

The Men Who United The States

America's Explorers, Inventors, Eccentrics, and Mavericks, and the Creation of One Nation, Indivisible

Simon Winchester

HARPER **LUXE**

An Imprint of HarperCollinsPublishers

Frontispiece: The "The Good Roads Train," courtesy of Project Gutenberg

HarperCollins books may be purchased for educational, business, or sales promotional use. For information, please e-mail the Special Markets Department at SPsales@harpercollins.com.

FIRST HARPERLUXE EDITION

HarperLuxe™ is a trademark of HarperCollins Publishers

Library of Congress Cataloging-in-Publication Data is available upon request.

ISBN: 978-0-06-227851-7

13 14 ID/RRD 10 9 8 7 6 5 4 3 2 1

February 23, 2012, was the eightieth birthday
of my mother-in-law,
MRS. AKIKO SATO.
Shortly before the family celebration, I told her of my
plan to structure my book around the five so-called
classical elements. She briefly left the room, returning
with this card on which she had handwritten this
aide-mémoire for me, the five elements rendered in
English, Chinese characters, and Japanese.

Wood	木	Moku
Fire	火	Ka
Earth	土	Do
Metal	金	gon
Water	水	sui

Three hours later, toward the end of her party, happy
and surrounded by friends and family, Mrs. Sato
collapsed and later died.
This card was thus the very last thing she ever wrote in
her life—one ample reason among many for me to offer
this book as dedication both to her daughter
SETSUKO
and, with gratitude and respect, to the memory of
MRS. AKIKO SATO.
Born, Tokyo, 1932. Died, New York, 2012.
May this small offering be her memorial.

Think of the United States today—the facts of these thirty-eight or forty empires solder'd in one— sixty or seventy millions of equals, with their lives, their passions, their future—these incalculable, modern, American, seething multitudes around us, of which we are inseparable parts!

—WALT WHITMAN, A Backward Glance o'er Travell'd Roads (PREFACE TO THE 1888 EDITION OF Leaves of Grass)

Contents

Maps and Illustrations

Author's Note

On Independence Day, July 4, 2011, I swore a solemn oath before a federal judge on the after-deck of the warship USS *Constitution* in Boston Harbor, and by doing so I became, after half a century of dreaming, a naturalized American citizen. The following day I acquired my voter's registration card; a week later I was issued my first American passport, a document on which I have traveled ever since. When I returned to Kennedy Airport after my first trip overseas as an American, I was little prepared for my reaction when the immigration officer remarked with casual warmth, "Welcome home." I felt almost overwhelmed by at last now being a part of *all of this*.

The most recent design of an American passport incorporates a series of declarative epigraphs at the top of each visa page. Samuel Adams: "What a glorious morning for our country." The inscription on the Golden Spike at Promontory Summit in Utah: "May God continue the unity of our country as the railroad unites the two great oceans of the world." And Jessamyn West's description of the railway as "A big iron needle stitching the country together."

But of all the quotations, the one I like most is a paragraph taken from Lyndon Johnson's inaugural address of January 20, 1965. The nation was at the time still shocked by the tragic shooting of President Kennedy—the event that elevated LBJ to the presidency. The country, still mired in Vietnam, was in a liverish mood, and many more tragedies were yet to come. But Johnson, seeking by his speech to help salve the country's wounds and to better the temper of the times, spoke in an optimistic vein:

> For this is what America is all about. It is the uncrossed desert and the unclimbed ridge. It is the star that is not reached and the harvest sleeping in the unplowed ground. Is our world gone? We say "Farewell." Is a new world coming? We welcome it—and we will bend it to the hopes of man.

The pages that follow are devoted in large part to those men who, in the overarching interests of welding the nation together, traversed those uncrossed deserts and scaled those unclimbed ridges, offering in their own times and their own ways the promise of a better place and of better times ahead.

Preface:
The Pure Physics of Union

E pluribus unum.
—SINCE 1782, THE MOTTO ON THE
OFFICIAL SEAL OF THE UNITED STATES

Early in the crisp small hours of November 7, 2012, a weary but exultant Barack Obama was thanking his countrymen for just handing him a second term as forty-fourth president of the United States. His speech was brief, but it rang with an eloquence that moved well beyond the platitudes of the pitiless election season that had mercifully ended in this culmination just moments before.

It was a speech that spelled out President Obama's unyieldingly optimistic belief in the future of a country that had allowed him, a young black man, to be

invested, now for a second term, as the most powerful human being on the planet. He had been given this
role, he said, with a new chance to perfect still further
the immense entity that is the American union, more
than two centuries after his country had declared its
independence from colonial rule.

Such was the crowd's exuberance that much of
what the president said was drowned in a cacophony
of cheering and frenzied delight. Sensing the mood,
he prudently kept what he had to say brief and to the
point. After no more than ten minutes of high rhetoric,
the tone of his voice fell and quieted—he was coming
to the end.

"I believe we can seize this future together," he
said, "because we are not as divided as our politics
suggests. We're not as cynical as the pundits believe.
We are greater than the sum of our individual ambitions, and we remain more than a collection of red
states and blue states. We are and forever will be"—
and here he paused for just a beat, to add solemn
emphasis to the adjective—"the *United* States of
America."

The *United* States. This unique national quality—of
first becoming and then remaining so decidedly
united—is a creation that, in spite of episodes of trial
and war and suffering and stress, has been sustained

for almost two and half centuries across the great magical confusion that is the American nation. The account that follows, then, is on one level a meditation on the nature of this American unity, a hymn to the creation of oneness, a parsing of the rich complexities that lie behind the country's so-simple-sounding motto: *E pluribus unum.*

America is, after all, a nation founded as a home for the single simple ideal of universal human freedom. The country was established as a grand experiment, with people invited from all over the world to take part, to help build a nation of free souls, each to be given an equal opportunity to seek as each saw best the greatest happiness for themselves. The question I try to address in the following chapters is: just how has it managed to adhere, to keep itself annealed into one for all the years and decades since?

Unity among peoples, in a country as complicated as America, is just not an organic thing. In countries with less convoluted pedigrees it might well be. By way of analogy, people in tribes tend toward natural unity— whether they are Kikuyu, Comanche, Wurundjeri, or Micmac, individuals within each tribe bond together tightly. Clans in Scotland are proud of being firm-welded entities of great antiquity—all McKenzies and MacNeils are one, Scots like to say, whether fortune or happenstance has led them to be dukes or dustmen. Elsewhere class and the tendency toward an intellectual aristocracy have magnified a sense of union— Etonians, graduates of Hotchkiss and Science Po, Harvard and Christ Church may all bond clubbably, as may most European marquesses and counts or their American equivalents, the Biddles, Lowells, Cabots, and Saltonstalls. Race likewise has an annealing affect: Harlem and Hough and Watts and a score of other places have long offered local concentrations of great resilience, strength, and pride.

But America as a whole, once its early Puritan settlement had been diluted by those who followed or those already there, became too much of a mongrel nation to enjoy the simpler organic benefits of union. Lacking the communal simplicities afforded in some other countries—Japan, say, or Norway—by the existence of one race, one ethnic group, or a single class or a dominant intellectual or spiritual tendency, the great experiment that is America has had to make a union for itself, not wish it to grow in the dark out of time and nothing. It has done so purposefully by the deliberate acts of its own people. Man has had to do the hard work in bringing America together, forging something that in other, less complex places has been accomplished much more simply.

And surely all must agree that man in America, bent to this single task, has done most creditably. Excepting of course the tragic period in the 1860s when the union was so cruelly tested by civil war, this work has been performed with a consummate degree of success. The states are now generally united, and as a body united, the nation has enjoyed a steady growth of prosperity and power known by no other country on earth. And all the while, the American people have managed to remain staunchly together while countries in so many rival regions around the world—in Europe, Russia, China, and India—have been plagued by bickering and struggling and division, and have been rendered much the lesser thereby.

But just how has America's uniquely stable union been achieved? What factors have ensured that, say, a Chinese migrant in rain-swept Seattle can find himself locked in some near-mystical concord with a Sephardic Jewish woman in Manhattan or a Cherokee student in Minnesota or a Latina stallholder in a market in Albuquerque—all of them being able to enjoy the same rights and aspirations, encapsulated in their shared ability to declare so simply, I am an American?

How did the notion of creating a more perfect union of such peoples and of such administrative entities—the now fifty states, comprising the 2,955 counties of forty-eight of them, the 64 Louisiana parishes, and the 18 Alaska boroughs—first come about? And how did this idea of union translate into the practical, physical, and concrete terms we know today, which have worn so well and lastingly?

The main purpose of the pages that follow is to consider what might be called the physiology and the physics of the country, the strands of connective tissue that have allowed it to achieve all it has, and yet to keep itself together while doing so.

For the ties that bind are most definitely, in their essence, practical and physical things. It would of course be idle to dismiss the adhesive nature of the ideas on which the nation was founded. It would be a grave mistake to forget that the guiding national concept is based on a set of common purposes, on ideals and constitutionally guaranteed freedoms that are so publicly cherished by all. But over the years, these inchoate things have all been of necessity underpinned by innumerable real, visible, tangible connections—by survey lines and marks; by roads; by canals; by railways, telephone lines, power grids; and, more recently, by submerged rivers of electrons—all of which have proved crucial both in maintaining the union and in preventing, or at least lessening the likelihood of, its fracturing and spinning into a thousand separate parts.

This book tells the story of making such connections as these and of the remarkable and visionary figures from the country's history who first made them.

Most of them were already Americans when they did so. Though we might nowadays wish it were otherwise, most—but not all—were men. Most of their achievements—but not all, most especially that which permitted the private ownership of land—were made after the Louisiana Purchase, which suddenly doubled the country's size into the truly transcontinental entity it is today. Most of their achievements—but not all— remain as vital to the nation's preservation as they were when first they were created.

From the very visible nineteenth-century explorations of the Lewis and Clark expedition, by way of the geological surveying expeditions and the highway-building ventures and waterway excavations, to the less easily describable twenty-first-century mystery makings of the Internet communications backbone— there are fully two centuries of inventive zeal that have left as legacy a nation now as comprehensively interconnected and as practically unified as it is possible to imagine.

But how best to organize the wealth of work that has brought about this unity? The sheer complications of it all—the overlappings of the work of road builders and survey makers, of the pioneers of flight and the makers of radio, of the work of those who dug canals and those who excavated the tunnels for the railroad lines—made it well-nigh impossible to narrate the story in purely chronological terms. By the same token, to list the characters who were involved in the forging of the union would lend the account the feel of a catalog or an encyclopedia. A device was needed, it seemed to me, that would link the achievements thematically and give the story some greater degree of structure and logic.

An idea came to me one morning when I was writing a letter to a friend in China.

Beginning in the mid-1970s, I had lived for many years on the far side of the world and had spent much time tramping the territories between Vladivostok and Vietnam, between Manchuria and Malaysia, and between Kashmir and the Khyber Pass. All the countries of Asia—as well as the ancient civilizations of the Mediterranean—had held for centuries a philosophical view that everything and everyone can be reduced to the barest essentials, the five so-called *classical elements*. While the ancient Greeks revered just four elements, most other civilizations, from India eastward, nominated five.

The various eastern countries in their histories have made subtle variations in just what these five elements are, but those most commonly selected are wood, earth, water, fire, and metal. While I was writing the letter to my friend in Shanghai that day and explaining the idea behind the book, it suddenly seemed to me that the five elements could be a logical way of placing into context the basic themes behind the making and joining together of the United States.

The earliest explorers of the country, for example— Lewis and Clark and all the others in the years immediately following—were confronted by endless stands of ancient forest. Despite the myths, these forests were seldom as impenetrable as those in Russia or the tropics: Native Americans regularly set fires to manage and to thin them, to create pasture and to make usable landscape. But they were woods nonetheless, and they were vast and ancient.

The early explorers paddled through them and up and along the various rivers of their expeditions in wooden boats. In winter and at night, they kept themselves warm by building fires of oak and ash wood. They framed their earliest houses of timbers of cedar and pine.

Wood, in other words, could be claimed as an abiding elemental theme of their voyage of discovery, and it would go on to be a dominant feature of every subsequent early voyage across the country. Wood, then, could provide an overarching theme for a chapter that considered these first explorers and settlers, an emblem of the frontier in the forested wilderness that was the American continent.

Once the basic geography of this continent had been established, there came the equally vital task of learning what riches might lie beneath the woodlands and the carpets of vegetation. Geologists—men who were quite unschooled at first but highly sophisticated in later years—began to probe for the mineral riches and determine the agricultural worth of the land, the value and potential of the earth. The vision of mineral treasures lying locked within those millions of acres, or the possibilities of fertile farmland for crops or livestock, and of livelihoods to be made from raising them, would in due course lure out the settlers and prompt their treks westward into a country that was now established to be blessed with the promise. The earth and its riches, in short, would offer a second theme well worthy of exploring.

And the remaining three elements—water, fire, and metal—prove equally suited to this broader organizing principle.

Water, for instance. There is no gainsaying the use of the country's rivers and streams as early highways and the later employment of these waterways for trade, for the making of power, for the creation of frontiers. Then, if the waterways were not wide enough or deep enough or straight enough, there came the making of artificial rivers—the canals—which might ease the passage of people and goods across mountain chains. For scores of decades, right through to today, there are stories to tell of figures who were prominent in such unifying endeavors, which could all be linked by the essential element of water.

After or overlapping with these stories, there came the invention of the engines and the concept of employing these engines as agents of motive power. The common physical feature of all such early engines was the employment of heat; whether they were powered by steam, gasoline, or aviation fuel, these engines would eventually allow the country to be journeyed across swiftly, expeditiously, and easily. The nation could now be intimately linked along roadways and highways traveled by a variety of contraptions, all powered by fire.

Finally: metal. The copper cable of the telegraph, the steel wire of the telephone, the iron mast of radio and television, the subterranean and aerial titanium and cadmium and platinum mysteries of the Internet—the elemental common denominators of the transmission of information might be varied indeed, but in the terms of the ancients, metal was the common factor. Metal was key.

Armed with this basic notion, I set off for several months of exploration. Like a mantra, the words *wood, earth, water, fire, and metal* became a phrase, repeated over and again, that lay always in the back of my mind as I traveled back and forth between the coasts and crossed the prairies and the mountain ranges of the United States.

I equipped myself for the journey with tent and compass and sleeping bag, as well as numberless maps, books of history, and novels by the classic writers of the American experience: Willa Cather, Wallace Stegner, John Williams, Theodore Dreiser, Sinclair Lewis, Sherwood Anderson. And as vade mecum, I also managed to collect all fifty volumes of the American Guide Series—the famous WPA Guides, still among the most thoughtfully composed and intelligently edited books about the individual states.

The books date back to the late 1930s and were each assembled, as part of President Roosevelt's New Deal, in a federal government effort—the Federal Writers' Project—to give work to unemployed authors, journalists, and photographers. Though as sources of precise travelers' information, they are long past their sell-by date, their essays still have a sustaining importance, and they offer wise counsel and a grand perspective for anyone wishing to venture into the great American hinterland.

The WPA guides—government-made books, it has to be remembered—offer a reminder of a highly divisive argument about the making of America: the role of government in the creation and sustenance of human society.

That is not to say that in these pages I wish to offer an uncritical apologia for the concept of big government. Far from it. There are all too many examples of unforgivable excesses. The savage and divisive melancholy of the Trail of Tears was, after all, a consequence of overzealous government behavior toward America's own native peoples, with results that ran entirely counter to the principal thesis of this book. The amassment of vast armories of atomic weapons, the involvement of the United States in scores of cruel and unnecessary foreign wars, the lunacies of Prohibition, of the Tuskegee experiment, of the infamous MK Ultra program, and of the fully legislated and half-century-long antipathy to Chinese immigration—all of these and more were the acts of a government that had simply become on occasion too big for its boots.

Yet there was much good done, too, and not a little of it was and still remains on display in the telling of this story. Without an engaged and functioning federal government, the development of these various strands of the country's connective tissue would probably have been either delayed or never achieved at all. That is why my reading of the WPA Guides provided me with a symbolic madeleine, a means of remembering a single sobering fact: while today's political hostility to big government is an understandable reality of contemporary life, the historic role of big government in the creation of the American nation is a reality, too, one that might as well be acknowledged and celebrated for its value and great worth.

The first two volumes of the series that I decided to use were published in 1940 and 1941, respectively: the first was devoted to Ohio; the other, to Missouri. I took them along because I had decided to travel first to a pair of places that seemed to me to have played crucial roles in the making of a united America.

Each town stands on the right bank of a great American river, and in each case the river gave its name to the state in which the town is situated. The first town was East Liverpool, which is both in Ohio and on the Ohio; the second was Saint Charles, which likewise is both in Missouri and on the Missouri. Neither place is especially well known, whatever its chamber of commerce might say. The importance of each has faded over the centuries. Neither seems to me lovely enough to attract many visitors.

But each town was once most important to the man who originated the idea of creating a properly *United States of America*, a Founding Father who would go on to be the country's third president, Thomas Jefferson. And each town now has a fine-looking memorial—in one case a plaque, in the other an obelisk—to the events that occurred there and helped make each community briefly famous.

Both memorials are now surrounded—and in the case of one of them, half hidden—by trees. A reminder, if any were needed, that at the time Thomas Jefferson gave these places their brief significance in making the physics of the union, America was a land swathed by long reaches of barely penetrable forest, most of its mysteries still half hidden by wood.

I never before knew the full value of trees. My house is entirely embossomed in high plane-trees, with good grass below; and under them I breakfast, dine, write, read, and receive my company. What would I not give that the trees planted nearest round the house at Monticello were full grown.

—THOMAS JEFFERSON,
LETTER TO MARTHA RANDOLPH, 1793

. . . I was not prepared to see the pine timber so valuable and heavy as it is above and about here. The trees are of large growth, straight and smooth. . . . With the exception of swamps, which are few and far between, the timber land has all the beauty of a sylvan grove. The entire absence of underbrush and decayed logs lends ornament and attraction to the woods. They are more like the groves around a mansion in their neat and cheerful appearance; and awaken reflection on the Muses and the dialogues of philosophers rather than apprehension of wild beasts and serpents.

—CHRISTOPHER COLUMBUS ANDREWS,
Minnesota and Dacotah, 1856

PART I

When America's Story Was Dominated by Wood

1785–1805

A View across the Ridge

Thomas Jefferson was a man with a lifelong fascination with trees. He thought of them as his favorite kind of plants, wrote of them as his pets, and went to much effort and expense to place those he liked best around the great west lawn of Monticello, the house he made for himself in the foothills of the mountains of Albemarle County, Virginia.

He was an extravagant man, given to extravagant visions—which Monticello's present-day garden conservators have done much to reproduce. So just as he wished when he first bought Monticello in 1768, there are today long allées of willows, great terraces of magnolias, and stands of sugar maples. There are linden trees, mulberries, and honey locusts; there are oaks and pines and pecans, catalpas and gingkoes and chestnuts, sycamores, walnuts, slippery elm, and Osage orange and border plantings of persimmons, black gums, and fruit-bearing peach trees.

Monticello's fortunes have fluctuated dramatically over the years—not least because the third president's excesses left the estate hopelessly mired in his legacy of debt. For many decades the house itself was a magnificent ruin, the estate gardens were left to run wild, and the surrounding forests were choked with underbrush. Some of the greater trees survived the rigors of time and neglect, however, and in recent years it became something of a sport to try to say with certainty which of these gnarled monsters Jefferson himself might have planted. It somehow made the country's best-beloved Founding Father ever more human to imagine him out in the garden on a summer's evening, digging the saplings deep into the hilltop's rich loamy soil, to think of him spreading mulch above their roots and then leaving the shoots to the soothing balms of warm Virginia rains.

An X-ray device invented by a Dutch arborist was recently brought in to work out the age of the nobler-looking trees, and there was much exultation on the mountain when four of them could be proved to be at least two hundred years old—and thus quite old enough to have been planted by Jefferson. But irony has no respect for antiquity: no sooner had these trees been identified as most probably the work of the man himself—or the small army of slaves he had working on his estate—than all four of them keeled over and died.

Two of them were massive but fragile tulip poplars, one of which was fully ten yards around at its base and had begun to pose a dire threat to the building beside it. The others were a larch and a copper beech, immense shade trees under which the aging Jefferson was said to have whiled away many of the afternoons of his latter days. To most there was a gentle poignancy in their passing, because it severed one certain and romantic connection with the man who, above all others, still stands today as the architect of most of the central ideas behind the making of the United States.

But one other connection, a small and little-noticed arboreal conceit at Monticello, also links the man and his vision—and quite literally his vision—with those who visit today. It is a small and cleverly created spy hole in the woods that surround Monticello, and it affords visitors a subtle view of their surroundings that in its own way is every bit as inspirational as that which Jefferson, in laying out the plans for his estate, had once designed for himself.

Monticello faces almost exactly to the west. Were it not for one low hillock in between, Jefferson would have been able to contemplate an uninterrupted panorama clear across to the Blue Ridge Mountains, thirty miles away. The 1,200-foot Montalto—on top of which he once planned to build an observation tower but never did—does slice off some of the ridge's more southerly aspect. But only a little. Otherwise the view was unobstructed. The trees Jefferson planted around his lawns had not in his lifetime grown tall enough to be much of a barrier, so that toward the end of his life he could sit on his porch and watch the sunset over the distant folds of hills, with only Montalto slightly in the way.

Today, however, this is no longer true. The trees have grown high, and someone sitting where the president liked to take his evening ease could no longer see in the summer his blue remembered hills. Instead he would be confronted by a mighty wall of green—or in October, when the trees take on their autumn colors, a tableau of brilliant yellows and reds and oranges. The view might well be chromatically beautiful, but because of the sheer number of fully mature trees that have sprung up today, it is not at all what Thomas Jefferson saw.

Those who run Monticello today have long sought to re-create the estate just as it was in the fifty-eight years it was his home, views and all. To help achieve this, they have cut a spy hole in the trees. By judiciously pruning and carefully planning, the foresters have cut in the faraway wall of oaks and hemlocks and

white pines what looks like a tiny eye-shaped rent—
though up close it is a hole probably measuring a good
twenty feet by ten, at least. By careful cutting back and
shaping, they have managed to keep it clear year after
year—with the result that it is quite possible to squint
through it and see, or at least to glimpse briefly, a frac-
tion of what Jefferson saw.

Because Jefferson was most especially proud of
having created the University of Virginia, the tree
cutters and spy-hole makers have managed to frame
its great rotunda, which Jefferson did not live to see
finished, in the dead center of the view. Behind,
though, are the soft, wood-smothered hills, with the
sinuous curves of the Skyline Drive and the Blue Ridge
Parkway marking their summit lines. These are hills
with a special significance in American history and in
the story of the eventual unification of the country. For
in Jefferson's time these hills toward which he gazed
marked the outer limits, the western edge, the border
of the pale of properly settled America. They were a
line of hills which Jefferson never managed to cross but
which intrigued him, pulled at him, and nagged at him
all his life.

Drawing a Line in the Sand

There are a great many aspects of Jefferson's character that led him to play so crucial a part in the physical creation of the eventual transcontinental republic. It is a commonplace to repeat that he was a man of contradictions. He was a scientist, first and foremost, as well as a learned aesthete and a slave-owning aristocrat with apparently profound feelings for the furtherance of human decency, kindness, and civilization. At thirty-two years old, he was described by the überbiographer James Parton as "a gentleman . . . who could calculate an eclipse, survey an estate, tie an artery, plan an edifice, try a cause, break a horse, dance a minuet and play the violin."

There was something else though. Thomas Jefferson may well have been a sophisticated foreign traveler—he had been minister to France, after all, and later for four years was the US secretary of state—but his travels within the republic were limited indeed. And yet for most of his life, he was quite enthralled by the concept of the American West. He suffered from a bewildering, almost uncanny, and romantic fascination with the continental Occident. He was obsessively interested in particular in just how its immense and generally unknown acreage could and should eventually be apportioned among his country's fast-growing citizenry.

To know its geography was a first imperative. As far back as 1783, while he was a Virginia member of the Continental Congress, he had formally suggested the mounting of a private expedition to the Pacific. "I have always had," he declared, "a peculiar confidence in men from the western side of the mountains."

But Jefferson did not mean by this the grand crystal crags of the Rockies or the Sierra Nevada (of which he, in common with most, knew precious little). Rather he meant the relatively modest ripples of the Appalachians, of which the Blue Ridge hills that he could see from Monticello were the easternmost. For these endless ridges of Devonian rock that rose out of the coastal plains from South Carolina up to New York essentially marked the edge of the United States proper, in Jefferson's time. Beyond them, America was barely known.

Five million people (a fifth of them black, mostly enslaved) lived within fifty miles of the Atlantic Ocean, hemmed in by these confusing swaths of mountains. Only four dirt roads pierced the passes between the hundreds of miles of ranges. Poor weather, frequent rockfalls and mudslides, or else the occasional understandable hostility of the Creek, the Iroquois, or the

Cherokee who once owned these lands (to the extent the ownership of land was a concept recognized by indigenous Americans) increased the difficulty for settlers wishing to travel across the hills, between South Carolina and Kentucky, say, or from Tennessee to Pennsylvania. As late as the middle of the nineteenth century, it could still take nine days of fitful journeying by railroad, canal, riverboat, and stage line to get from New York across to Pittsburgh, because these Appalachian ranges were so ruggedly impenetrable. In Jefferson's time, travel across them was for the fainthearted all but impossible.

Those scattered few who then lived permanently beyond this cordillera, those whose homes were south of the Great Lakes and down in the Ohio Valley, turned out to be so decisively cut off from the American mainstream that there was serious talk of secession, though in the end it came to little more than campfire grumblings. Meanwhile those who lived farther out still, in the wilds beyond the Ohio River and in the American-possessed (but hitherto Indian-reserved) lands that were then designated as the Northwest Territories, were as scarce as they were brave. They may have enjoyed Jefferson's "peculiar confidence," but they were initially outnumbered ten to one by Native Americans; they had extremely limited opportunity for work, mainly in the fur trade; they were protected, but only somewhat, by outpost contingents of American soldiers; and they lived under a scrappily benign version of martial law. It was only when their numbers reached five thousand—in 1798, two years before Jefferson became president—that they were given a representative government with a proper little parliament sited in the first instance in Marietta, the tiny town that was their territorial capital.

It was these doughty western settlers who were to become, however, the eventual first beneficiaries of one of the greatest and most revolutionary ideas put forward by Jefferson, as the territory in which they eked out their rugged existences became its initial testing ground. Jefferson's great notion was that Americans could and should have the right to do something that was hitherto quite unimaginable to the Indian tribes: they should be able to *own* the land of which their territory was made.

Land ownership was an unfamiliar and almost alien concept. Bands of the Iroquois, the Creek, the Shawnee, the Delaware, the Miami, and their native kin had certainly passed through these lands for hundreds of years, had hunted it, settled it, raised families on it. But they had never imagined it as something that could be possessed. It was much the same for the early white settlers: they may not have had the same nomadic urges as the Shawnee and the Iroquois, but land ownership was conceptually well beyond their ken also. It might have seemed possible and reasonable for a settler to own a canoe or a cow or a cottage—or back in those unenlightened times in the Americas, even a slave. But land, an immovable part of the eternal fabric of our celestial body—that seemed somehow to be an entity beyond ownership, the possession of it lying perhaps within the divine prerogatives of kings but certainly not ever in the name of ordinary citizens.

Thomas Jefferson thought quite otherwise. He had developed this thinking long before completing his work on the Declaration of Independence. It is spelled out in a sulfurous pamphlet, published in 1774, in which he denounced King George III's plan that American colonists on the far side of the Appalachians should live in a feudal arrangement, with the king owning the land and his tenants obliged to pay their feu to his court.

The thirty-one-year-old Jefferson denounced this as a barbarism. It was a formula for studied inequity, based on a fiction of kingly and ecclesiastical privilege that had been developed in Britain a thousand years before by the Norman conquerors and believed by many in the homeland ever since. Jefferson declared that such a concept—that only kings, churches, and the aristocratic mighty could own land—would not be allowed to infect the vast tracts of real estate that he suspected lay on the far side of his new country's mountain ranges. All men should have, in his opinion, the right to own land there as they see fit—to buy, sell, or borrow against its value and to hand it down over the generations. And to pay taxes on it, moreover, from which good governance might be purchased and paid for.

This belief had helped propel Jefferson into his seat in the Continental Congress. And his unwavering devotion to its principles led to his sponsorship, a decade later, of a new law, the practical effect of which can be seen nowadays in just about every American town and city beyond the East Coast and on just about every field west of the Ohio River. This was Jefferson's Land Ordinance of 1785—An Ordinance for Ascertaining the Mode of Disposing of Lands in the Western Territory—a piece of legislation that laid down the rules for how the immense tracts of new American countryside, at the time neither owned nor properly known, were to be described, divided, and eventually distributed.

Though a dreamy Jeffersonian idealism lay at the legislation's heart, this prescient and profoundly significant piece of legislation not only provided land for those who wished to own it, but also raised money for a new government that was financially exhausted and depleted by the war with the British. The western lands were the new nation's greatest physical asset—albeit an asset taken without regard for the Native Americans who inhabited them. The new government could sell these lands, in parcels, to anyone who had the wherewithal to buy them. So Jefferson's ordinance set out principles for creating the parcels. Most crucially, it laid down the requirements for a survey, for the creation of a grid of meridians and baselines from which to create these parcels.

To start the process, there also had to be established a place where the surveys of western America would be formally begun, a place that was then touchingly named, as it remains named today, the Point of Beginning.

The honor of locating this point went to Ohio—or what would later become Ohio, the crucible of the Old Northwest. The point can still be seen today, just. It is on the outskirts of a grimy industrial town called East Liverpool, close to a family firm named S. H. Bell, which processes, crushes, and screens, as well as stores and ships, many of the basic materials of the country's industrial lifeblood—bricks, wire, cement, oil-fracking sand, pig iron, steel billets, fertilizer, and limestone.

Here, at the point where Pennsylvania becomes Ohio—and 1,112 feet north of where, a few score yards out into the river, a slim tongue of West Virginia licks its way between—is the monument which, though it doesn't exactly say so, truly is a memorial to these two most Jeffersonian ideas, private land ownership and public westward expansion.

From near this unlovely spot, beside a railroad line and an industrial storage yard outside East Liverpool, Ohio, all of western America is still measured. The obelisk marks the Point of Beginning, the origination site for the meridian and baseline used since the first surveys of the nation.

It is a cement stele, about chest high, sitting on a circular stone mat on which are engraved the four cardinal compass points. The monument is a four-sided obelisk, not unlike the very top of the Washington Monument, with suitably portentous inscriptions on each side. Few of the motorists hurtling by on state highway 39 bother to stop to read, even though the obelisk is surrounded by a small copse of other cast-iron markers and stone boundary posts, and by rights it should be most alluring. It is indisputably one of the more historically significant sites in the nation, a place that should have tour buses and fountains of cool drinking water, even a souvenir stall. Instead it sports a scruffy parking space, one forlorn utility pole, and a scattering of litter.

Jefferson's name is not there; instead the marker notes that "On September 30, 1785, Thomas Hutchins, first Geographer of the United States, began the Geographer's Line of the Seven Ranges." Mr. Hutchins was very much Thomas Jefferson's man, a keen supporter of the distant vision of the American West as an immense "empire of liberty." He was a soldier, a cartographer, and the architect of a system of surveying that continues to be employed in America to this day.

Taking this arbitrary spot as his starting point, he drew lines—one north and south, the meridian; another at right angles to it, the baseline. Once having determined, with the use of sextants and star charts and chronometers, the precise longitude and latitude of the site—40° 38' 33" North of the equator, 80° 31' 10" West of Greenwich—he then set off with his rolls of twenty-two-yard-long iron Gunter's survey chains,* then later with his theodolites and compasses and plane tables, and his party of army-protected cartographers, to survey America.

And by *America*, Hutchins meant the entire continent, though at that time the nation extended only to the Mississippi River, the boundary with the lands then owned by Spain. For the baseline, that magical arrow-straight line at 40° 38' 33" North, known to this day as the Geographer's Line, was by law decreed to extend westward through "the whole territory," all the way to the Pacific Ocean. America might not yet have title to all of the lands between the Ohio River and the Pacific, but now that it had a baseline computed, it was not entirely fantastic to imagine that one day it might.

This was Jefferson's dream, after all. Now that his ordinance was firmly a part of the nation's law and the

---

* Edmund Gunter's chain—invented by this Welsh preacher-astronomer in 1620—became a vital helpmeet for the early physical unification of America. It was a set of a hundred iron links, marked off in groups of ten by brass rings. Its total length was an exact sixty-six feet. Ten square chains, easily measured out, make an acre. Six hundred forty acres—equally easily measured, by the slow and steady process of what is still called chaining—make a square mile.

survey well under way, he made a famous remark: that despite his young country being hemmed in by lands in the north still belonging to Britain, by lands in the south belonging to Spain, by territories in the near west under the vague control of often hostile aboriginals, and by lands in the farther west controlled by France, "it was impossible not to look forward to distant times, when our rapid multiplication will expand itself beyond those limits, and cover the whole northern, if not the southern continent, with a people with similar laws."

It was certainly not fully anticipated that a cash-strapped Napoléon would ever actually sell the land he called Louisiana—let alone that he would sell all of it and all at once. At the time of the survey's beginnings, no one except Jefferson thought much beyond the coming months. Settler life was precarious, and even policy makers tended to think in seasons, not decades, their business more concerned with planning for harvest than for history.

Some say Mr. Hutchins invented the survey system under which he worked, which has endured as a model for many of the world's great surveys. It called for the creation of townships, six miles square, stacked north and south in what were called ranges. Each township was divided into thirty-six numbered sections of one square mile each (640 acres). The sections were divided into half sections (320 acres), quarter sections (160 acres), and quarter-quarter sections (40 acres), which led to such phrases as "the lower forty" and "forty acres and a mule." This system—ranges, townships, sections, and subsections—is now woven deep into the fabric of modern American life, the basis for everything, a systematically numbered* design for almost the entire nation.

It was intended that the distribution of the territory begin at a great clip. Sales offices were promptly set up—the main center being in the nation's capital, New York City—where petitioners put down their money (a dollar an acre minimum, no land sold on credit) and walked away with a title document. The results of the plan and the purchases can be seen today on any map—Western farm after Western farm regularly spaced and perfectly aligned beside undeviatingly die-straight roads spearing east and west, north and south; the country towns with their impeccable grid patterns

---

* Precisely how the 36 section squares in a township are numbered has tended to vary with fashion. Most commonly section number 1 is sited at the southeast corner. Often sections 1–6 are then numbered in a westward direction, sections 7–12 going eastward, 13–18 back again, and so on. This back and forth is known as boustrophedon numbering, after the plowing pattern of an ancient Greek farmer behind the oxen. It is a system often used by flight attendants delivering food and drink to first class passengers.

of streets laid out from North Dakota to Arizona, from Oregon to Alabama; the siting in each township of schools (usually in section 16, with one in section 36 added later), town halls, courts, and railway stations; and the government's retention of some sections (8, 11, 26, and 29) for future sale, the lawmakers in the capital believing, optimistic always, that once the township had been developed, the value of that land would skyrocket.

Matters in fact began rather hesitantly. The ordinance came into formal effect on May 27, 1785, and Hutchins began his survey of the first seven ranges of Ohio—the tract of land spanning the first forty-two miles west of the meridian—a scant three months later. But then scouts reported that a Delaware war band had attacked settlers some miles ahead—a trading post had been sacked and a migrant American murdered, his doorway smeared with red paint as a warning. Already the local indigenes—Shawnee especially— had expressed reservations at Hutchins's plans. To add to their quite understandably cynical attitude toward white men's treaties, and to their pervasive and quite reasonable fear of dispossession, they felt little sympathy for the settlers' apparent need to draw straight lines through lands across which they had been content to meander for centuries, following the routes of animals and streambeds and other natural features. They welcomed the haphazard and felt slighted by the straight.

Hutchins's surveying team was understandably spooked by the killing, and all pulled back to Pittsburgh. It took them nine more months—and a guarantee of cavalrymen's protection—before they recovered their nerve. They then picked up their chains and came back, extended their 40° 38' 33" baseline to a town called Magnolia, and soon managed to survey with a fair degree of accuracy four of the ranges in 1787. By the next year, they had completed all seven.

The surveys were done hastily and often quite imperfectly; each section was merely marked with a white stone at its corners, and at first no surveys at all were performed inside the sections themselves. But it was a start. Congress was formally notified. Maps were then published, and the selling of America began, formally and in earnest. Tacked to office doorways and trees and published in such local newspapers as then existed were advertisements displaying the beginnings of a new phase in American history. Each showed a map

of the Seven Ranges of Townships, being part of the Territory of the United States, NW of the River Ohio, which by a late Act of Congress are directed to be sold.

That part which is divided into sections or tracts
of a mile square will be sold in small tracts at public
auction in Pitsburg the residue will be sold in quar-
ters of Townships at the seat of Government.

During the next few years, all of the rest of the
Northwest Territory was surveyed, and the salable sections were duly disposed of. Towns and villages and
hamlets were born where the land was deemed most suitable for settlement, and thousands of hopeful migrants
streamed westward. Within fifteen years, the quarter-
million square miles of the old territory had been transmuted into the five-and-a-bit states that now occupy its
immensity: Ohio, Indiana, Illinois, Michigan, Wisconsin,
and a part of Minnesota. Independent America had
started out with thirteen states; now the constellation of
stars that was symbolizing them on the nation's newly
adopted flag was starting to swell, and fast.

The more westerly of these states (Minnesota aside)
stopped short where the territory had stopped short:
the east bank of the Mississippi River. Beyond the cliffs
and mudflats of the river, the writ of America did not
run. West of the Mississippi, the land still belonged to
France, and in theory and law, it was no business of
Americans to travel there, either to survey or to settle.

Except that everything changed on April 30, 1803, when the American government—in what many consider the most prescient triumph of Thomas Jefferson's presidency—bought from France all 820,000 square miles of its possessions on the American continent.

With a few strokes of a pen and the payment to Napoléon, America overnight doubled her size. With the acquisition—for $3 million in gold and $12 million in bonds—the country turned herself into what seers of the time recognized as a potential world power and even at that very moment a force to be reckoned with. The Louisiana Purchase, as the transfer was known, suddenly untangled America from the most pernicious of colonial snares and guaranteed her (because the port of New Orleans—Louisiana's third and final capital city, after Mobile and Biloxi—was naturally a part of the sale) unencumbered access to the Gulf of Mexico. Included were seemingly unending stretches of real estate for the settlement of many more millions of yeoman stock— men and women who could be encouraged to undertake their advance from the rigors of respectable Eastern poverty to the nobility of hard-won Western wealth.

If this great tract of land earmarked for Jefferson's imagined millions of settlers was to be properly incorporated into a United States of America—if, in short, it was to be united with all the existing rest—it all now needed to be surveyed and sold, just as the Northwest Territory had been surveyed and sold in the years before. To be surveyed and sold it needed also to be known. To be known, it needed to be crossed.

Peering through the Trees

The decision to cross the continent had actually been made some months earlier, on a late-summer afternoon in 1802 at Monticello. Thomas Jefferson was sitting in rapt attention, poring over a thick, heavy book of more than four hundred uncut and untrimmed pages, bound in blue paper and published by the firm of Cadell and Davies in London. *Voyages from Montreal to the Frozen and Pacific Ocean* had been written by a Scottish fur trader, from Stornoway in the Scottish Outer Hebrides, named Alexander Mackenzie. Or more accurately, Sir Alexander Mackenzie—since King George III had awarded him a knighthood for becoming the first white man ever to cross the entirety of North America.

Mackenzie had completed his voyage almost nine years earlier. He suspected that his seven-month over-land journey to the Pacific was probably of historic moment, and so he had left a memorial. He had created what he hoped would be a lasting inscription on a tiny sea-washed rock near the present-day British Columbia fishing village of Bella Coola: "Alex. MacKenzie, from Canada by land. 22nd July, 1793." He had inscribed the message with his finger, using an old trappers' trick for long-duration messages, dipping it into a poultice made of bear grease mixed with vermilion powder and smearing out words that he hoped would survive the cold and lashing rains for which the Pacific coast is notorious.

One can imagine Thomas Jefferson's reaction as he read the closing pages of the blue-bound volume, on learning that a mere Canadian, a British loyalist, had been the man to first traverse the American continent. It was unseemly. It was an affront. It was a claim that simply should not be allowed to stand.

As legend has it, he put down the book, his face purple with apoplectic annoyance, and turned to his twenty-eight-year-old secretary, Meriwether Lewis, who was sitting beside him. He told Lewis in no uncertain terms promptly to organize an expedition to cross his own country—an expedition that would, among other things, trump the ill-considered wanderings of a forelock-touching Scotsman. He had little reason to suppose the venture would in time become and remain, to all Americans, the most noted of all the many exploratory journeys through and around their now mighty, sprawling, and mysteriously alluring new nation.

This Monticello moment was the birth of the so-called Corps of Discovery, the United States Army's expedition best known today by the names of its joint leaders, Lewis and Clark.

Meriwether Lewis himself needed little by way of convincing. He was not just curious and eager to match and better Mackenzie's achievement, but he, like Jefferson, wanted to ensure that the detested British in no way preempted any potential American claims to the distant West. Lewis also had a highly personal motive to mount a new expedition, stemming principally from having been thwarted by Jefferson in a bid a decade earlier to try such a venture.*

---

* This was in 1792, when Lewis had applied to join a cross-country expedition planned by a Frenchman named André Michaux—a botanist best known now for having brought the ginkgo tree to America. Jefferson refused to let Lewis join his party, saying he was too young. Lewis had made no attempt to join an even earlier venture, in 1785, when a Connecticut opportunist named James Ledyard had the entirely eccentric idea of trying to cross the country from *west to east*, but only after first traversing the entirety of Russia. He was arrested and kicked out of Siberia, was found weeping beside a road in Poland, died of an aneurysm in Egypt, and was buried in a sand dune. Jefferson had met the man and laconically noted in a diary that "thus failed the first attempt to explore the west."

Now, ten years on, Lewis was being asked to go. And to go, moreover, as the expedition leader—Jefferson being by now fully familiar with his assistant's practical abilities as a trapper and hunter, a man who could travel far away in extreme discomfort and not come whining home. The president also admired him for being an exceptionally quick study—his "luminous and discriminating intellect" being one of the many reasons that prompted Jefferson to employ Lewis at the White House in March 1801.

Another reason Jefferson decided to order him out west was Lewis's unusually sympathetic awareness of America's aboriginal people. Lewis already had firsthand knowledge of various Indian tribes: the Cherokee in Georgia, where he had lived as a young man; the Chickasaw and the Shawnee when he was stationed as a soldier near modern-day Memphis; the Miami when he was involved in the mighty Battle of Fallen Timbers in far western Ohio; and later the Potawatomi near his army camp in lands close to today's Detroit.

At the time of Jefferson's pouting decision, the Louisiana Purchase had not yet been consummated. The *drapeau tricolore* still flew on the far side of the Mississippi. There was also evidence that maybe Britain was about to make some kind of claim on the territory too, leading to a certain urgency. Lewis had to leave, it was decided, and in double-quick time.

The depth of ignorance of the soon-to-be-acquired territory was profound. Its precise borders were unknown, for a start—and the French had made it abundantly clear they were not going to give Americans any information about them. Lewis and the party he would choose would have to start essentially from scratch. Where were the land's natural frontiers, where were the mountain ranges, and how exactly did its rivers twist and turn down from the hilltops to the sea? They would have to find out. Moreover, were there truly, as stories of the time suggested, great peaks out in the vastness that were made entirely out of salt? Where were the territory's snowfields, its deserts, the pastures and the prairies? What kinds of flowers and trees grew there, and what species of animals, which types of birds?

And who exactly were the peoples—the Indians, as Columbus had supposed—who had belonged to the land before? Was it true, as some said, that many of them were Welsh? Or, according to others, the Lost Tribes of Israel? And whoever they might be, where did they now live and have their being?

By March 1803 the necessary congressional authorization for the venture was in hand. A sum of $2,500 was appropriated—with $696 set aside for gifts to the natives. A month later and the transfer papers were formally signed in Paris, and the land that had so intrigued Jefferson was now fully American owned and so could be legally explored.

Lewis, now certain the expedition was to begin, procured a note of limitless credit signed by President Jefferson himself as a guarantee, just in case they ran out of cash. He then began assembling his gear. He found his rifles at an army arsenal in Virginia. He found builders for an iron-framed fifty-five-foot wooden keelboat in Pittsburgh. He found his ammunition, his trinket gifts, and his comestibles in Philadelphia. He had to imagine what else he might need: mosquito nets, waterproof lead tubes for holding ammunition, various bibelots and silver medallions struck with Jefferson's profile to be handed out as marks of amity to the encountered Indians, large quantities of powdered ink, 193 pounds of dried soup, twenty-five axes, and four gross of fish-hooks. He also took instruction in field medicine— mainly from a doctor of somewhat crabbed views who believed most ailments could be cured by powerful laxatives, especially one made of mercuric chloride and a ground-up Mexican purgative root named jalap.

With the gear assembled, it was now time to gather the men. In June, shortly before the secretary of war gave formal authorization on July 2 for the corps to select volunteers for the expedition from any of his army garrisons along the Ohio, Lewis wrote to his old army friend William Clark, offering him the position of joint leader of the expedition. The latter accepted cheerfully: "This is an undertaking fraighted with many difeculties, but My friend I do assure you that no man lives whith whome I would prefur to undertake Such a Trip."

Clark was four years older and, in their previous encounter, had been the senior officer. Clark was more rough-hewn and both less literate and by all accounts less given to dark moods than Lewis. Clark had been tested in battle with Indians, while Lewis had not. And Lewis was very much Jefferson's protégé, while Clark had barely a nodding acquaintance with the president. Nevertheless, the pair—who joined up in the Ohio River town of Clarksville, Indiana, in October to begin their formal collaboration—got on famously well on just about every day of the 856 they would spend away together.

In Clarksville they assembled their full team, the soldiers chosen from the scores of fort-weary volunteers ready for an opportunity of real excitement. In the end some twenty-nine men, including Clark's slave, York, were sworn in for the duties ahead.

There were ten weeks of training and preparation before the team was prepared to start. History records with some precision the formal beginning of the expedition: three thirty in the afternoon of Monday, May 21, 1804.

By now the winter was well over; the ice was gone, and the rivers were brimming with snowmelt. Having crossed the Mississippi separately, Lewis rejoined Clark some slight way along its principal tributary, the Missouri. The place he chose was the village of Saint Charles, a threadbare settlement on the river's north bank with a population of about four hundred, most of them French Canadians.

There was a simple topographic reason for the choice of the expedition's starting point. Close to the junction of the two streams, there was a mess of fluvial indecision, with the tributary rivers swiveling direction at the behest of their conjoined currents, leaving a maze of swamps and oxbow lakes and blind-alley bays all across the landscape. But in Saint Charles, the Missouri seemed at last to start pulling itself firmly away to the west—the direction in which the expedition wanted to go.

The river's course was directed by the local geology—the same geology that also enticed the first settlers. There was a low bluff of Devonian sandstone hills on the river's northern bank, the first elevated ground west of its junction with the Mississippi, which would both keep any settlers safe from floods and, in case of attack, offer their pickets a good view of the waters downstream. So a cluster of buildings was built along the bluff—a Catholic chapel, a hundred poorly made houses, a few shops. All of them looked southward across the deep brown stream—the Big Muddy, as it would later be widely called (Clark claimed to find a wineglassful of ooze in every pint of Missouri river water)—toward the scattering of houses in distant Saint Louis, toward the familiar and the known.

Behind, beyond their village pale, was the true unknown—a terra incognita of brown Indian hills, expanses of lands unfamiliar and potentially hostile. Hunters and trappers ventured there—but no settlers, not yet. Saint Charles was thus for many years the most westerly European settlement, the last bastion of immigrant civilization, a town that lay at the very point of intersection between settled America and untamed native lands of the frontier. It could scarcely have been more appropriate as a departure point.

A thunderstorm was raging when Lewis arrived from Saint Louis. He took what churchly men still charmingly called a cold collation—a snack, allowed on fast days—and then crossed the river, where he found

Lieutenant Clark and his party encamped for the evening. Most of the party (except for one member, who the night before had received fifty lashes for going AWOL and then displaying "behavior unbecoming" at a party) were "in good health and sperits." Small wonder: Clark had been royally looked after during his four-day stay: the local Gallic swells offered far better food and wine than had ever been available back east, together with invitations to balls and visits to his boats by numbers of ladies of the town.

One could imagine that Clark would have rather liked to stay, but just after lunch the next afternoon, they set off—"under three Cheers," wrote Clark, "from the gentlemen on the bank."

They headed first directly toward the west, toiling against a slow river current and the whirling of the deadly water-boils they would endure for the next fifteen months, and until they eventually crossed the unknown, unimagined wilderness of the Continental Divide after more than three thousand miles of travel.

They spent the first six weeks journeying easily enough through what is now the state of Missouri. During the early miles, a number of limestone cliffs and sandstone bluffs rise up beside the stream—indeed, Clark fell from one three-hundred-foot pinnacle early in the trip, saving himself only by digging his knife into a crevice and dangling there until he felt brave enough to clamber back up. But generally the countryside here is more floodplain than valley, more prairie than canyon, and the river winds and wanders irrationally, all over the place.

The party found they were making only minimal forward progress, even though their daily distances turned out to be wearyingly long. Today the highways and the Union Pacific rail lines follow much the same exhausting path along the riverbank. They do so not because contour lines compel them to, but because if they tried to go straight where the river winds—with every single bend given a name, Bushwhacker Bend, Bootlegger Bend, Cranberry Bend—far too many costly bridges would be required. It is more prudent and economical to follow the stream than to fight it, today just as it was back in the expedition's time.

The Frontier and the Thesis

After some weeks of sailing and rowing and poling along a willow-banked river, the party reached a junction, with a river they called the Kaw, today the Kansas River. The leaders were at last quite impressed with the landscape—"the countrey about the mouth of this river is verry fine," wrote Clark, and said it would be a good site for a future army fort.

The army must not have agreed, but civilian settlers eventually did, in their thousands, for they later turned the spit of land between the two streams into an enormous campsite, a base for the long and heroic westward treks along the Oregon Trail, the Santa Fe Trail, and the California Trail. And the metropolis that some of their number then stayed behind to build, Kansas City, has become a classic of frontier America.

I had been here before, some thirty years earlier. It was shortly before the bicentennial celebration of 1976, when I spent six months traveling through the Midwest, trying to understand the importance of that uniquely American phenomenon, the frontier. Along the way I had met many people and had seen many things: two of the more memorable happened to be right here, where Lewis and Clark were pressing westward through the very frontier I was studying.

The first encounter was of rather lesser importance, though it still had some poignancy. I had been invited to visit a marble memorial to an enormous white Charolais bull. He was named Sam 951, and until 1972 he had lived on a ranch in the town of Chillicothe and had been famous for miles around as an example of bovine excellence. Sam's frozen semen, once produced in exuberant gallons by what all agreed was an excessively jolly creature, was worth millions, and was packaged in nitrogen-cooled vials to be sent off from Chillicothe to eager customers all over the world.

The Litton Charolais Ranch was in consequence once perhaps the most profitable cattle-breeding outfit in America. Sam 951 was primus inter pares of the large and carefully managed herd. Each bull—the best of them lived in air-conditioned barns kitted out with red carpets—weighed a ton or more, had ears the size of dinner plates, had a vast muscular body joined necklessly to an appropriately immense head, and sported dewlaps that would take two strong hands to move.

Cattle like Sam had made a great fortune for the ranch owner, Jerry Litton, and had now brought him within a hair's breadth of true fame. I spent two happy summer days with him—a handsome and engaging man who had married a former Miss Chillicothe (and a runner-up in the Miss Missouri pageant) and who for the previous four years had been a member of the US Congress, a Democrat. His home at the time I stayed was abuzz with political excitement: in two weeks voters were due to decide whether or not to elect him a US senator. Many, indeed, thought he would and should run for national office—President Jimmy Carter was a supporter—and in early 1976 he was sufficiently intrigued to announce that he would indeed take this obvious next step along the political glide path.

When I turned up, his work was nearly all wrapped up. He was in the closing stages of what all agreed had been an impeccably nuanced and well-funded campaign for the primary election. And two weeks after I left, he did indeed triumph, leading a stunning upset in a twelve-man primary race. Jerry Litton was on the verge, I have long since believed, of well-deserved political greatness.

On the night of his victory, he was to be flown back to Kansas City for a celebration. But then came calamity. The crankshaft in one of the engines sheared in half; the little plane lost power and crashed on takeoff; and Litton, his wife and children, his Beech Baron's pilot, and the pilot's son were all killed. Jerry Litton had been born in a house without electric power, in 1937, when this part of Missouri still had the feel of the frontier about it. He would have brought something of this spirit to Washington had fate permitted it. He was a figure of whom it can rightly be said, He could have been a contender. But fate saw to it that he never got the chance.

My second excursion of that 1976 adventure concerned the polar opposite of a cattle farm. I spent time touring a sprawling Minuteman nuclear missile site, centered at the Whiteman US Air Force Base, an immense complex of men and their flying machines set close by a village just south of the river with the engaging name of Knob Noster. Back in the 1970s, it was quite possible to visit the immense missiles and to descend deep into the bunkers where clean-cut young officers—curiously decked out in uniforms that included starched white ascot collars—sat beside their pairs of launch switches, enduring a bleak shift of existence in air-conditioned subterranean silence, waiting to execute a world-destroying command that, mercifully, never came.

The Cold War is now over, but America still has deployed around the country three wings of Minuteman missiles, all nuclear tipped and more powerful than ever, as ready to go as ever they were before. They are, however, no longer deployed in Missouri but in more distant and protected wilderness bases in North Dakota, Wyoming, and Montana. One can still try walking across lonely Montana meadows up to the edge of the wire-protected silo where a missile lurks beneath its concrete blast doors, to test how long it takes before a security jeep or a helicopter, with flashing blue lights and a crew of soldiers with full authority to maim, comes to find out what you are up to. Back in Missouri in the 1970s, I was invited to try and found it took no more than a couple of minutes for my breach of security to be discovered and repulsed. But it is no longer possible to play such a game at the Whiteman Base there since the men have all been stood down and their missiles dismantled and destroyed under the terms of the various treaties signed with a Russia that is no longer the Soviet Union, no longer seen as quite the threat it once was.

Yet Whiteman Air Force Base itself still exists, and if not armed with missiles today, it still sports a title and wields an ability that sends chills down spines. It is part of a terrifying arsenal of weaponry that is now called— after numerous organizational changes and semantic alterations of title—the United States Air Force Global Strike Command.

The command has its headquarters in Louisiana, from where it controls all of America's air-launched atomic weapons—the three Minuteman missile wings in the northwest and a large number of B-2 stealth bombers, all of them designed to drop thermonuclear bombs. The bulk of these bombers happen to be based at Whiteman—at a site a short way from that point on the river where, in 1804, William Clark recorded hearing an "emence snake" that inhabits a small lake nearby "and which gobbles like a Turkey & may be herd several miles."

The planes belong to a US Air Force wing, the 509th, that proudly reminds visitors that it is the direct descendant of the only unit in history that has ever dropped live atomic bombs in wartime, on Hiroshima and Nagasaki in 1945. Today it is a battle-proven assemblage of aircraft and crew that, its commander says, can now bring massive firepower to bear, in a short time, anywhere on the globe.

B-2 stealth aircraft of the 509th Bomb Wing, a twenty-strong nuclear-armed operational component of Global Strike Command, at Whiteman Air Force Base, Missouri. The wing is a direct descendant of the group that dropped atom bombs on Hiroshima and Nagasaki.

There seems a certain irony in this nuclear firebase being located so very close to the Lewis and Clark expedition route, not least because what Jefferson's explorers were seeking to do, even if unknowingly at the time, was tied to the unique American concept of the frontier and to the development of what to this day is known—and argued over—as the frontier thesis. The irony stems from the argument that the frontier mentality, if such a thing truly exists, still plays a nourishing—and controversial—role at the intellectual roots of much of today's American foreign policy.

The famous argument, put forward in an 1895 paper by a University of Wisconsin history professor, Frederick Jackson Turner, held that there was an immense social significance in the simple existence of the frontier—that ever-westward-shifting margin between civilized society in the East and the untamed savagery and wilderness to the West. Kansas City, the city that rose from one of the campsites of the Lewis and Clark Expedition, became a classic, if momentary, point of frontier contact: on its eastern side were traders, trappers, farmers, settlers, surveyors, villages, and towns; on its western side were empty prairies, nomads, lawlessness, and an unprotected and shelterless void of stony plains, tornadoes, and starvation. Between them lay the line of contact, division, and separation: the frontier.

The rolling clash between these two extremes gave rise, Turner argued, to a peculiarly American set of character traits. The experiences suffered or enjoyed on the frontier left Americans inherently different from what they and their antecedents had been in their homelands. Those tested in the borderlands were by comparison more violent, more informal, more democratic, more imbued with personal initiative, and less hamstrung by tradition, class, and elegance. More *American*, Turner suggested. Strength, power, might—the ability to tame rather than to persuade, the tendency to demand rather than request, the tendency to shoot rather than to talk—these were all tendencies compounded by the frontier experience, uniquely different building blocks employed in the making of the modern American. The Western myth, the legends of the cowboy, the cinematic and entertainment-park allure of concepts like Frontierland—all of these were born from this single simple (some would say simplistic) thesis offered by Frederick Jackson Turner.

In the century since the publication and promulgation of his views, Turner has been attacked roundly and mercilessly for ignoring such matters as race, gender, and regionalism. Yet what has gone essentially unanswered still remains: just why do Americans believe they are so different, so exceptional? Why the persistent belief in the idea of America as the "shining city on a hill"? Why the notion of Manifest Destiny?

And why, indeed, did Thomas Jefferson believe so keenly in the idea that America should and could and in time would extend herself from sea to shining sea, and accordingly dispatch Lewis and Clark to see if and how this could be achieved? Was all of this, as Frederick Jackson Turner would later argue, rooted in that same peculiar experience, shared by all, born in the process of the steady closing of the frontier?

Some may consider it injudicious to conflate, on the one hand, John Winthrop's inspirational city-on-a-hill sermon of 1630 and the tenets of the frontier thesis with, on the other, the notion of conducting Manifest Destiny at home and so many interfering adventures abroad. And yet viewed from some perspectives it does seem right and proper to ask, particularly here in Missouri: why does America still believe, as the slogan of Whiteman Air Force Base has it—why did it ever believe, in fact—that it has a right and a duty to be able to deliver "massive firepower, in a short time, anywhere on the globe"? Why America? And if such a belief is somehow rooted in a deep-seated conviction that it should, that it needs to, and if called upon, that it must—then was not this all born, as Frederick Jackson Turner and his supporters would also argue, from the experiences gained by early Americans in their closing of the frontier? Isn't this determination to extend itself across the planet simply a reflection of the strength and crudity and informal decisiveness of the pioneer Americans, brought up to date, made global, and now writ large for all the world to see?

Does the mission of the huge atomic firebase, sited so close to where William Clark first heard of *the snake that gobbles like a turkey*, have its intellectual origins in this very same tiny, brave expedition that first crossed the frontier and in the consequent development of the huge city now lying just a short drive away to the east, which once so vividly encapsulated the notion of the frontier, two centuries before?

These days it is by no means easy to see the inner workings of Whiteman. In the 1970s it was simplicity itself to win an invitation for a tour. The air force was only too proud then to show off its wares and its weaponry, reasoning that doing so helped display to the Soviets its perpetual readiness to strike. Today, terrorism has introduced a new reality: long scimitar glints of razor wire, battalions of ever-scanning cameras, and heavily armed sentries at the entrance gates all stand guard to protect the planes and their weapons from the innocently curious and the ill intentioned alike. Tours still happen, but application lines are long, details demanded weeks in advance, cameras forbidden.

Once in a while, though, along this steady reach of the wide Missouri, there will come a distinctively huge and quite unexpected rumbling sound, a thunder of jet engines that shakes the willows and the stillness of the stream. Then from its lair behind the wire, rising from an invisible runway folded among the cornfields, a great gray bomber will slowly lumber upward and hoist itself into the skies.

It is always an awesome sight—all the more so if other planes follow and the singleton becomes part of an airborne armada, a squadron of unimaginable power bound on an unannounced mission to a place no one will ever disclose for a purpose never to be known.* As the craft vanish into the clouds and the thunder ebbs away over the woods, it is tempting to wonder just what corner of the planet might soon be basking under the unasked-for invigilation of these nuclear-tipped watchers from the skies.

It is at moments like this the irony of history presents itself. For it seems not too much of a stretch to suppose

---

* In late March 2013, two B-2s from Whiteman were ordered to fly seven thousand miles nonstop to the Korean peninsula and to drop test bombs on a local training range—all to display American readiness to a North Korean leader who was behaving somewhat tiresomely.

that America's present-day global reach, insisted upon as a right and represented by weapons like this, is a concept that actually enjoyed its infancy here, more than two centuries ago. This was when two young soldiers, on orders from their president, were engaged on a mission to extend the reach of their young country, not then clear across the world, but from just one gray ocean clear across to another. The world would come later, when canoes became bombers and wooden paddles jet engines.

The Wood Was Become Grass

Beyond Kansas City the river turns northward, and William Clark offered his views about the kind of terrain that he believed now lay on its western bank. His spelling and grammar were never exemplary: on the evening of June 21, 1804, when he wrote this simple observation, he was probably quite weary:

Supplied with water the Small runs of (which losees themselves in the bottom land) and are covered with a variety of timber such as Oake of different Kinds Blue ash, walnut &c. &c. as far as

the Praries, which I am informed lie back from the
river at some places near & others a great distance.

The *Praries*, as he had it, were indeed nearby, and
they were of a landscape very different from what had
gone before.

Until this point in the journey, the expedition had
been quite overwhelmed by trees, by forests, by glades,
by copses—by wood. The valleys through which the
men traveled and the hills they saw from the water
were usually thick with trees. They were burned in
some places by Indians, who needed places to conduct
their agriculture, but otherwise they seemed totally to
carpet the land. Red and white pine forests; oak and
chestnut forests, copses of hickory and cottonwood;
groves of aspen, birch, maple, and cedar; stands of
balsam fir, oak, ash, and walnut—all these and more
make their way into the journals of Lewis and Clark,
for whom scarcely a day went by without some mention
of a tree or a wood or the worrisome absence of wood-
land where the explorers had believed it should be.

Though Lewis had some scant botanical training, the two were focused primarily on the commercial possibilities of timbering, not forest science per
se. Early America ran on wood. People had an urgent
need of it for every aspect of life, from fuel to housing,

from boatbuilding to the making of crude paper and the construction of that most esteemed emblem of pioneer life, the log cabin. And in those settled parts of the country, wood was abundant. From the white pine forests of Maine to the magnolias of South Carolina and the elms and chestnuts, the cottonwoods and willows of Missouri Territory, the stripling America was bristling with trees.

Except, as William Clark was aware, this suddenly was not so anymore, on the west bank of the upper reaches of the Missouri River. Up on these riverbanks, sometimes close to the river, on other occasions some great distance away, and first seen in the long reaches upstream from where the river makes its directional shift from the west to the north, there appeared glimpses of a landscape now in a state of arboreal undress, much changed from what had gone before.

What Clark glimpsed was a relatively treeless brown-green country, stretching away into a violet horizon that was longer and flatter than any that these hill-born Easterners had ever witnessed or imagined before. It was landscape laid out, flatly undulating, beneath a sky so big it was overwhelming. It was a new kind of prairie, a limitless tableland of grass, a huge grazing-plain, with a wind that soughed near-constantly above the vegetation, the temperature of the drifting air the only clue to the season. Its sky was flecked with mare's tails of clouds, where lightning could be seen a hundred miles distant and you could watch the black storms chewing their way toward you, the sky suddenly darkening overhead as the squalls arrived and smashed down wafting curtains of hail until the earth was quite white and crunched underfoot, though within moments the reappearing sun then melted it away, steam suddenly began to rise from the grass, and you could almost hear the plants bursting upward in the newly made and richly damp sauna of heat. America was someday to be a united nation, for sure, but in places its newly seen landscape evidently comprised the greatest imaginable differences.

The explorers had reached the eastern edge of that immense, hitherto mostly unseen and uniquely American geographical phenomenon: the Great Plains. Uniquely American, but not unique: there is no shortage of vast midcontinental expanses elsewhere—the Russian steppes, the African veldt, the Argentine pampas, and even some African savannas all offer much the same confluence of flattened topography, pitiless windblown climate, and endlessly unvarying botanical covering. But in America, the Great Plains have been sintered into what is now a cultural, as well as a geographic, entity—a tract of thinly settled grassland of between half a million and a million and a half square miles, depending on the chosen boundaries, a place and an entity that is now an essential component of what America has made of herself, part of the country's shared triumph and, for many years, part of the narrative of her shared national tragedy, too.

The Great Plains boundaries are fugitive, vague lines that shift from year to year, drift from climate to climate, or wander and wobble like the polar axis. The sudden upsurge of the Rocky Mountains more or less marks their western limit. In the east, where Lewis and Clark became the first confirmed American explorers to encounter them,* their boundary is ill defined at best. Some like to suggest that the Missouri itself provides the line. The land on the river's eastern side is thick with lush vegetation, the soil so Russian black and damp and rich that some have remarked that it might

---

* Spanish conquistadores crossed portions of the Great Plains' southern tier in the sixteenth century. Fur traders—most of them French—knew something of the region, too, had built small trading camps, and had made maps and left diaries. David Thompson, a Welsh-born Canadian explorer, also reached the plains in what is now North Dakota in 1797, almost a decade before the Americans. There is still understandable vexation among Canadians that the reputation of Thompson—who went on to become the first white man to travel the entire length of the Columbia River—has been overshadowed by that of Lewis and Clark, whose explorations were scarcely any more arduous.

as well be eaten without any need to pass vegetables through it. The lands on the far side, by contrast, are said to be parched and dusty, their grasses scrawny and patchy, and such meadows as exist having a persistent brown and sun-scorched look about them. But this is all a fancy; scarcely anywhere along the river is the division ever so neat and clear-cut. In fact, seldom can a traveler from the east be entirely sure he has truly entered the plains proper until their presence, after miles of slow and subtle alterations, becomes fully— and to some stunningly, even alarmingly—obvious. And that has little to do with the changing nature of soils or vegetation: it is generally when all the visible world around seems sky and endless curved horizons, where nothing else seems to exist before or behind or on either side but an apparently limitless, wind-hissing emptiness.

Though geology and glacial history have deter- mined the extent and topography of the plains, it is quite simply rainfall—or rather, its lack—that is the real key to their existence. The climate patterns here are so classic that they might be lifted from a text- book. The huge, moisture-laden weather systems that trundle relentlessly eastward across the continent from the Pacific Ocean are forced upward as they pass over the Sierra Nevada and the Rockies; this ascent cools the air, reducing its capacity to hold water. Gravity then insists it fall as rain or snow on the crags below.

What happens next determines the fate of the plains; by the time the weather systems are done with the mountains and swish downward from the heights on their eastward drift, they are exhausted, wrung out, and bone-dry. They roll on for hundreds of airborne miles without immediate purpose, without maturing clouds, and without the will or ability to deposit any further moisture on the grounds below.

The flatlands beyond the Rockies thus lie in a rain shadow, and the vegetation that grows or clings to life within it is peculiar and appropriate to the waterlessness it imposes. And since the vegetation is almost always the key to both animal and human settlement, the role of these flatlands in at least the beginnings of the American story was as fully determined by it as in any other settled corner of the planet. Just as the Inuit and the polar bear inhabit the northern snow country, just as the Tuareg and the camel make their own very different kind of living in the hot African deserts, and just as the San and the Yamana and the Ainu and the Kazakh all adapt to their own unique habitats according to climate, topography, and the local flora and fauna, so too in these prairie parts one finds people and creatures uniquely suited to the conditions: the Comanche and the prairie dog, the Sioux and the rattlesnake, and all of the other Plains Indians—the Blackfoot, the Cheyenne, the Arapaho, and the Crow—together with uncountable millions of the single species of animal that once so dominated and symbolized the grass-covered landscapes here, the American bison.

The plains grasses from which these bison fashion their cud are of very different kinds and appearance, depending largely on the rainfall, the mean temperatures, and the thickness of the soil. Latitude plays its own part, of course; but longitude has the greater role in dividing each from each. Generally speaking, the Great Plains extend between the 95th and 105th meridians— with the midline, the line marking 100 degrees west of Greenwich, denoting by tidy coincidence the approximate limit of twenty inches of rainfall a year: west of the line is drier, true rain-shadow country, while to its east the rainfall becomes ever more abundant and more steady. Altitude plays a part also: because the plains generally slope downward from west to east, from the Rockies to the Missouri River, the western plains are higher, the made-for-movies hardscrabble country of the High Plains, indeed.

This is the great Dust-Bowl-to-be country, rarely much good for agriculture, where otherwise munificent bankers were traditionally reluctant to lend to settlers who were proposing to live there and farm. In this western dry country, the plains are dominated by very short tufted grasses like fescue and needlegrass, and later by hardy imports like crested wheatgrass.

On the 100th meridian itself, in the midplains, there is more of a mix. In what is now Nebraska, say, with its wide, empty farm fields, Willa Cather's famous "shaggy coat of the prairie" has a pile six feet high at least, made of deep big bluestem, *Panicum* witchgrass, wild rye, perennial tussock grasses like yellow Indiangrass, and a weave of flowering timothy and blue grama. (The last is a prairie grass that currently displays its own limitations, for it manages at once to be sufficiently abundant to be the official state grass of Colorado and yet is classified as endangered only five hundred miles east in Illinois, whose western counties, if not quite the Great Plains, are very much a part of the tallgrass prairie.)

Lewis and Clark saw all of these grasses—even timothy, the only non-native of the group, which had been introduced from Europe more than a century before and had spread across the nation with astonishing speed. But one plant they would not have seen, despite its now being a near-legendary symbol of the plains—was tumbleweed.

The image of tumbleweed—a ghostly botanical thing looking like a bouffant hairpiece, bouncing steadily across a dusty road before a cold and gritty wind, lodging itself eventually in a barbed-wire fence—is persistent, emblematic, frequently adopted by Hollywood, and generally best viewed on the screen in black-and-white. In most cases, the plant involved is the Russian thistle, *Salsola tragus*, a pest of a weed, loathed by farmers. The reason Lewis and Clark never reported seeing it is that they arrived too early on the scene by many decades: the vanguard of the tumbleweed invasion came with the accidental importation of thistle seeds in a sack of flax brought to the Dakotas by settlers in the 1870s, six decades after the Corps of Discovery had passed by. It is now just about everywhere, occurring clear across the middle country, from the dusty American West to the lush soils of the Missouri Valley.

It was the eastern tallgrass prairie that Lewis and Clark would have first glimpsed when they made initial contact with the plains during their gentle upriver paddle through what is now Kansas, Nebraska, and the Dakotas. On July 4, for instance, when the party was near Leavenworth, Kansas, Clark wrote:

The Plains of this countrey are covered with a Leek Green Grass, well calculated for the sweetest and most nourishing hay—interspersed with cops

of trees. Spreding their lofty branches over Pools
Springs or Brooks of fine water. Groops of shrubs
covered with the most delicious froot is to be seen
in every direction, and nature appears to have
exerted herself to butify the Senery by the variety
of flours . . . raised above the Grass, which Strikes
and profumes the Sensation, and amuses the
mind.

Clark's "Leek Green Grass" of 1804 is simply today's big bluestem, the classic of the tall grasses. Its appearance among the scattered copses here hints at the borderline between prairie proper and Great Plains. And Clark is prescient indeed in remarking on its "sweetness" and on the "froot." This tract of countryside, with its two-foot-deep soil that once gave support to these long grasses, would (once John Deere had perfected his steel plow blade in the 1830s to create a splendid tilth) become America's present-day granary, with section after section laid to the endless acres of wheat and corn of the richest and most productive grain belt in the world.

But that is the eastern edge, where the soils are rich and fertile. Just six scant weeks and seven hundred miles later, when the expedition had come to what is now Fort Thompson, South Dakota, all had changed.

The men had by now passed the mouth of the sand-laden Platte River. Frontiersmen scorned this long and wandering stream, "a mile wide and six inches deep," as "too thick to drink, too thin to plow," and held that passing northward over it held a symbolism similar to crossing the equator. The explorers were now in a harsher, drier territory, a wilderness of small braided streams, alkaline flats, and immense buffalo herds, where small cottonwood groves grew only in the deeper stream valleys and where the rich planting soil had given way to rougher grazing land, as Lewis himself noted:

> I found the country in every direction for about three miles intersected with deep revenes and steep irregular hills of 100 to 200 feet high; at the tops of these hills the country breaks of as usual into a fine leavel plain extending as far as the eye can reach. . . . [T]he surrounding country had been birnt about a month before and young grass had now sprung up to a hight of 4 Inches presenting the live green of the spring. . . . [T]his scenery already rich pleasing and beautiful, was still hightened by immence herds of Buffaloe deer Elk and Antelopes which we saw on every direction feeding on the hills and plains.

North of the Platte, they had now passed into the true short-grass prairie, and they would have made it farther west, perhaps into the High Plains themselves, had not winter intervened. The first snows came in October. The Missouri was by now substantially shallower, slower moving, and, free of the Platte sands, clearer and purer. As destined by its hydrodynamics it soon started to freeze, and the expedition was obliged to set down its planned winter camp.

By now the expedition had already begun to encounter scatterings of Indians—and in November would meet with one group of Native Americans—and one Native American in particular—who would profoundly change the tempo, the temper, and the reputation of their adventure.

Encounters with the Sioux

The smallest commercial nuclear power station in America is in Nebraska, standing slightly sheepishly on the west bank of the Missouri. Since 1973 it has supplied with a fair degree of constancy (though lately interrupted by flooding) a modest amount of electric power to the city of Omaha, forty miles to its south. It is officially the Fort Calhoun Nuclear Generating Station, and it stands more or less on the very point, just around river milepost 645, where Lewis and Clark had their first official meeting with a delegation of Indian chiefs.

The encounter took place on August 3, 1804, three months out from Saint Charles. The chiefs were from one of the country's lesser seminomadic peoples, the Otoe tribe. They were not the first Indians the explorers had seen. Every so often, Clark noted in his diary having passed by trapper boats, with usually a Frenchman and a native client aboard, but these Otoe were the first to be properly and formally met. And the explorers were fully prepared for them, backed as they were by the full authority of the White House, with dozens of preprinted forms on stiff white card ready to hand out when appropriate.

"Know ye," the opening of each card declared, that the United States government "will be at all times extended to (your protection), so long as (you) acknowledge the authority of the same." In other words: enter into a treaty of peace and amity with Washington and the white men, and expect protection, harmony, and good neighborliness. Refuse, and face the consequences.

It is not entirely clear from the diaries that the Otoe people were either at first given or were thought to entirely merit the gift of this handsomely printed peace offering. They were certainly not to be offered the very highest quality of the three kinds of peace medals, each with President Jefferson's profile on one side, which the party carried with them. The Otoe were, after all, regarded as something of a second-rate tribe. They were seen as a small group of interlopers from Lake Superior. Though they may well have adopted the modus vivendi of the Plains Indians and so had once (smallpox had drastically reduced their numbers) been given to riding horses,* hunting buffalo, carrying their goods behind them on a pair of parallel ground-scraping sticks called a travois, and living in small villages of tepees, they were not, in fact, considered quite the real thing. Such ceremonial as they might be offered would be little more than a rehearsal for the bigger events to come.

But whether giving adequate gifts or not, Lewis and Clark nonetheless made impressive-sounding speeches

---

* Horses were an unintended gift from Europe. The wild horse populations of North America had long since been hunted to extinction, and when sixteenth-century Spanish conquistadores like Coronado imported the animals, they found great favor with the native population, whose nomadic habits were quickly and profoundly changed—the buffalo being a special victim of their hunters' expanded range. Of the thirty-odd Plains tribes, it was the Comanche of the southern plains who took to the horse with the greatest enthusiasm, a man often owning as many as thirty. The Sioux in the north had a rather more moderate equine appetite, though they were invariably to be seen mounted.

to their six visiting chiefs, making each side feel diplomatically important. Lewis, a dour and introspective man at the best of times, delivered a gloomy and foreboding address that would prove the model for almost all of his future speeches: it was perhaps not the kind of address to suggest amity and cooperation. "Children," he told the assembled indigenes, "obey . . . the great Chief the President who is now your only great father . . . he is the only friend to whome you can now look for protection. . . . He has sent us out to clear the road, remove every obstruction . . . lest by one false step you should bring upon your nation the displeasure of your greater father, who could consume you as the fire consumes the grass of the plains."

Yet there were as many carrots as sticks. The men handed out packages, some for the arrivals, others of greater worth to be delivered to the absent seniors. Included in the gift boxes, besides presidential medallions of the second and third class, was a jackdaw clutch of beads, tomahawks, scissors, a comb, some mirrors, and American flags. For good measure, Lewis offered a bottle of whiskey and then fired his rifle into the air, astonishing the visitors and underlining the power and potential authority of these boat-borne strangers.

It was not necessarily the most auspicious meeting, but it was important enough for the party to name the place Council Bluff. Today there is a bright steel memorial marking the site, with a peace pipe and a shiny steel arrow shaft above the inscription, which records the event. The nuclear power plant hums just a few miles away.

(The important-sounding name of the place has since, however, been shifted both across the river, into another state, and a dozen miles downstream. Council Bluff, Nebraska, has become recast and pluralized as Council Bluffs, Iowa, a sprawling riverside city of railway trains and gambling casinos, which is now the better-known memorial to the meeting. When I visited Lewis and Clark Overlook here, a senior manager of the Federal Reserve Bank's Omaha branch was offering at full volume an expansive history of the Corps of Discovery's route. He seemed not to be aware—most aren't—that the crucial first meeting with Indians actually took place upstream, where instead of this overlook there is the rather less impressive metal monument, of just the peace pipe and the arrow.)

There was actually another meeting with Indians from the same tribe two weeks later. But by then the explorers were consumed with misery over the sickness of one of their own, Sergeant Charles Floyd, who died of a ruptured appendix during the talks. He was buried nearby; his grave, a miniature Washington

Monument–like affair near Sioux City, still stands. He was the only member of the party to die during the expedition, was the first American soldier to die west of the Missouri, and most probably also was the first to die west of the Mississippi.

This time they did hand one of Jefferson's peace-and-amity cards to a quite naked Indian chief, only to be mightily offended that he handed it right back and said he preferred to have more of the enticing-looking goods the Corps had lodged in their canoes. He had to be told off, and sharply. Through an interpreter named Mr. Fairfong, *words were spoken,* and the Indian left with a flea in his ear.

If the Otoe were not quite the genuine article, the next native inhabitants to be formally encountered most definitely were. It was at the end of August, after the Corps had crossed what is now the James River, near the town of Yankton, South Dakota, when they encountered a third and rather more important group of native inhabitants. By this time they were becoming fascinated by the astonishing abundance of wildlife on the Plains—huge gatherings of buffalo, antelope (which they called goats, as some locals still do), prairie dogs, jackrabbits, magpies, bull snakes, mule deer, elks, coyotes. To men who had spent their years in the eastern woodlands, where wildlife was quite scarce, this was beyond belief: only the French trappers who traveled with them as hired interpreters exhibited (typically, one might say) an unimpressed sangfroid.

But late one Monday afternoon at the end of the month, a young Indian boy swam fearlessly out to their boats, and the expedition made its first encounter with the tribe for whom President Jefferson had most especially instructed the soldiers to watch out: the Sioux. Once others had gathered to supervise the youngster's meeting, William Clark took a long look at them and declared himself mightily impressed:

> *The Souix [sic]* is a Stout bold-looking people (the young men hand Som) & well made. The Warriors are Verry much deckerated with Porcupin quils & feathers, large leagins and mockersons, all with buffalow roabs of Different Colours. The Squars wore Peticoats and white Buffalow roabs.*

Whatever the Otoe had been, these men at last were true Plains Indians, most certainly. They were a people of great number and power, and most assuredly not

---

* Scholars who study such matters claim that Clark employed no fewer than twenty-six other spellings of the tribe's name.

to be trifled with. Yet the white man did trifle with them from the very beginning—by first calling them something they did not call themselves. They had long termed themselves the Dakota. The name Sioux is a complicated French corruption of a much more complex Ojibwa word and, so far as is known, has been employed since the mid-1700s: the Irish-born colonial official Sir William Johnson, who traded with the Indians from his home in New York, wrote in his diary for 1761, "I picked up a pair of shoes made by the Sioux Ind[in] to the westward."

Properly the Sioux formed a part—an extremely large part—of the Plains Indians. The Sioux linguistic group (the easiest means of classification, ethnologists say) enfolded an immense area that arched from the upper Mississippi River in Minnesota's Thousand Lakes region clear across to the Rocky Mountain foothills in Montana and Wyoming, down in the east to Texas, and down in the west to parts of western South Dakota. Confusingly, several Plains tribes were not members of the Great Sioux Nation—the Blackfoot and the Gros Ventre tribes to their west were not, nor were the Cheyenne, the Arapaho, and the Pawnee to their south. (The Ojibwa, the Kickapoo, and the Illinois beyond to the east were not part of the Sioux Nation, nor were they Plains Indians at all.)

Within the Sioux Nation there were three main groups, based on subtle differences in their language. In the west were the Lakota and Teton Sioux; toward the east, such groups as the Santee and the Osage; and here where Lewis and Clark first met them, the Yankton. Each—together with their many subgroups, most of these more sedentary than the endlessly nomadic Sioux proper—had a reputation for power, determination, and utter fearlessness.

The best-known of their number, Crazy Horse—the leader who in 1876 oversaw the defeat and death of George Custer at the Battle of Little Bighorn—remains their most vivid exemplar. Sitting Bull, who did much to unite the various Plains Indian tribes to resist the depredations of the whites and whose spirit oversaw the same battle, is another; he was one of the historical figures chosen (if somewhat controversially) by President Obama in a book published in 2009 as a role model for his young daughters.

Both men seemed tougher than tungsten. Sitting Bull, bowlegged from a life in the saddle, seemed to have had an unlucky left side: he limped because he had been shot in the left foot by a Crow Indian, he had a wound in his left hip after being shot there by a soldier, and he had taken an arrow in his left forearm after a tussle with a posse of Flatheads. Before his backstage role at Little Bighorn, he offered a sacred pledge of a hundred pieces of his own flesh and sat with bovine stoicism while his brother carved fifty tiny morsels out of each of his arms. Small wonder that the Lewis and Clark Expedition diaries offer similar tales of Indian grace under pressure: of Sioux warriors who walked unflinching into any battle, unprotected; and of a group marching on ice who disregarded cracks and fell through and drowned, with those following disdaining the idea of walking around, but marching ahead regardless.

Matters might have turned out more peaceably if Lewis and Clark had realized from the start the immense pride of these peoples and the significance of the Sioux's samurai-like code. For although the meetings in the autumn of 1804 between those first Yankton Sioux, and then on a later occasion in September with the much more belligerent Teton Sioux, both went well enough, the encounters in hindsight turned out to be the starting points in a spiral of hostility between the ever-westward-moving whites and a people—an enemy, in time—who turned out to be case-hardened, imperturbable, and initially well-nigh undefeatable.

The explorers might have suspected something from the uneasiness of their meeting with the Teton on September 23. For although it did end well, there was a potentially dangerous row—the Teton chiefs wanted tobacco and wouldn't let the boats pass upstream until they were given some. Lewis lost his temper, cast off his fleet, and contemptuously threw a number of carrots of tobacco onto the bank. The Teton, on a hair trigger, might have slaughtered the expedition members there and then—but accepted the tobacco without the slight and let the ropes go.

It was a small enough event. But even though over time white Americans and some Indian tribes developed a degree of mutual understanding and friendship, in general there grew a deep and pervasive mutual loathing between them, a hatred that metastasized during the rest of the century, marked by attacks, skirmishes, battles, and eventually in 1876 by an all-out nation-enfolding war—with Custer's famous Last Stand at Little Bighorn its most especially savage episode.

Savage from the white perspective, that is. Fourteen years later a welling-up of white revenge led to an even greater tragedy, one never to be forgotten by any Native American. It was in the winter of 1890 that US cavalrymen, many legatees of the Little Bighorn battle, descended en masse on a group of 120 Lakota Sioux, all members of the mystical and mysterious and much-feared group known as the Ghost Dancers. The soldiers herded them, together with more than 230 women and children, along the banks of the Wounded Knee

Creek on the Pine Ridge Reservation in what is now South Dakota. And there, on the bitterly cold, snow-dusted final Monday morning of the year, and after a brief altercation that acted as a tragic tripwire, the soldiers opened fire on them—shooting with their rifles and, most notably, with four newly bought rapid-fire Hotchkiss cannon, which in a matter of minutes mowed down the trapped Indians by the score, the detonations of their enormous shells creating a true bloodbath.

At the very least, 150 Sioux and their families died in the chaos of the shooting. Once the cavalrymen had lowered their weapons, nature conspired to render the scene more permanent, as frigid weather rolled in from the west to consolidate and harden the day's terrible handiwork. It snowed a full Dakota blizzard, and when it eased the bodies were left frozen in grotesque and unforgettable contortions. There is a famous, shameful photograph of the leader Spotted Elk, his body etched with snow, his arms frozen by cold or rigor, seemingly trying to get up from the ground, pinioned in icebound pain, his face the picture of purest agony.

The Massacre at Wounded Knee left a panorama of memory that of course Lewis and Clark can never have imagined—yet some may say that their occasionally high-handed behavior toward those who had inhabited the lands over which they ventured must have played some part in sowing the seeds of ill will, and which culminated in so much eventual misery. The intent of the men and their president may have been noble; national unity may have been their distant aim; and yet division, in later years, was to be at least one unintended consequence.

First Lady of the Plains

Yet not all of their encounters with American Indians were so fraught. It was some few weeks later, in November, when the winter chill had begun to freeze the rivers and farther upstream travel was proving difficult, then impossible, that they first met up with a middle-aged French fur trapper named Toussaint Charbonneau and his two wives—one of whom, most memorably, was a heavily pregnant fifteen-year old Shoshone girl named Sacagawea. A captive youngster from an Indian tribe based in the distant western mountains, she would become in time an unforgettable, romantic American heroine and perhaps one of the better-remembered human legacies that the Great Plains would bestow upon President Jefferson's great unifying expedition.

By now the men had moved beyond the main Sioux lands and had reached the territory of three of the lesser Indian tribes, the Arikara, the Mandan, and the Hidatsa (the latter also for some reason known by the French who met them as the Gros Ventre, or Big Belly), which were all affiliated with and linguistically part of the Sioux. But unlike the nomadic Sioux proper, these tribes were in the main sedentary farmers, who raised crops (developing a strain of maize still planted today) and kept dogs and livestock, and (long after their encounter with Lewis and Clark) who died in massive numbers of a smallpox epidemic.

But in 1804 they were healthy, numbered in the thousands, and lived in large circular earthen lodges arranged in villages, in groups of twenty or thirty. They were a people who had not entirely abandoned travel: on occasion their hunters set off on horseback to bag buffalo. But the Mandan in particular were generally more homebodies and quite amiably disposed to all. The Hidatsa people by contrast were still wanderers and frequently took off westward for the distant mountains, to hunt not only for food but to seize new horses, once in a while to collect Indian slaves, and from time to time to give a few old enemies a bit of a hiding.

On Sunday, November 4, while the expedition team was building its heart-shaped stockade (a fort "so strong to be almost cannonball proof," it was noted), the French Canadian trader named Toussaint Charbonneau arrived from his home in a nearby Hidatsa village, asked for work as an interpreter, and was hired more or less on the spot.

Charbonneau had worked for the North West Company for some years and had lived with the Hidatsa most of that time. We know from the expedition diaries just a little of his appearance—that he was small and dark—and a little more of his character; he was said to have been cowardly and aggressive by turn, valued initially only as a translator, though later found to be indispensable for expedition morale as a talented *maître de cuisine*. But though his early worth may have been trifling, that of the younger of his two wives, Sacagawea, has since become inestimable—even if her value may have been magnified and driven by the popular demand for compelling narrative, and maybe also by a need to introduce a decisively female personality into the largely male-dominated world of the Western story.

No one has the slightest idea of what Sacagawea looked like, though there has been much speculation and invention. Fanciful images of her—in oils, water-colors, mosaics, pastels, cartoons—are plentiful. The Iowa-born white all-American beauty Donna Reed played her in one older movie (with Charlton Heston playing Clark and Fred MacMurray as Lewis), and more recently a Japanese Cherokee actress named Mizuo Peck did so in two others. Sacagawea stamps, mountain peaks, and rivers abound. The eighteen best-known American statues* of Sacagawea usually display a tall, robed woman of classically noble bearing, invariably carrying, papoose-style, the boy child she bore in camp in February 1805. His name was Jean-Baptiste, but the expedition members more familiarly called him Pomp. (Clark could never get his tongue around Sacagawea's proper name and called her Janey instead.)

But while the artistic world might allow some license, the United States Mint is more severe in its demands. To present as accurately imagined a profile as possible for the gold-tinted Sacagawea dollar coin required some rather more intelligent speculation. The coin artist chosen by the Treasury eventually chose as her model a twenty-two-year-old college student named Randy'L He-dow Teton, a Shoshone from Idaho who won some lasting fame as the girl on the coin and has since become a motivational speaker for the American Indian cause.

---

* Sculptures are scattered along the expedition's route, from the Dakotas by way of Montana and Idaho out to the Oregon coast, though there is one at the Cowgirl Hall of Fame in Texas, another at the old International Fur Exchange building in Saint Louis, and one with Lewis and Clark by Monticello, where the expedition was conceived.

By general consensus Sacagawea is thought to have been a Shoshone from the mountains of Montana or Idaho captured by a Hidatsa raiding party. Many others claim evidence that suggests quite the opposite: that she was a Plains Indian, a member of the Hidatsa tribe all along, who had actually been captured by Shoshone raiders and spirited back to their lair in the Rockies, from which she escaped and managed to get all the way back to the Plains with the help of (the story at this point somewhat straining belief) a party of sympathetic wolves.

To judge from the expedition diaries, in which Sacagawea is seldom named as anything but "the Squaw," there is little evidence she did a vast amount for the explorers. There is no suggestion, in particular, that she enjoyed an affair with Clark, as Donna Reed so luridly did with Charlton Heston, though the fact she gave him two dozen white weasel tails as a Christmas present was apparently enough to set Hollywood's imagination going. Her quick thinking may once have saved some of the expedition papers from getting waterlogged when one of the pirogues tipped over. She was helpful in recognizing parts of the Western landscape from her childhood memories, and she was able to nudge the scouts to cross a particular mountain pass she knew. She was helpful as a translator and interpreter of the Shoshone language and of those other tongues that were its linguistic kin, and she knew a little French, as did Captain Lewis.

Probably her most valuable contribution was her simple presence. She was a Native American, a woman, and a mother. She traveled with her child. Any expedition that included so innocent a member could not—at least to the many Native Americans who might be encountered—pose a threat of any kind. Sacagawea thus became, unwittingly if not unwillingly, the key that opened the gates of the West and allowed the white men through.

High Plains Rafters

When the winter broke and the prairie ice had begun to melt, the party set off again. They sent their big iron boat back downriver, laden with reports, specimens,* and booty for the White House. But in their two original pirogues, together with half a dozen newly made cottonwood canoes, Lewis and Clark, along with most

---

* Including four magpies, a prairie hen, and what Clark called a "barking squirrel"—a prairie dog. Three of the magpies and the hen failed to survive the rigors of the trip.

of their soldiery, their new interpreters, and Sacagawea and Pomp, set off upstream. It was April 7, 1805.

There was a general mood of excitement and no little regret at leaving, even among the men. During the previous six months, there had been plenty of sexual activity—the Mandan Indians were generous in offering their wives' favors to the visitors, and the irritations of syphilis (with which many of the locals had been infected, reputedly by the French trappers) and the frequent need for mercury-ointment treatments were getting frankly tiresome. "We were now about to penetrate a country at least two thousand miles in width," Meriwether Lewis wrote later, apparently without any intended punning, "on which the foot of civilized man had never trodden."

At first there were many miles of loneliness and heartache—the plains desolate, the rivers shallow and fast with snowmelt, the winds "violent" and incessant, and the breath of Canada—for the territory they explored was just a few miles south of the present boundaries of Manitoba and Saskatchewan—intense and unpleasantly chill. Initially they did not encounter any Indians, other than a dead man they found on a specially built coffin-platform, offerings to the gods scattered beside him. But they did see a good deal of new wildlife: the terrifically dangerous and nearly unkillable grizzly bears most notably, as well as gophers, bald eagles, mule deer, bighorn sheep, prairie rattlesnakes, a kind of avocet, and a snipelike bird now called a willet.

There were new plants, too—such as the just-about-edible white-apple-like prairie turnip (which Sacagawea munched and seemed to enjoy) and in the drier plains the prickly pear, which painfully abraded the soldiers' feet during the ever-more-necessary portages. And there were minerals, most especially long outcrops of the coal that is so important a part of the economy of the western plains today. The immense Union Pacific and BNSF coal trains that rumble along the horizons on their way out of the region today provide a visually appropriate kind of mobile legacy, a memorial to the expedition that first noted the wealth underground.

A month later the men, still on the ever-narrowing Missouri, were past the confluence with the Yellowstone River and two weeks later still were gliding through a peculiarly harsh landscape that is nowadays known as the Missouri Breaks, lying within the boundaries of the giant state of Montana. Lewis loved the place. His writings here were more cheerful and lyrical than at almost any other place in the odyssey. He even sees a peculiar beauty in the strange bleakness of the landscape, where the river twisted through the loneliness, with its white canyon cliffs speared through with dark patches of volcanic rocks that hint at the mountains that we now know are ahead. That the Breaks, a region so far away from the main western trails, a territory that is sparsely settled and has been largely unpopulated and unfarmed for most of its history, would later become infamous as a hiding place for outlaws and brigands only adds to an allure that Lewis can never have imagined.

Both Lewis and Clark believed that an immense range of mountains lay some way before them, and Clark first spied them from the Missouri Breaks. The distance had unrolled furiously in the weeks before: they had now done 2,387 miles since leaving Saint Louis. It was Sunday, May 26:

> . . . *assended a part of the plain elevated much higher . . . from this point I beheld the Rocky Mountains for the first time with Certainty.*[*]

---

[*] Academic parsing of the Lewis and Clark papers is still a major industry, with assertions like this—Clark's "certainty" at having seen the Rockies here—being endlessly analyzed. One writer insists that the mountains Clark saw that day were in fact the Bear Paw range, outliers of the Rockies to the north of the Missouri (a direction that hardly squares with the southwest bearing that Clark mentions here). Most analysts, however, seem content to accept his vision, and to share in his excitement at seeing for the first time mountains of which only the Indians knew.

I could only discover a fiew of the most elevated
points above the horizon. The most remarkable of
which by my pocket Compas I found bore S.60W.
Those points of the rocky Mountain were
Covered with Snow and the Sun Shown on it in
Such a manner as to give me a most plain and
Satisfactory view. Whilst I viewed those
Mountains I felt a Secret pleasure in find myself
So near the head of the heretofore Conceived
boundless Missouri.

Before they could reach the mountains, there were still many more days of difficult slogging—none more so than when they came to the Great Falls of the Missouri and had to portage around the rapids for the better part of a month. But by now the landscape was dramatically different from anything they had seen since coming through the Appalachians: up on the foothills there were trees again, and before long the foothills gave way to mountains, grander mountains than they had ever seen before. They would soon slice deep into them, and pass the Continental Divide, and begin their steady drop down to the ocean they knew on the far side.

Passing the Gateway

There is something indescribably magical about Montana, and every experience I have had in the state, in more than thirty years of visiting, has been a good one. Most of the events on which I now look back so fondly took place along the same long, scimitar-curved route Lewis and Clark took when they first came through and spent their twenty weeks there. Here they are, recalled not in chronological order but according to their location along the line that the Corpsmen followed as they paddled upstream and then finally as they lifted their boats out of the ever smaller and shallower rivers and walked across the great Divide.

The men—led by Lewis only; Clark and a small group had gone a different way—entered the Rockies by way of a narrow rock-walled defile that Lewis named the Gates of the Mountains. It is still called that, and for good reason. A curious optical illusion confronts anyone who boats upstream toward the towering line of cliffs, more than a thousand feet high, that marks the leading edge of the range. The river initially seems to vanish into the rock itself until, just a few score yards before you are dashed against the cliff face, an opening appears, an opening that, as your boat moves left and right with the current, seems to open and close, as if with sliding doors. Or gates. The river is little more than a hundred yards across—the entrance to the Rockies, the river's exit, is spectacularly slender, half hidden, secret.

I had never seen the Gates; and on the day I arrived by car from the state capital, Helena, ten miles away, it looked unlikely that I would see them. I was upstream, above the defile; it was a Sunday, in early spring, and there were no boats to hire. It was more or less impossible to walk along the cliff edge, and much of the land was in any case private. I was glum indeed—until a fisherman pulled his car down to the water's edge and began to ease his aluminum boat off the trailer and into the lake. He introduced himself—Jeff Key—and his ten-year-old son, Jason. They were spending the day trolling for trout. When I explained who I was and what I was doing, there was no hesitation. Hop in, Jeff said; Jason won't mind a few hours' delay. There'll be plenty of fish.

And so downstream we went, driving gently down the twists and turns of the canyon, the water slapping happily against the hull, the sun glinting on the water. We had to crane our necks and squint into the sky to see the tops of the peaks, each crowned with lodgepole pine, balsam, and aspen.

At one stage, on the west side of the river, there were some Indian pictographs, tricked out in black and ocher, and then a scar of landscape where greenery seemed a little newer, the trees a little shorter. This was the valley, the Mann Gulch, where there had been a terrible uncontrollable forest fire in 1949—the most fearsome kind, known as a crown fire—in which thirteen men had died. Norman Maclean had written a classic book: *Young Men and Fire*, which told the saga of the smoke jumpers who had been dropped in and who, when the fire suddenly boiled and turned, had been burned alive or had suffocated that day. It told of how an escape fire had been burned that might have offered them a way out, but the men's radio, which might have told them about it, had been smashed when its parachute failed to open. The Mann Gulch Fire is a legendary episode, a lesson in how not to fight infernos, which forest firefighters use in classrooms still.

And though the event occurred more than sixty years ago, it still resonates. A short while after my visit, a relative of one of the dead men—who was Jewish—came to these hills with a Star of David to replace the cross that had memorialized him. The other tiny monuments can be seen from the river, a scattering of white against the fresh green of the undergrowth, dotted up the impossibly steep hillside. Fire can rage uphill with astonishing speed. A man can hardly run up such a slope at all. Such was the core of the tragedy, all those years ago.

But finally we came out of this gloomy canyon with its macabre memories and out into a burst of sunlight: we were back in the flatlands all of a sudden, the river now coursing through the Big Sky country that gives Montana its current nickname. Jeff turned his boat around—and as he did so, we were able to see just what Meriwether Lewis had seen more than two centuries before: the immense black gates of the Rocky Mountains, opening and closing slowly before us as the boat pirouetted in the water. It was mesmerizing; small wonder Lewis was so enthralled. Since the beginning of the adventure, his world had been dominated by the horizontal. Now it had been upended, and the dominance belonged entirely to the vertical.

We stayed for half an hour, admiring, remembering; and then my companions remembered that they were bent on fishing. So Jeff then gunned his motor and sped back upriver, finally depositing us on the dock where we had started, by the very place where Lewis camped on that celebrated night of July 19, 1805. The expedition leader had been overjoyed at getting to the mountains, but when he heard a single shot ring out, he suddenly imagined a Blackfoot war party bearing down on them. It turned out to be a signal from Clark, telling his colleagues that he was over the mountains, too.

I tried to thank Jeff and to apologize for taking his time and spoiling his son's holiday fishing. But he said it was nothing, that it had been his pleasure. At least let me pay for your gasoline, I said.

"No," he replied, quite firmly. "Remember: *this is Montana!*"

Later that day, when I was in Helena, I decided to buy a copy of one of my more recent books and send it to Jeff and Jason as a gift for their kindness. Jeff had given me his address. There was just one bookstore open on this April Sunday. By good fortune it had what I wanted. I signed the back suitably, and asked the elderly lady behind the counter if she would gift-wrap it and mail it. I then paid, said my farewells—only just as I was leaving, I realized to my shame that I hadn't paid for postage. I turned back to the desk.

The lady looked at his address and smiled. "I'll drop it by his house on my way home tonight," she said. "It'll be no problem."

I thanked her, effusively. She shushed me.

"I said it's no problem," she repeated. "You have to remember: *this is Montana!*"

The party had to deal with a river that was now fast diminishing, in width, in depth, in strength. It was a river that would soon cease to be and instead would split into what would be recognized as its three main tributaries. Sacagawea had already told the leaders what they could expect, and she had already recognized the Gates. She also knew that the three forks, as she called the place, were only a few miles distant.

And it was just eight days after entering the Gates—on July 27—that they reached this point of the great divergence—a watery plain, with groves of willow, box elder, and cottonwood, towering mountains on all sides in the distance. They had come 2,833 miles upriver: the Missouri had turned out to be a mighty long stream indeed. But now, close to its source, it was quite something else; and Lewis and Clark gave its three feeder rivers the names they retain today: the Gallatin, the Madison, and the Jefferson—named for the secretaries of war and state and for the president. There is nowadays a town of 1,700 or so at the junction: Three Forks, Montana.

The expedition party went through some small agonies of indecision at Three Forks. The choice was which of the tributary streams to follow. All looked of similar size and flow and navigational complication; all seemed to head down from the highest of the snow-topped ranges. In the end, they agreed to follow the Jefferson. It was the right choice, because within days they were high up in the clear cool air of the hills, paddling as best they could through streams only inches deep, getting themselves lost, having their notes to one another eaten by beavers, losing one of the men (a soldier named Shannon, who seemed to have a penchant for losing himself, as he had earlier gone missing for two weeks, and had lived for nine of those days entirely on wild grapes), sinking their canoes, and having to deal with men who were becoming ever more exhausted by the fetching and carrying they were having to do.

And then Sacagawea spotted a prominent rock, which she said her Shoshone people had named the Beaver's Head, since from some angles it resembles such a thing. But she had actually made a mistake, so excited was she to give the good news. The real Beaver's Head rock was another day's passage upstream. Nonetheless, she was right to exclaim that they were now deep in her own remembered tribal territory. And so they were now also very close to the Continental Divide, the ridgeline that separates the streams that flow down eventually into the Gulf of Mexico or the Atlantic Ocean—like the Jefferson, its tributaries, and all the rivers below—from those that flow down eventually into the Pacific.

They were close, in other words, to the expedition's topographic and spiritual tipping point.

Lewis, who had gone on ahead, was the first to cross. Coming up from the streamside, he had seen an Indian on horseback standing in the trees and tried to make small talk. But the man had looked down in silence at all of Lewis's attempts at friendship—unrolling a blanket, scattering gifts on it, offering his rifle, spreading his hands and showing he meant no harm—and then turned away, took off at a smart canter, and vanished into the brush. In doing so the man inadvertently led Lewis toward a hitherto unseen trail—a well-used Indian path that headed up to a mountain pass; and on Monday, August 12, Lewis and his small party of scouts plodded up it:

> . . . at the distance of 4 miles further the road took us to the most distant fountain of the waters of the mighty Missouri in surch of which we have spent so many toilsome days and wristless nights . . . the mountains are high on either hand [and] leave this gap at the head of this rivulet through which the road passes . . . here I halted a few minutes and rested myself . . . we proceeded on to the top of the dividing range from which I discovered immence ranges of high mountains still to the

West of us with their tops partially covered with snow. I now descended the mountain about ¾ of a mile which I found much steeper than on the opposite side to a handsome bold running Creek of cold Clear water. Here I tasted the water of the great Columbia river.

He was exactly right. Geographers today judge that first stream to be Horseshoe Bend Creek, which flows into the Lemhi River, thence to the Salmon and the Snake Rivers, and finally into the waters of the ever-westward-rolling great Columbia.

The pass he had crossed, which the rest of the party would traverse two weeks later, is now called the Lemhi, named for a figure in the Book of Mormon. It has never achieved commercial prominence: there was a stagecoach route for a few years, but when the railroad was built, it crossed the divide some miles south, over the Bannock Pass, and the main highway crossing was at Chief Joseph Pass, a few miles to the north. There is a rough grass track today, strewn with bluebells, lupines, and wild strawberries—looking not too different from the time when Lewis, Clark, and in later years the Blackfeet Indians crossed—a lonely mountain memorial to the Corps of Discovery Expedition.

The Lemhi Pass was not altogether wanting in importance, though, for it and the entire Continental Divide marked what would be for the next forty years the western boundary of the United States of America. Lewis and Clark strode across that grassy hilltop, out of America and into what was still then a foreign entity— and not even an organized country.

Getting this corner of the continent organized and alloyed into the Union proved a mammoth task. In six years' time, this immense tract of land, extending from the Divide to the Pacific, would have its first formal name—the Columbia District. As such, it would be a formally organized fur-trading region of the North West Company, one of the two major beaver-fur suppliers in Canada. When ten years later the North West merged with its rival, the Hudson's Bay Company, the region became known as the Columbia Department— although the Americans, who claimed free and open access to the territory along with Britain, preferred to call it Oregon Country. In 1818 the northern boundary of the country was set—by agreement, it should pass along the forty-ninth parallel, which went back east to the Great Lakes, en route traversing what the treaty documents called the Stony Mountains. In 1846, with yet another treaty, the federal government finally wrested total control of the lands away from London and named its new possession Oregon Territory. Last of all, in 1859, the most southwesterly quadrant was awarded formal statehood and named what it remains today: Oregon.

But of course none of that was in place when Lewis first breasted the ridge. What lay before Lewis that August was territory that had in fact been explored—though only to a very minimal extent and almost entirely from the distant Pacific—by sailor-explorers, a few of whom had been bold and curious enough to take their boats upriver along the Columbia. No one nation had initially claimed the land for any particular purpose—not Britain, Spain, France, or the United States, although a company operating from Montréal, the North West Company, was generally considered to have supervision. So what Lewis saw—the far snow-dusted mountain peaks and the rivers he thought would lead to the Columbia, were still Indian territories, still a confection of places to be brought into the federal fold.

In the mid-1990s, by which time Montana and Idaho had each been states for almost exactly a century, I spent $40,000 for a small tract of land in this corner of Montana, just to the north of the Chief Joseph Pass. Its fate tells much about one corner of the economy of the modern American West.

The land consisted of eighty acres, in the valley of the Bitterroot River between the towns of Hamilton and Darby. Lewis and Clark passed directly along this same valley, heading north through easy, beautiful country, before turning to the west and crossing over the snow-filled Lolo Pass into the headwaters of the Columbia. They remarked only casually on its beauty; I was captivated by it—by the views of the great jagged mountains to the west, by the chuckling of the waters of the impossibly clear and cold trout streams, by the green of the lush grasslands, by the smell of balsam firs, by the fugitive presence of bears and mountain lions and great stags, and by woods filled with birdsong. The small towns, too, had an easy, late-Victorian charm to them, and people still left their doors open and the keys in their cars and their children quite free to roam as wild as they wished. The big city of Missoula to the north was a friendly place, with a good university, fine bookshops—all that was needed for civilized life. My plan was to build a small log cabin on the land, to write there, and to live out an imagined Western dream.

Two things rapidly became apparent. The first was that others of far greater resources had much the same idea. Hollywood was starting to embrace the Bitterroot Valley. My immediate neighbors all turned out to be famous: a rock star named Huey Lewis and two actors,

Christopher Lloyd and Andie MacDowell. Then there was talk that bankers and great figures of the financial world—Charles Schwab most notably—were considering buying ranches nearby. This led to the second realization: that the style of life I envisioned was something I could ill afford. Besides, I lived at the time in Hong Kong, eight thousand miles away across the Pacific Ocean. Montana might be heart-stoppingly beautiful, but it was beauty I was going to have to live without.

So a year after I bought the land, I did the dull and responsible thing: I sold it—this time for $80,000. My melancholy was somewhat assuaged by having made a tidy profit. Yet the decision saddened me. It rankled. Montana had long been central to my dream, and it was trying to have to accept that it was not to be.

It must have been twenty years later that I returned. I stayed with the realtor who had sold me the land and who had then sold it for me. She had prospered, mightily. She and her husband lived in a stupendous mansion, had property on the Pacific coast of Canada, and lived, by their own admission, tremendously well. The land boom I first noticed had been sustained, had become overwhelming. Huge houses were being built high up in the hills, expensive restaurants were everywhere, the local airstrip was busy with private planes, and the road through the valley—the very track Lewis and

Clark had taken two centuries before—was jammed with shiny four-wheel-drive trucks. People were complaining about the difficulty of getting help, because for working people there was suddenly nowhere affordable to live locally, and in cafés I heard wealthy newcomers expressing amazement that their gardeners or pool boys had to drive sixty miles each way to get to work.

For four days the realtor and I explored the valley, fascinated, mouths agape at the way everything had changed, so dramatically and so very quickly. But the rich outsiders who had bought into the Bitterroot Valley were never there, someone remarked: they spent just a few days, then went off to some other equally opulent corner of the world—and, my friend remarked, by doing so they created a kind of absence, a kind of poverty. The sense of community that had made the valley towns so special had evaporated. The beauty and solitude of the place, the kind of world that Norman Maclean and Wallace Stegner had so loved, was fast vanishing. It had much more to do with money.

And with that, my friend drove me down to my land. She had been waiting to tell me about it. She hadn't wanted to depress me further, she said. I wasn't quite certain what she meant.

So we drove down the old road, bumped across the stream, and came to a small paved highway that hadn't been there before. We breasted the ridge where there was a fence and a "Private, No Trespassing" sign. We parked the car. The air was heavy with the smell of pine needles and horse dust. Everything began to look and feel familiar, and then, as we rounded a bend in the track, there at last was my land—a long sloping parcel of yellowing grass and rock, and on it, a house of such appalling vulgarity as quite beggared belief. Eaves and arches, wings and columns and a mighty porte cochere, all done in white and ocher stucco, with a long black Escalade parked outside.

The house must have been unimaginably expensive. But what of the land, the eighty acres I had briefly cherished? My friend coughed discreetly and looked at her feet before replying. It had last been put on the market, she said quietly, for *one and a quarter million dollars.*

There is indeed something—for some—quite magical about Montana.

Shoreline Passage

From here it was downhill all the way for the explorers. Sacagawea was in her element here: this was Shoshone country, and she knew the language, remembered friends, and could and did persuade the local people to supply packhorses for the difficult trek down-hill. As soon as the expedition members discovered among the forests and the crags the most navigable of the swarms of westbound streams—the frighteningly all-whitewater Lochsa, and then what they called the Kooskooskee, but which is now the Clearwater—they began their descent in earnest.

They built themselves another clutch of canoes by hollowing out ponderosa pine trunks with hot embers, then set off, screaming down mountain rivers that had a combined length of no more than 120 miles (in a straight line less than 80) until the hills flattened out and the rapids became ever more sluggish and steady waters.

When they had left the Bitterroot Mountain head-waters of the Lochsa River, they were at 7,000 feet. When they reached "the leavel pine Countrey" at the end of the Clearwater River, which coincides with the western edge of the Rockies, they were only 740 feet above sea level. The party had thus descended almost a hundred feet with every westward mile of travel, reaching with stark suddenness the bone-dry grasslands of what is now eastern Washington State. The Snake River joins and takes over the Clearwater here, with a river-bisected pair of towns once colorfully known as

Ragtown and Jawbone Flats but now called the more respectful and anodyne Lewiston and Clarkston.

Down on these waterless and treeless flats, the men's moods seemed changed. They were now more like stable-scenting horses, creatures who were beginning the run for home and could scarcely be persuaded otherwise. They began to chew up remarkable daily distances—the now placid nature of the river helping, of course—and the team plowed across the prairies like men possessed.

The sea now tugging them west was still some hundreds of miles distant, but there was growing evidence that it did indeed lie not too far away now, just below the western horizon. One of the local Indian parties showed the explorers a sailor's jacket, another a red-and-blue blanket made of cloth—both from one of the maritime expeditions that had already explored and charted the West Coast. They then saw sea otters in the river one day; and then, crucially for history, they glimpsed far away the snow-capped summit of one of the volcanoes of the coastal ranges known as the Cascades.

This moment—it was Saturday, October 19, and by now they had joined the great flow of the Columbia River—is of great importance because in his diary William Clark gives this mountain peak a name:

I discovered a high mountain of emence hight
cover with Snow, this must be one of the
mountains laid down by Vancouver, as seen from
the mouth of the Columbia River, from the Course
which it bears which is West I take it to be
Mt. St. Helens.

Not unsurprisingly, there is dispute. Some historical geographers insist that Clark could not possibly have seen this particular peak from his reported location—and that the mountain he saw was actually the then unnamed Mount Adams. The distinction is important. For if the mountain he saw was Saint Helens, then he was noting without fanfare something that was transcendentally *intercontinental.* For the Royal Navy explorer Captain George Vancouver had already seen this mountain, by chance on exactly this October day thirteen years before, in 1792. He had been the one to name it. He had done so in homage to his great friend Alleyne Fitzherbert, who had just been made British ambassador to Spain and had been created Lord St. Helens to add dignity to his posting.

Vancouver had seen the mountain from its western side. Now William Clark reported seeing its eastern side, and in doing so he was also seeing the first far-side-of-the-continent entity *that had already been seen and named by another non-native discoverer.* The circle of unveiling had thus now been closed. A great blow had been struck for the geographic and topographic unification of America, for the making of trans-America, and for uniting what would in due course become, even out here, the United States of America.

Mount Saint Helens is a volcano known these days for its devastating and lethal eruptions (the latest in May 1980). Perhaps now it could be more suitably memorialized as the capstone for this first-ever attempt to unite the American states. It could be seen, if a little fancifully, as the capstone of the idea itself, or more prosaically as the fastening that finally closed and secured the fabric of human knowledge and imperial adventure that now covered the whole breadth of America. If, that is, it was the mountain that William Clark professed to have spotted from his vantage point on the high Columbia.

But no memorial to the moment stands on the banks of the river. Nothing stands to say, Here was America first United. Instead there are two less agreeable monuments, if you will, to modern American life.

One, at a place named Umatilla, is a secret and highly secure army base that was built specifically to destroy the nation's stocks of nerve gas. The troops deployed here started work in the 1990s, and so numerous were the warheads filled with sarin and VX and mustard gas that they are still hard at it twenty years later.

The other monument, if such it deserves to be called, is an enormous silvery-looking factory—just as secret and secure in its own way as the Umatilla Army Depot—owned by the giant agribusiness corporation Con Agra. It is called Lamb Weston, and though it looks more like a steam-belching power station, it does make food, all of it out of potatoes. Its owners wouldn't allow me access but instead referred me to a press release, which said in part:

> *Potato products are the most profitable food item on foodservice menus today. And no other product is so universally loved, so broadly versatile and available in so many styles, cuts and flavor profiles.*

Local employees said that their plant makes french fries, one of the most popular of the humble potato's "styles, cuts and flavor profiles," for McDonald's.

From here matters for the Lewis and Clark expedition changed fast, climatically and topographically. The dry plains gave way with startling suddenness to forest—rain forest, in fact, with low clouds and dripping moss. The river picked up speed as it squeezed through the Cascade ranges. There were rapids and small waterfalls—nowadays all smoothed and calmed by a succession of great dams, the Bonneville most notable among them. And then, once past the rapids, it seemed that in the ever-increasing risings and fallings that the team noticed each day, it might well now be affected by tides, from the sea. Sea frets—thick wet fogs smelling of fish and seaweed—began to trouble the scouts in the party, canoeists who were now having to pick their course carefully as they passed along on an ever-widening, shoal-rich estuary.

On November 6 it seemed that they might have attained their goal. "Ocian in view! O! the joy!" Clark's line is often quoted. But he was wrong. Though they had done "4,212 miles from the Mouth of the Missouri R," they were still in the Columbia estuary. It seemed so unfair: ocean waves were breaking into the bay, setting their craft rocking with an intensity as if they had been offshore. But it would be two more weeks of foul weather and disappointment before, at last, Lewis was able to disembark at a spot in full and undeniable view of the true Pacific and carve into a tree, just as Alexander Mackenzie had daubed onto that stone off Bella Coola twelve years before, a simple inscription: "By Land from the U. States in 1804 & 1805."

They built a camp on the left bank of the estuary and called it Fort Clatsop out of respect for the friendly local tribe. They spent the winter there, hoping in vain that a ship might come and take them back home by way of Panama, thus sparing them another long trek across the country. In the end, of course, they opted to walk home and reached Saint Louis in late September 1806. They had not found a water route across the country; they had not found the Northwest Passage; but they had forged some kind of relationship with almost twenty distinct Native American tribes, though to what ultimate benefit remained uncertain. They had unified the nation in a purely geographic sense; they had achieved in the very crudest sense what Thomas Jefferson had expected of them. And they had gained a formidable amount of information, thousands of pages of fascination and wonder for all America to pore over for decades to come.

And Fort Clatsop would go on to become Astoria, after John Jacob Astor, a butcher's son and flute maker of Walldorf, near Heidelberg, established just to its north the headquarters of the great fur-trading empire that made his one of the wealthiest families in America. The names Astor and occasionally Walldorf are now memorialized almost everywhere—in New York at the Public Library, the Waldorf-Astoria Hotel,

Astor Place, and Astoria in Queens; in a novel, *Astoria*, by Washington Irving; in four American towns called Astor and three others called Astoria; in Britain in both Houses of Parliament; at Cliveden and Mackinac Island; in Waldorf salad; and in scandals aplenty—the catalog of achievement and memorial and fortune is as endless as the family's present fecundity and its former (since the family's star is now slightly dimmed) celebrity. There is also, on a hill outside the Oregon terminus town, a marble column of great height built by the Astor family in the 1920s, with an inner staircase that allows visitors to clamber up and see unimpeded the view that Lewis and Clark would have seen in that early winter of 1805.

Beyond where Fort Clatsop stood and where the city of Astoria now straggles, there was only ocean—the wide, gray, slow-moving, and entirely open Pacific Ocean—to be seen ahead. There was no farther point of land to the west. With their arrival at the mouth of the river and their crossing of the bar, America had been crossed and the continent physically unified by the travels and the travails of a party of newly made American men. President Thomas Jefferson's intention had perhaps not been fully realized—his men had not opened a water route across the country, for the Rocky Mountains had proved to be an impenetrable barrier—but they now had accomplished something of unimaginable courage and determination. They could now declare that they knew—and America knew as well—just where America was.

The basic shape and size and topography of the continent now being satisfactorily established, all that was needed next, at least in the short term, was to find out just *what* America was. How had America's land been made, what was it made of, and how could it best be settled and turned to American use and enjoyment? Or because America would in time become a nation built by peoples from all over the rest of the world, how could the land be employed for the use and enjoyment of all the rest of the planet?

The explorers had come first, as they always should. The scientists, bent on answering the questions that the explorers had posed, would inevitably come next. And then, guided by what these explorers told of their findings, would come the settlers, who would plant their flags and shovels deep in this hitherto untouched soil, deep in the virgin American earth.

. . . we pass each other alternately until we emerge from the fissure, out on the summit of a rock. And what a world of grandeur is spread before us! Below is the canyon through which the Colorado runs. We can trace its course for miles, and at points catch glimpses of the river. From the northwest comes the Green in a narrow winding gorge. From the northeast comes the Grand, through a canyon that seems bottomless from where we stand. Away to the west are lines of cliffs and ledges of rock—not such ledges as the reader may have seen where the quarryman splits his rocks, but ledges from which the gods might quarry mountains.

—JOHN WESLEY POWELL, ON FIRST SEEING
THE GRAND CANYON, JULY 1869

At Pacific Springs, one of the crossroads of the western trail, a pile of gold-bearing quartz marked the road to California; the other road had a sign bearing the words "To Oregon." Those who could read took the trail to Oregon.

—DOROTHY JOHANSEN, "A WORKING HYPOTHESIS
FOR THE STUDY OF MIGRATIONS," 1967

PART II

When America's Story Went Beneath the Earth

1809–1901

The Lasting Benefit of Harmony

The small southwestern Indiana town of New Harmony is not much to see these days—just a clutch of frame houses on the banks of the slow-drifting Wabash River. It is neat and tidy, quiet and peaceful. Nine hundred or so people live there, deep in the lush countryside where Kentucky, Illinois, and Indiana meet, down in the broad alluvial farmlands of the Ohio and the Mississippi Rivers that sidle past, not too far away.

A closer look at the town will hint at links with an interesting past. There is a museum designed by Richard Meier; a roofless church of curious design, which turns out to have been built by Philip Johnson; and a pair of spectacular gates designed by the cubist sculptor Jacques Lipchitz. All were created to memorialize that brief time in the early nineteenth century when New Harmony, as its name might now suggest, was founded to be the spiritual center of a great social experiment.

It was among the earliest of a scattering of similarly optimistic, hope-filled communities that sprang up in the early days of the United States, but New Harmony enjoyed a peculiar connection with that most elemental and unifying aspect of the stripling American nation: its geology.

To understand the geology of a country is to understand and then to realize all of its possibilities—its wealth, its strengths, the nature and kinds and value of its resources. Geology, after all, and without any intended pun, underlies everything. Human settlement on an unknown landscape perforce depends on a deep knowledge of what and where is potentially being settled—on whether the geology of this region suits it to farming, to mining, or to industry heavy or light; whether this range of hills is traversable, this cold prairie is cultivable, this wide river is fordable.

There can be no gainsaying the importance of the first crude discoveries made by the geological pioneers of early America: their findings, surveys, maps, and forecasts were the guides and lures that tempted and then scattered millions of people across the country. Eighteenth-century geology, infant science though it still may have been, offered the keys to unlock the country's promise, bringing men out to inhabit the farther reaches of this country and create their nation.

And the town of New Harmony, Indiana, was where this realization of geology's importance was born.

The town, first simply named Harmonie, was settled initially by early-nineteenth-century Germans, men and women fleeing to America much as the Pilgrim Fathers had fled two centuries before, to escape religious restrictions back home. Their piety and hard work paid off quickly, and they eventually moved on to larger quarters, selling their tiny settlement to another idealist adventurer, the campaigning Welsh socialist Robert Owen. He, flushed with the success of a millworkers' commune that he had organized outside Edinburgh, planned to establish a utopian beachhead in America, based on socialist ideals. He renamed the former German village New Harmony; and once he had settled during the winter of 1825, he invited like-minded idealists to join him.

Such was the educational reputation of Owen's earlier Scottish experiment that New Harmony became an immediate magnet for intellectuals, philosophers, teachers—and, in particular, scientists. Geologists, most notably, pitched in with a special enthusiasm, such that at the peak of New Harmony's fortunes, no fewer than seven geologists of considerable later distinction could be counted among the inhabitants.

This tiny town briefly became "a scientific center of national significance," as the University of Southern Indiana describes it today. New Harmony can fairly be regarded more specifically as the birthplace of American geology—not least because Robert Owen's closest colleague and ideological soul mate, an equally eccentric visionary who came to join him on the banks of the Wabash River, is generally acknowledged today to be American geology's founding father. He, too, was a foreigner, a middle-aged Scotsman of wealth and distinction whose fortune was in no way connected to the science of the earth—William Maclure.

The Science That Changed America

Robert Owen and William Maclure were both strange and remarkable men. Owen was a social reformer of lasting repute—though his fame remains largely in his home country, to which he eventually returned. But when it comes to the story of geology as a unifying force in the making of the United States, the person of William Maclure is the one to be remembered preeminently—even though, ironically, he was not really a geologist at all.

William Maclure, the wealthy Scotsman whose immense fortune allowed him time to indulge an early passion for American geology, and who in 1809 drew the country's—and the world's—first geological map.

Maclure, born in southern Scotland in 1763, was by his early thirties already a very rich man. He had amassed a fortune as a trader, helped by the annuity from his equally successful Ayrshire father, in whose mercantile footsteps he followed. He had come to the American colonies when he was a teenager, had set up his own import-export business when he was only nineteen, and soon afterward, profoundly influenced by the revolutionary events of 1776 and 1789 in America and France, moved to Philadelphia, throwing in his lot with a society that seemed to him to embrace his own beliefs in fairness and equality. He assumed American citizenship in 1796 and promptly established himself as a fully paid-up member of Pennsylvania's fast-growing patrician society.

Except that being merely patrician held little interest for a man of such a restless nature. No more than a year after becoming a citizen, when he might otherwise have started to enjoy the sedate comforts of early middle age, he made two important decisions.

He first decided to devote much of his remaining life to promoting educational reforms among America's working classes. He vowed, as Robert Owen had already vowed, unbeknownst to Maclure, that the farmer, the miller, and the forge master would each have the same access to society's potential as he and his wealthy peers had already been granted. It might take him years, but he would at least now begin to make plans.

At thirty-six, he believed himself young enough and fit enough to take on such a challenge. He had already achieved great eminence among the East Coast thoughtful: he was a leading light in the fiercely intellectual American Philosophical Society, founded by Benjamin Franklin, and would later go on to help found and run the American Academy of Natural Sciences, the oldest such institution in the country today. He believed deeply in the unifying powers of democratized science.

His second idea was more precise, as he explained in a letter to a friend. He "adopted rock-hunting as an amusement." Geology, he declared, was far preferable to the conventional bourgeois pursuits of hunting and fishing, not least because it was "most applicable to useful practical purposes." Moreover, "it has always appeared to me that the science of geology was one of the simplest and easiest to acquire: the number of names to be learned is small, and the present nomenclature, although rather generic than specific, is not difficult."

He first toyed with the science during a brief stay in Europe, delving into the small mysteries of its nomenclature at the very time when the numerous schisms that plagued the calling were at their most dramatic. Perhaps no science has ever been caught up in such turmoil. On the one hand, geologists were busily unleashing themselves from centuries of dogmatic interference from the church. The less pious of their number were no longer content to believe unquestioningly in such literal Bible-based teachings as the creation of the earth on a precise October date in 4004 BC, for instance.

There were also continuing disputes raging within the science—between the plutonists and the neptunists, for example, or between the catastrophists and the calmer-minded supporters of uniformitarianism. And the aristocrats who had claimed the science for themselves—rich men who amassed gaudy collections of minerals and fossils to decorate the drawing rooms of their mansions—were also at the time beginning to yield to a wider public sense of inquiry, with every farmer and walker and settler keeping an eye on the land, curious to know what it was made of and why.

But for some, the academic din in Europe proved perhaps just a little too much. William Maclure soon came back to America, admitting to being overwhelmed by the topographic complexity of the European landscape. He decided he would instead be more suitably self-employed discovering the geology of his newly adopted homeland. He would find out what America was made of, he decided. And he would draw a map of it.

Drawing the Colors of Rocks

This was in 1799. For the next ten years, Maclure tramped and stagecoached relentlessly up and down and across the narrow swath of territory that lay between the Appalachians and the Atlantic—the country's most known and settled region. He wandered on the far side of the Alleghenies, too, through what is now Arkansas and Mississippi, though it is not certain if he managed to get as far north as the sparsely settled territories of Ohio and Indiana. He managed to get himself all the way down to Georgia. We cannot be certain that he got all the way up to Maine, but he did claim to have crossed the hills and valleys of the Appalachians at least twenty-two times—which, given the condition of the roads at the time, was no mean achievement.

But however he did it, wherever he happened to go, however many miles he walked, rode, or went in greater comfort by carriage and diligence and cart, he achieved something truly memorable, with a significance that went beyond what might have seemed its purely American relevance. He announced it in an address before an evening meeting of the American Philosophical Society in 1809 after offering his first thoughts on the geology of the nation. Crucially, he included with his seventeen pages of explanation a hand-colored map—the first geological map in the world, some say, and certainly the first recognizable geological map of the United States.

It is a document of curious beauty, even though its simple innocence rendered it of little real use. It is based on a topographic map engraved by Samuel Lewis, a Philadelphia mapmaker who was at the time perhaps the country's preeminent cartographer.* Maclure used vivid watercolors—yellow, red, blue, pink, and green—to paint onto the base map's eastern half the five main divisions, as early geology was inclined to see them, of America's rocks.

The swaths of color he used—to show different *kinds* of rock, not different ages, as maps do today—all trend across the states from the southwest up to the northeast. They sweep along in approximate parallel to the lines of the Appalachian hills—which on the map sheet are picked out in caterpillarlike lines of ugly

---

* Samuel Lewis was the man chosen in 1806 by William Clark, postexpedition, to produce the first-ever map of America that showed both coasts and a reasonably accurate interior—the first map of the true United States. He had a business connection with the great British cartographer Aaron Arrowsmith, whose great maps of the world, of Canada, and of India remain classics of their kind. Samuel Lewis—no connection to Meriwether Lewis—remained preeminent on the west side of the Atlantic.

fuzziness; that was the device Lewis had employed to depict chains of mountains on all of his maps.

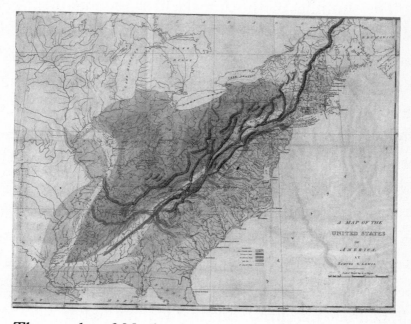

The results of Maclure's estimated fifty survey journeys across the Appalachians led to the publication in 1809 of this crude but memorable first geological map of the then United States. It predated by six years the much more famous British map by William Smith.

Four basic types of rock make up the geology of America, as it was described in 1809. On the western side of the mountains, everything on Maclure's map is colored pale blue, indicating the presence of so-called secondary rocks, fossil-bearing sediments, by and large. On the eastern side, the ocean side, all is by contrast hand-colored yellow, indicating alluvial rocks, gravels and sands and fresh-from-the-ocean clays.

The summits of the Appalachian mountain ranges Maclure showed to be quite geologically different and painted them in vivid streaks of deep blue and deep red denoting what early geological theorists called transition rocks. He also noted the presence of hard granitic outcrops of what he called primitive rocks; these he colored in pink. And for good measure, he also invented a fifth category for deposits of rock salt.

As art, Maclure's maps—he made a revised version in 1818—are undeniably pretty; but as science, they were confusing and, in truth, pretty useless. Had they been Maclure's sole achievement, he might then have slipped into obscurity. But that was not to be.

The Wellspring of Knowledge

Maclure's eagerness to instill in American working-class youth a love for the practical—for the skills of farming; for a knowledge of geography; for the learning of natural history, statistics, biology—remained for years little more than an unrealized dream. But all changed in 1824, when he traveled to Scotland and had his first meeting with Robert Owen. That was when he was first seized with the idea of joining a utopian commune, transforming himself from a mapmaker into a missionary, and becoming America's first geological messiah.

Owen was a Welshman who had made his fortune from the spinning of cotton in Scotland. He had carefully created in New Lanark a showpiece of social engineering for his mill workers—a near-ideal industrial environment, as he saw it, a community that was clean, healthy, well paid, disciplined, and morally sound, its children better educated than those in the finest paid schools in the land. So successful and admired had been New Lanark that Owen decided to expand. In the winter of 1824, he took his millennial dreams and blueprints for popular communal perfection across to America and started the process without delay by buying all of the land and real estate that the departing German settlers had created for themselves in New Harmony.

He reasoned that two thousand or so people could live together around an immense quadrangle he would build in the town. They would govern themselves, farm the land collectively and intelligently, live congenially without money, commune among themselves in the gardens within the buildings, and discipline themselves to hard work and moderate celibacy. His ideals were to all intents and purposes the ideals of the early Soviets, with communities to be run according to the familiar Marxist precept of fifty years later: "From each according to his ability, to each according to his needs."

After settling his purchase of New Harmony, he came back east on a whirlwind recruiting mission. The fame he had won from his Scottish experiment preceded him, and as a successful industrialist, he found immediate and ready acceptance everywhere. At least, he did at first. He was able to meet without difficulty all of the privileged and the progressive figures of the Philadelphia Main Line, as well as chiefs of two Indian tribes. He won an audience with President Monroe, took tea with Thomas Jefferson and James Madison, and gave two public lectures in the Capitol. John Quincy Adams, the president-elect, came to both talks, and was so taken that he had Owen build a scale model of his proposed New Harmony building and display it at the White House.

It was while he was in Pennsylvania that Owen achieved his greatest coup, the one whose effects would linger longest, in managing to persuade William Maclure to come on board.

At the time, Maclure, his mapmaking success well in his past, had won fresh fame as a campaigning education reformer; and as president of the American Academy of Natural Sciences, he was seen not just as one of the preeminent scientists of his time, but as a great educational theorist, too. At their first meeting, Owen lost no time in reminding Maclure of his own, rather similar credentials. He assured him that what Maclure had seen of his success back in Scotland just a matter of months earlier could and should now be re-created in America.

What followed was an epiphany. After an initial bout of dithering—he was shrewdly wary of Owen's eccentricities and shortcomings, even then—William Maclure finally and decisively bought into the revolutionary plans. He agreed. He would uproot himself from the comforts of his Pennsylvania life, move the eight hundred miles across and down to New Harmony, and throw in his lot with Owen's strange new settlement.

Moreover, and more important still, he persuaded a number of his scientific colleagues to come along with him. They were a die-hard group, young men and women, also largely from Pennsylvania, who thought the idea of going off to live in Owen's eccentric new commune was both worthy and noble. Most of those who volunteered were younger than Maclure. All were as eager as he was to teach youngsters the knowledge they had accrued. All were dreamy and impractical idealists.

So he made the journey a suitably impractical adventure. Rather than have the party travel down to Indiana in the comfort of the stagecoach, Maclure had them all go down on a boat. It was a shallow-draft keelboat, with barely room for forty, rowed by six oarsmen. Officially it was named the *Philanthropist,* but Owen proclaimed that "it contained more learning than ever was before contained in a boat," so it was and still is informally known as the *Boatload of Knowledge.*

The vessel took off down the Ohio River from Pittsburgh on a bitter Sunday in late November 1825. After punching its way though the ice for the next seven weeks—its passengers listening to the onboard piano, taking off for skating ventures, reciting poetry and reading, reading—everyone arrived at New Harmony on a bitter cold day in late January. Fifty tons of books and what was termed "philosophic apparatus" joined them a few days later, whereupon the team promptly began— under the supervision of Maclure and Owen (who had come down at ease, on the stagecoach)—a hyperactive program of teaching to all and any of the youngsters from the towns nearby, just what they had to offer.

But there was more hyperactivity than most had bargained for. The furious energy of Owen's New

Harmony experiment barely survived Maclure's arrival. The enthusiasm sputtered out within weeks, and the community soon began to fail, and it did so miserably and quickly. As is so often the way with utopias, factions developed—no fewer than ten had formed within just two years of Owen's arrival, and all began bickering, squabbling, and arguing for various rewritings of the commune rules, each splinter group jostling for ideological supremacy. In the end, a demoralized and disillusioned Owen, shocked at a brand of waywardness he had never experienced back home among the Scots, returned to Britain. His confidence was sorely shaken: his ideas for the universal betterment of the working classes began slowly to evaporate, and he became steadily ever more marginalized and ridiculed a figure.*

---

* Robert Owen's final grand gesture was the creation of an immense and ruinously expensive cooperative community in Hampshire called Queenswood, in which seven hundred people lived, their inner quadrangle illuminated by "koniophostic light," with conveyor belts bringing food from central kitchens to their dining halls. Couples moved in. A first baby born at Queenswood was formally named Primo Communist Flitworth. But the community never really prospered and closed after only a short while. Owen then changed gear once again and provided valuable help in settling the rival US-Canadian claims to Oregon territory, then tried in vain once more to sell his socialist ideas in Paris, where he died in 1858 and was buried in the grounds of a deconsecrated church.

But William Maclure did not immediately leave New Harmony. He remained behind to use the community as a base to preach the benefits of science and science education—and most especially the value of geology, the science that had first anchored him to America.

And in that sole regard, New Harmony was to become in this fresh incarnation something of a success. Maclure saw to it that the leaders of the more quarrelsome factions were persuaded to leave, that houses were now bought and sold and rents were expected and paid, that new shops were opened, and that the vigor of commercial life replaced the rigor of communal life. A printing shop was set up, and produce from the gardens was sent down to be sold in New Orleans.

Most of all, Maclure began to plan and finance his revolutionary education system, preaching and then practicing in town his long-held beliefs in the gift of free education for the American working youth. He gave his superb personal library to the town and opened it for the benefit of all. The young scientists—botanists, physicians, geologists—who had come down with him on the *Boatload of Knowledge* were to be the first teachers in the schools that were opened, and soon students came from towns and villages both nearby and far away. The town began to flourish again, and soon began to win a reputation—which spread nationwide—as a center of educational excellence.

Members of the community began to write books: there would soon be definitive multivolume works on fish, insects, the shells of mollusks, and the trees of North America. There was a resident engraver and color printer in New Harmony, too—and finely wrought monographs soon began to appear for sale at nearby fairs and bookstalls.

But William Maclure was beginning to feel his age. The Indiana winters were settling their cold deep into his bones. He started to take off on southerly explorations, finding himself eventually in Mexico, declaring a liking for it and settling on a new ambition to create progressive schools there. By 1830, when he was sixty-seven, he decided finally to cut loose from the winter cold of Indiana and stay put in the soothing balms of Mexico. He would for a while continue to finance New Harmony, but now only from afar.

The Tapestry of Underneath

The presiding intellectual genius who then ran New Harmony in his place was Robert Owen's youngest son, David Dale Owen. who would become one of America's leading geologists and a key player in the surveying of the nation as it expanded all the way westward across to the Pacific. William Maclure certainly started it all and is revered as the father of American geology in consequence. David Dale Owen, apprenticed in New Harmony, set in train the practical tasks that proved necessary for finding out what America was made of. Maclure had the vision and led the way; Robert Owen's son went the distance and did the work.

When David Dale Owen was born, in 1807, there had been almost no geological maps made of anywhere. That soon began to change very rapidly, in response partly to Maclure's American map of 1809 but more to William Smith's map of England and Wales published in London in 1815, which demonstrated decisively how a proper stratigraphic map should be made. Not for nothing is Smith's cartographic achievement still regarded as "the map that changed the world." His revolutionary idea of illustrating the rocks *according to their relative ages* allowed for extrapolation and prediction: armed with a Smith map one could forecast with some certainty where a plunging coal seam might lead or where iron or copper—or one day, oil—could be found deep below the surface. By the time David Dale Owen assumed control of New Harmony, such mapping was standard practice in Europe, and both the federal government and state governments soon saw a pressing need to bring America similarly into order.

The first regions to be properly and systematically examined were in the Eastern states. The capital of New York State, Albany, was mapped in 1820 by Amos Eaton, a blacksmith's son who two years earlier had published a cross section of America from the Catskill Mountains through Massachusetts to Boston and the Atlantic Ocean—a thing of sinuous curves and colors, showing the rock layers rising and falling in great subterranean swoops of blue and yellow that perhaps owe more to art than to science.

The practical demands of commerce soon introduced more scientific rigor to the mapmakers' efforts: in 1832, Massachusetts became the first state to be systematically surveyed for its invisible underneath. The driving force behind the design was nakedly mercantile, the state's governor demanding that the survey show "valuable ores . . . quarries . . . coal and lime formations . . . for the advancement of domestic prosperity." Such imperatives would soon produce a torrent of new surveys and maps, invaluable guides to an America that was by now quickly evolving into an overwhelmingly industrialized nation.

The country's mills, smelters, and forges were demanding iron and coal and copper—while wealthy city dwellers were demanding other precious metals and stones to be brought out from underground, too. Agriculture was expanding westward with the settlers: fertilizers were needed, and beds of phosphate and marl needed to be identified by a cadre of elite scientists who were now all of a sudden being seen as ever more vital to the national interest. The maps they made—not entirely comprehensible to most, true—were becoming popular items, in vogue at least among those eager to be able to forecast where needed treasures might be found.

David Dale Owen was a key player during this ebullient period in America's expansionist history. His first duties involved helping with the geological survey of the state of Tennessee, which was begun in 1833. He was appointed assistant to a Dutchman, Gerard Troost, whom Tennessee had appointed to be its first-ever state geologist and who, as a passenger on the *Boatload of Knowledge*, had been a keen member of the utopian community. The men knew each other: both were legatees of Maclure's ever-spreading influence, both were graduates of the New Harmony schools.

But there were to be many more. The United States Congress was at the time making certain that all American public land that held proven or suspected reserves of minerals—lead, iron, and coal in particular—be sold in an organized manner, without either favoritism or fraud. Owen, his skills honed in Tennessee, was next appointed an official in the General Land Office, the body that made both the rules and the sale, and in 1840 the agency demanded that he survey eleven thousand square miles of the ore-rich corners of Iowa, Wisconsin, and Illinois.

He achieved this survey with remarkable dispatch. Within two years, he had finished and had turned in to his superiors a report that encompassed "161 printed octavo pages, 25 plates and maps, including a colored geological map and several colored sections." He had had help—no fewer than 139 assistants, every last one of them drawn from the schools in New Harmony, all of the young men trained by him and Maclure. According to an official history, Owen's organization of the survey "was a feat of generalship which has never been equaled in American geological history . . . one more illustration of the energy, persistence, and virility of the Scotch emigrants and their descendants in

America." It was a testament also to the enduring role of New Harmony in the making of early America.

By the time Owen died, in 1860, at least twenty-eight states had organized well-established geological surveys. Scores of maps were being published from all sides. Moreover, geologists who had arrived by sea on the West Coast had looked carefully at California and Oregon and had declared that it was likely that great mineral wealth existed there, and that discoveries of great value, such as the one made at Sutter's Mill in 1848, were likely to be repeated.

All of the land between the coasts was also soon about to yield to squadrons of men who were equipped just as Owen and the Eastern, Midwestern, Californian, and Oregonian explorers had been. American scientists would in short order offer up thousands of detailed and very beautiful cartographic images of how the entire country had been constructed. So the knowing of the country was now well under way, and with this knowledge came the pressing urge to settle those places now being revealed map by map, survey by survey. To settle places deemed suitable for living, for farming, for mining, or for the birth and nurturing of an unbridled frontier optimism—a territory that was fast being fashioned and united into something that for millions of settlers could soon be called a homeland.

Setting the Lures

It is surely a universal truth that men and women who choose voluntarily to pack up everything, acquire a wagon, set off down a rutted track into the sunset, and then endure weeks and months of privation, misery, and real danger in order to create new homes for themselves countless miles elsewhere must have a powerful reason for doing so. Modern America's very existence is based on the awe-inspiring reality that thousands upon thousands made this very choice. And at first blush it appears they had just as many thousands of reasons for setting forth toward the sunset.

Nearly all were going off to the West because they imagined a better and more congenial life there. Perhaps some were afflicted by a goading restlessness, but only a few went out on a whim. Some were drawn by reasons religious, others were compelled by a need to escape—to get away from political persecution, from the hand of the law, the clutch of a pestilence,* the misery of a failed romance, or the stench of an unsavory past. A number in America's Eastern and Southern

---

* A mid-nineteenth-century malaria epidemic in the Mississippi Valley acted as a powerful recruiting sergeant for the settler movement.

states found the whole business of segregation and slavery unpalatable, and imagined that out west they might encounter a more tolerant and liberal atmosphere.

But for most, the West was simply the Promised Land. "Eastward I go only by force," said Thoreau, "but westward I go free." And the pioneers who were bold enough to head in that direction did so, generally speaking, imbued with a spirit of ambition and adventure and an unyieldingly optimistic sense of enterprise.

And yet—what was it, more specifically than all of their stated reasons, that truly provided the lure? What intelligence was it that had produced the necessary temptation—the impetus, the final trigger, that decided a hitherto settled Easterner to obtain a wagon, to pack up all his belongings, to say farewell to scores, and then to head off for thousands of miles into the Western unknown?

The answer almost always had something to do with the land. People went in multitudes because of what they knew, what they had heard told, or what they suspected about the very earth of which the West was made.

They learned that the far reaches of the country held places that sported a variety of temptations. There might be vast acreage of thick black soils. There might be cliffs rich with exotic ores. Some might have found rivers running over gold-specked gravel beds. Wanderers might have returned with news of coal seams, tar pits, strangely glinting mineral crystals, deposits of marble, beds of rich red sandstones, or prairie tablelands and valleys covered with a wealth of grasses and flowers that, once tamed and watered, could be farmed and persuaded to yield measureless wealth. These were lands well suited to those who had the necessary ambition, vision, and endurance, for those possessed of the true grit.

It was from the 1820s onward that those in the East were first being told about the remarkable qualities of the Western lands, and were being told of them in great and fascinating detail. The information was contained in the often breathlessly excited reports of the men who had made the journey already. Some of these were American Fur Company trappers; some others were missionaries—two missionary women, Narcissa Whitman and Eliza Spalding, crossed the Continental Divide in 1836, and the husband of one came back east to tell of their adventures and to plead with others to come follow her example. Most reports, however, were sent back from soldiers or scientists accompanying those soldiers, who had been sent out officially by the United States government, charged with exploring the full extent of the trackless continent. Such men—and there were so many, a roll call risks becoming a blur— would turn out to be the vanguard of all the great migrations that followed soon after.

They were men like Edwin James, in 1820, who conducted a geological survey of the Appalachians, the Ozarks, and the foothills of the Rockies on an expedition run by a Major Long, United States Army. Or like Henry Schoolcraft, who went with one Major Cass to the headwaters of the Mississippi, also in 1820, and there found substantial deposits of copper, lead, and gypsum. In 1823 a geologist named William Keating found copper in West Virginia. In 1824 the heroic explorer, trapper, and mapmaker Jedediah Smith rediscovered the low and easy South Pass* through the Rockies, and Benjamin Bonneville, who took a wagon train through it eight years later, wrote of his discovering the famous salt flats in Utah in 1832. Two years later still, the first-ever official United States geologist, George Featherstonhaugh, drew a remarkably accurate

---

* The South Pass, through which scores of thousands of migrants on the California, Oregon, and Mormon trails passed westward, was actually first discovered by Robert Stuart and his party of Astorians going eastward, heading from John Jacob Astor's Pacific headquarters in 1812. The first town on the western side of the pass was Pacific Springs; here the trails parted ways, a pile of quartz marking the way to California, a written sign showing the direction of Oregon.

cross section of the country from Texas to the Atlantic Ocean, noting the presence of interesting-looking mineral deposits along the way.

In 1841 the eminent mineralogist James Dana—his *Manual of Mineralogy* was still a classic when I studied more than a century later, and its twenty-third edition was published in 2007—explored the Sierra Nevada and wrote extensively and temptingly of the mineral possibilities of the Far West. John Charles Frémont made a remarkable series of explorations of the West. One trip was made with the great frontiersman Kit Carson; on another he discovered Lake Tahoe and mapped Mount Saint Helens, then wrote what turned out to be the definitive map and guide for anyone thinking of traveling overland to Oregon or California; it remained in print for years. Howard Stansbury reconnoitered the near-empty territories between Salt Lake City and the Sierra Nevada in 1849; Marcy and McLellan wrote about finding coal and many other mineral treasures in the valley of the Red River in Louisiana and Arkansas in 1851; and Mr. W. P. Blake described "auriferous gravels" in California in 1853. The following year, seemingly to place a capstone on all these furious endeavors, Josiah Whitney—of Mount Whitney, the highest peak in the contiguous forty-eight states, and of Mount Shasta's Whitney Glacier—wrote *The Metallic Wealth of the United States,* which for many years served as vade mecum for the legions who dreamed of traveling westward, striking it lucky, and making a fortune.

Such temptations! All that scenery, all that gold, free farmland, open space, political freedom, copper, coal, abundance. And all, or almost all, of it was reported by those geologists who had gone out exploring, with hammer and magnifying glass and compass and acid bottle. Their reports, which would prove to be catnip to a restless generation, played also into a swelling current of official opinion, which John Quincy Adams had expressed so succinctly in his famous letter to his father, the former president, in 1811:

> *The whole continent of North America appears to be destined by Divine Providence to be peopled by one nation, speaking one language, professing one general system of religious and political principles, and accustomed to one general tenor of social usages and customs. For the common happiness of them all, for their peace and prosperity, I believe it is indispensable that they should be associated in one federal Union.*

Fourteen years after he wrote this most prescient and persuasive passage, Adams was himself elected president. And fifteen years after that, in 1840, an otherwise unknown Midwesterner named Joel P. Walker, together with his family and three missionary couples, decided that the drumbeat, the pressing need to move west was now, for him at least, too powerful to resist.

The allure of all that land, space, and possibility had been fully spelled out for Mr. and Mrs. Walker. The noble role his family might play in the creation of a national ideal had been made clear to him. The decision, bolstered by a logic that must have seemed all but inescapable, was now up to him.

So Mr. Walker bought tickets on a steamboat to one of the trailheads along the Missouri River. There he found himself a suitable wagon (for Jedediah Smith and others had reported that the South Pass was indeed suitable for the passage of wheeled vehicles); he yoked up a sturdy team of oxen; he piled up such possessions as he felt he needed for his new life; and in the late spring of 1840, he set out for the Green River staging post and rendezvous in what is now Wyoming and headed out to complete a two-thousand-mile journey into the West.

He traveled on the vague and rutted route that was already being called, by the fur trappers, traders, and missionaries who had already used it, the Oregon Trail. Joel Walker—whether he was real or a mere mythical symbol seems to matter little now—was the very first of a quarter of a million men, women, and children who would now follow him out west, as the great period of American migration and nation building got itself ponderously under way.

For the previous twenty years, geology had been paying out the lines, casting out the nets. Now at last it was reeling in the catch.

Off to See the Elephant

Starting in the 1840s, there were three principal westbound trails: one for those bound for California, a second that turned southward toward Santa Fe, and then the Oregon Trail, which was initially the busiest and, thanks to Francis Parkman's celebrated book of the same name, the best-known. The Mormon Trail, which was established for altogether different reasons, geology barely among them, took off six years later.

Those setting out on each of these trails chose as their first jumping-off point the town of Independence, Missouri. The Mormons, however, decided on Omaha, farther north upstream along the Missouri River—and though the set of westbound trails went initially parallel for some hundreds of miles through the prairies, the folk who were looking for somewhere to plant their beleaguered religion (doing so eventually in Salt Lake City) traveled on the northern side of the Platte River valley, while the more mercantile were on the southern side. They would not meet until Wyoming.

This being America, bustling centers of commerce began to pop up at the trailheads. "A multitude of shops had sprung up to furnish the emigrants . . . with necessaries for their journey," wrote Francis Parkman, "and there was incessant hammering and banging from a dozen blacksmiths' sheds, where heavy wagons were being repaired, and the horses and oxen shod. The streets were thronged with men, horses and mules."

Parkman didn't care much for the migrants, or indeed for such Indians as he met along the way. He was an indefatigable snob, a New England swell with money, ambition, courage, and a Harvard education. His academic brilliance (he was a fine horticulturist) won him a professorship; his historical writings won him prizes, standing, a great monument, a school with his name on it, and his face on a postage stamp. But his account of the Oregon Trail is wanting, in all too many aspects.

Even the title is scarcely true: he completed only a third of the journey, and then only the easy bits. *A*

Summertime Trip to Laramie would have been a more suitable title—not only geographically accurate (that is the farthest point he reached, scarcely a third of the way out from Omaha) but also underscoring the fact that his relatively simple eight-week journey, involving more mountain sighting than crossing, was for him a mere wheeze, "a summer jaunt," as he put it. His disdainful take on the expedition left him scarcely able to grasp the true historical importance of what he was seeing. He claimed that the true motives of most of the emigrants were a source of great puzzlement to him; they had in common only that they were "some of the vilest outcasts of the country," many of whom eventually "repent[ed] of the journey," and were "happy enough to escape from it."

The quality of Parkman's prose disguises the dubious quality of the facts; his style, a poor stand-in for substance. In fact we do know rather well why the emigrants went. We do know how many—or more properly, how few—repented of their adventure and went back home: no more than 10 percent, the figure falling steadily as the trail became more familiar. Moreover, we know that almost none of these turn-arounds or go-backs, as they were called, were happy with their choice. More often they simply reset themselves, pulled themselves together, and tried again.

In the year of 1840 that saw the departure of the first true pioneers, Joel Walker and his family—and their successful arrival six months later in Oregon—a total of just thirteen people made the journey. The following year, it was twice that number; three years later, nearly three thousand went. Soon so many tens of thousands of pioneers were going, so long were the trains of wagons, that perplexed Indians in Wyoming said they might themselves head off to the East, believing it to be fast emptying of all white people. The ruts left by the little white-canvas-sided prairie schooners—or more rarely by the three-ton Conestoga wagons, with their iron-rimmed monster wheels, ten-oxen teams, and wickedly large turning circles—were ground so deep into the prairie earth that they can still be seen today.

In places, the Oregon Trail was fully ten miles wide, with wagons veering wildly away from one another as the steersmen took different tacks to divert around the obstacles ahead. In others, it narrowed sharply, the ruts all commingling, incised ever more deeply into the earth. The Bureau of Land Management, which looks after the public lands of the American West, has seen to it that in many places these gatherings of ruts are preserved: they are easily visible at the great historic site of South Pass, in western Wyoming, for instance.

They can be seen just a few yards off State Route 28, which crosses the Continental Divide here. You step away from the pavement and onto the sagebrush, and there is the track, with its two parallel lines of hard-packed yellow dirt, a wagon-width apart, fading into the horizon. It is 7,500 feet up here, and you might take a few moments to catch your breath. But then if you stand here for a few moments, looking west as a chill breeze from ahead whistles through the short grass, making your eyes water, and the new snow on the peaks of the Wind River Range to the north glints in the evening sun, two things are worth pondering.

The first is the simple political importance of the place. Jedediah Smith, the great trailblazer of the Victorian West, was fully aware of its value. He reported that this was the one certain way through which wagons could cross the Rockies, the one certain road along which people in limitless numbers might one day cross the Great Divide. By urging settlers to cross here, creaking their wagons up from the valley head of the Sweetwater River (a stream whose waters were bound eventually for the Gulf of Mexico and the Atlantic by way of the Platte, the Missouri, and the Mississippi) and then down the far side to the head-waters of the Green River (which joins the Colorado and passes through the Grand Canyon to the Gulf of

California and the Pacific Ocean), he left an indelible imprint on the human geography of America.

For until that moment of discovery, places like California and the Oregon Country were months away, reachable for Americans only by boat up the West Coast, and then only by way of foreign countries like Panama or Argentina, Chile, and Peru. But now, thanks to the happy presence of the South Pass—with its slope "no more toilsome than the ascent of the Capitol Hill from the avenue, at Washington," as John Frémont had it—Americans could get to the West Coast territories directly. In time, they could get there quickly.

Continentalism—the notion that had been so eloquently advocated by John Quincy Adams in his 1811 letter home—was then swiftly realized. America's Manifest Destiny became a sure reality; the Pacific coast became America's sole remaining frontier; and in time, and for a while, the Pacific Ocean became an essentially American ocean.

And one can go further. The many transpacific ventures in which America has been subsequently involved—the colonization of the Philippines, the annexation of Guam and the Marianas and Micronesia, the assistance given to China, the war with Japan, the conflict in Vietnam, the bombing of Cambodia— all have their roots in America's own continental ambitions, which the discovery of the South Pass in western Wyoming ultimately made possible. A Bureau of Land Management official who takes visitors to see the rutted roadstead beside Route 28 says that more than a few veterans of the Vietnam War become visibly distressed on understanding this history, on realizing a connection far less tenuous than it first might seem. A great deal of recent world history had its origins in this wide, windy, and featureless pass.

In the first years of the trails' existence, most of those who went out west had guides. They took guides in part because the route was still uncertain, but they also took them—at no small expense—as guardians and morale boosters. For the pessimism of many Easterners was widespread: to try to cross the country by land, it was said by many, was the sheerest folly. There would be mountains suitable only for goats or Sherpas, deserts that only the Bedouin could manage, hostile Indians,* unavoidable and incurable diseases, unbridgeable rivers, sour waters, the risk of starvation, the onslaught of terrible storms, the wrath of various gods. It was a

---

* British papers wrote extensively of the migration. In 1843 the *Edinburgh Review*, which took a keen interest in American expansion, warned that anyone thinking of heading west could expect to meet Indians "of more than Scythian savageness and endurance, who cannot be tracked, overtaken or conciliated."

six-month ticket to hell, they said: far better either to go west by ship or to stay home, safe and dry.

For those who chose to ignore the Cassandras and go, a strong and able mountain man acting as guide could act as guardian angel of sorts and shepherd the weaklings through. In the early days, it might cost a dollar a person for such a man to accompany the trip to Fort Hall, the outpost in eastern Idaho beyond which the going got especially rough. Eighty dollars in all could then buy you a guide onward from Idaho through the mountains and to the coast. A very few mountain men might be persuaded do the entire journey with you, Independence to Chimney Rock to forts Laramie and Bridger, then on through the South Pass to Fort Hall and across the ranges to Fort Walla Walla and the Pacific— and all for $500 in advance asked, $250 actually given.

But that was in the early days: as the trail ruts deepened and gave the road a look almost of permanence, as the bridges became sturdier and more numerous, as the ferries became more reliable (avoiding the need for fordings took a month off the normal six-month journey), and as the en route campsites took on the look of small villages, with outfitters and blacksmiths and wagon shops and hardware stores, so the need for guides started to diminish: a man who could get a dollar per person for a crossing in 1840 could barely expect twenty-five cents for each wagonload five years on.

The right wagon was essential. The Conestogas were generally disliked, being too big and heavy and requiring more oxen or mules to haul them than most could afford. The smaller linseeded-cottonwood-sided schooners, made by wagon-and-wheelbarrow firms like Studebaker and sold in the jumping-off towns, were better suited to the rough roads. To haul them, most migrants used oxen: slow though the animals might be, they wouldn't wander far at night and weren't of much interest to the occasional inquisitive Indians. Mules were a second choice; horses became popular only when there were depots selling oats and bran along the way.

Everyone, the stock driver included, walked alongside the wagon; the vehicles were reserved for cargo. And despite their pygmy size, their design allowed them to carry quite enough for a family of four or five. They could certainly carry enough dry food— flour, coffee, beans, cornmeal, dried bacon, rice—for a typical journey. There were so many buckets, lamps, washtubs, spare bridles and tack, replacement wheels, saucepans, chains, toolboxes, tin plates and cups, and cages of chickens that the wagons rattled and clanged like a mobile scrap heap, each jolt against a rock setting off a terrific cannonade of sound. Those who loaded their wagons with family heirlooms—spinning wheels, finely carved beds and chairs, canteens of chased silver, and elegant lanterns that swung from the ceiling ridgepoles—found they soon had to discard them and dump everything on the trailside.

It became customary to have cattle, sheep, and goats walking alongside the wagon as a mobile food supply, with the family keeping them all in line. Usually someone carried a rifle, not so much for protection from the Indians, who in the early years were generally less troublesome than the doomsayers foretold, as to shoot game. There were plenty of buffalo in the early years, before they were quite wiped out; there were antelope, deer, geese, and elks, and even the occasional bear.

There were fish to catch, and in the Western rivers there were salmon, which the Indians were happy to sell, as they were also happy on the farther plains to sell the potatoes and vegetables that visiting trappers and clergymen had taught them how to grow.

And yet, withal, in a spirit of rough camaraderie, with campfires and sing-alongs and Good Samaritanism more notable than sparring and violence and competition, they made it. Most—more than 90 percent—of them reached their destinations—usually farmlands in the Willamette Valley, close to where Portland is today.

Between 1840 and 1869, between the Walkers' leaving and the rail-joining ceremony at Promontory Summit, as many as four hundred thousand went. There had been a pause for the Civil War, which disrupted this progress, even from so far away. But along all of the trails—those to Oregon itself, to San Francisco, to Salt Lake City, to Santa Fe—men and women and children walked and rode, and they did so with one vision in mind: to employ the saying that became so common at the time, they went off "to see the elephant." They went off to have that transcendental emotional experience that it sometimes seems is available only to those who take part in such a raw and extraordinary and draining experience. They all went west to get somewhere, true; but they went also for the greater reasons, like getting *to see all there is to see*. Most of them survived. Only a few turned back—and even those few who did played their part, too, bringing the mails with them to let those back east know how their more courageous compatriots were doing, allowing their wagons to be scavenged for spare parts, handing over supplies for the onward crowds.

The countryside over which these armies of migrants crossed is some of the most desolate and beautiful in all creation. But even in their crossing it, the travelers did not really come to know it. To almost all of them, the

American West was a vastness that had to be crossed, not considered. The plains, passes, and peaks were all to be endured, not analyzed. The destinations were what counted most.

Yet unknown to them and to everyone, the territory over which they journeyed was filled with an unimagined and unimaginable wealth. It was a wealth waiting to discovered, to be revealed and then related to the nation as a whole. And as the first phase of travel across its immensity wound down, and after the nation had for four wretched years halted all self-inquiry while convulsed in its terrible Civil War, then the United States government made its official decision. The West had now been won. But what *was* the West, exactly? America would now begin the business of finding out.

The West, Revealed

There were still immense blank spaces to fill on every map. During what today's government historians like to call the country's Great Reconnaissance, a mighty effort was put into unraveling the mysteries of the lands that lay between the hundredth meridian and the Pacific Coast. Detachments of elite and scientifically trained soldiers, formally known as topographical engineers, were sent into the most remote nooks and crannies of the nation, prizing out and chasing the details, noting and mapping all the arroyos and creeks and tarns and overlooked hillocks and canyons that made up the landscape, suggesting routes for roads and sites for the piers of bridges, and building the forts from which all these newly found regions might be further surveyed, organized, and policed.

These were men who were dispatched across the country by their superiors based at the West Point Military Academy on the west bank of the Hudson—which then was not a school for teaching warfare, tactics, and leadership, but rather the country's foremost engineering college. Within America's nineteenth-century military community, engineers were marked out as the elite. It was thought far nobler to discover a mountain pass or to climb a peak and map its height than ever it was to fire a gun and storm a rampart.

The orders sent out to the engineers in the field, whether they were sent to the snow-covered northern plains, to the swamps and deserts of the south, or into the rugged ranges of the Rockies and the Sierra Nevada, invariably concluded with a command, majestic in its simplicity, to "permit nothing worthy of notice to escape your attention." This the men did, with dedication and determination, and in doing so produced endless tonnages of reports and maps to show what they had found. And yet, as with those builders who laid the foundations rather than provided the finishing flourishes of a great structure, few of them achieved much lasting fame.

Stephen Long of the Yellowstone expedition is better known than many; Lieutenant Eliakim Scammon, preparer of the first accurate map of the upper Missouri Valley, rather less so. But of them all, John Frémont owns the name perhaps most recognized, not least because he was also one of the first two senators from the newly formed state of California, and then in 1856 ran as the first presidential candidate of the newly formed Republican party.* It was to be his exploration and survey work that remained his greater legacy. Not for nothing was he known as the Great Pathfinder, and his expeditions did much to open up the untracked corners of the country. "The occupation of my prime of life," he once declared, was to be "among Indians and in waste places."

Frémont's explorations and foragings took him and his fellow topographers into regions that were still not

---

* Despite his rousingly corny slogan—"Free land, free silver, free men, Frémont"—he lost the race for the presidency, losing even California.

yet part of the United States. Texas, for instance, had been an independent nation for a decade until the United States annexed it in 1845. Oregon had long been tussled over by both Washington and London until Britain gave up and gave in the year after. And the biggest tract of all—half a million square miles of territory stretching west from present-day Wyoming to the Pacific and south from the Columbia River to Tijuana, and which had been ruled from Mexico City, first by the Viceroyalty of New Spain and then by the newly independent Mexico—was handed to the United States by treaty in 1848.

The engineers of the corps were men who vanished for months and years at a time into the wilderness, content with their anonymity, happy to be forgotten. Their experiences were as varied as the countryside they surveyed. In Texas and the deserts of the Southwest, the topographers used camels, eighty of which were captured by the Confederacy during the Civil War, became officially prisoners of war, and escaped into the wilderness and died.

One expedition into the San Juan Mountains, near the headwaters of the Rio Grande, got itself lost and stranded in deep snow; the troopers watched as their mules dropped dead one by frozen one; then they ran out of food, tried in vain to trap mice, and finally lost eleven colleagues from their party of thirty. When the survivors returned to base, charges of cannibalism went around, and John Frémont said of the expedition leader, a man named Williams, that "in starving times no man who ever knew him walked in front of Bill Williams."

Another party, who took an iron boat named the *Explorer* into the Black Canyon of the lower Colorado River, came across an Indian of what they considered such staggering ugliness that one of their number, a German visitor attached to the party, voted to kill him, pickle him in alcohol as a zoological specimen, and take him back to New York for forensic inspection. The proposal was rejected, however, and the hapless man lived.

Relations with Indians were often very poor and, more often than not, lethal. But if the skirmishes with Native Americans did not prove lethal enough, then drinking a libation known as trader whiskey probably did: a party of engineers looking for dinosaur bones near the Nebraska-Wyoming line reported being plied frequently with this foul-looking brew, made of goodness knows what and by goodness knows whom, which they reported had been spiked with especially rank fillers, including chewing tobacco, red-hot peppers, and, in some instances, the heads of dead rattlesnakes.

The 1,800-mile boundary with Mexico was fixed by teams of these remarkable men—one of whom, William Emory, had previously performed similar duties along the very much longer western part of the 49th parallel that formed the frontier with Canada. He, like so many of his brother "topogs," as they were amiably known, was very much more than a mapmaker: he collected and named scores of plant specimens and kept detailed notes of the more peaceable Indians he met along the way, notes that are valued still as anthropological studies. He had a Texas mountain named after him, and more recently an army fort, but like so many of his colleague officers, he is today otherwise precious little remembered.

There are a scattering of memorials. There is a stone in Utah marking the spot where the army topographer John Gunnison and seven of his party were killed by Ute Indians while surveying a route for a westbound railroad. Emory and five others had mountains named for them, commemorating the days of the Great Reconnaissance, and Frémont, Gunnison, and less-well-known topographical corps figures with names like Nicollet, Warner, Stansbury, and Raynolds are marked today in the names of rivers, towns, counties, and valleys, all west of the Mississippi in the lands they were charged with exploring.

By the beginning of the Civil War, the existence of a separate corps of topographer-engineers essentially ceased, and civilian scientists were taking over as delineators of the nation's landmass. But the culminating triumph of the topographers' brief ascendancy was a map. A great and, for its time, quite memorable map.

It was an elegant triumph of cartography that still reigns supreme in the intellectual memory of America's making. It was drawn in 1857 by a curious and somewhat tragically luckless soldier named Gouverneur Kemble Warren, a man "with the look of an Indian" and clearly "of a nervous temperament," whose shaky military reputation was determined by two very contradictory events that befell him during the Civil War.

After the first, he was declared a hero. His topographically trained eagle eye had spotted a major flaw in the defenses of the Union troops as they lined up at Gettysburg. He managed to rustle up enough soldiery to plug the gap and then went on to save the day, though he was lightly wounded in the fight. In tribute, he was called, for the hill where he fought, the Savior of Little Round Top.

All changed a year later, when he became an instant pariah. A rival saw to it that Warren was summarily removed from his command during the heat of a battle for having been absent from the field. His removal was humiliating, and though the decision was later revoked and Warren was exonerated, it devastated his career and blackened his name.

His work on the map, however, had been performed some six years before the Civil War broke out, long before his military career was so engulfed in turmoil. It was in 1854, back while he was out surveying the Mississippi Delta, that he was summoned home to West Point and ordered to join a new government body to be called the Office of Pacific Railroad Explorations and Surveys. His job was to travel out west and to collect and collate all of the possible information about the American land that stretched to the west of the hundredth meridian.

This map would cover the entirety of the West. It was to be a defining snapshot of Manifest Destiny and a finely turned portrait of the much-vaunted policy of continentalism. It was to be designed to show how the nation could in time be united, coast to coast, as Warren was ordered to make special cartographic note of four main routes that could be taken by a railroad linking the Pacific coast ports with the terminals on the Mississippi River.

Such a construction project, of vast ambition, expense, and duration, was still a dream, but the map would show how a thing so insanely great might one day be accomplished—and thus how all of America's states might one day become fully connected with one another.

Gouverneur Warren worked four years, and the map that eventually resulted, the Warren Map of 1858, is quite without equal. It remained the definitive map of the western United States for years to come, the aggregate of all that was known about the West of America at that moment in history. The map has a beauty and an accuracy that entirely transcend the fuss over its maker's later disputed battlefield behavior.

And yet his map is far from complete. In the so-called Plateau Province, encompassing 130,000 square miles of the most spectacular countryside of Colorado, New Mexico, Utah, and Arizona, centered on the bronze marker that shows the only place in America where four states meet, the Warren map has a curious vagueness about it, a fuzziness of line and contour that suggests inadequate information. Elsewhere it is worse. The word *Unexplored* yawns across eight quite vacant degrees of longitude.

It would take the later efforts of four great civilians, in fact, to parse and determine and discern it all, to get it all right, to render Unexplored into Explored and Fully Known.

So far as the West was concerned, there were still too many myths to clear up, scores of fantasies to deny, clouds of bubbles to burst. Fantasists aplenty were feeding an insatiable public's lust for lurid tales about the nation's vast unknown. William Gilpin was one such: a papermaker's son from Delaware who resigned a commission in the army to join settlers on the Oregon Trail, he became notorious for his Messianic, eye-gleaming boosterism, quite detached from reality.

The need to sketch out the possible routes for a trans-continental railway led the Civil War hero Gouverneur Warren to draw one of the finest and most accurate early maps of the American West.

Two *billion* people could be accommodated with ease in the western territories, he would claim in his many public speeches. Scores of millions of cattle could be farmed in the prairies. "The destiny of the American people is to *subdue* the continent," he declared—and anyone and anything who stood in the way, be they Sioux or be they buffalo, could be swept aside. He was a firm believer in the climatologically nonsensical theory that "rains follow the plow."* And with this as his principal sales pitch, he peddled huge acreages of Western real estate—planning cities like Gilpintown and Centropolis, which in fact never got built—and made millions out of the gullible and the hopeful. He died in Denver, a very rich man.

Then there was a sometime imposter and mountebank named Samuel Adams, who almost managed to wheedle $20,000 from the US Congress as compensation for the hazards of an expedition he supposedly took along the Colorado, from which he sent back reports filled with blatant absurdities. He claimed, for example, that stern-wheeled steamships would be able to navigate up the Colorado for some hundreds of miles from its mouth and that such dangers and rapids as others had seen from a distance "almost disappeared at the approach of the steamer."

It was all the sheerest nonsense, of course, the bragging of a fraud and a charlatan. It displayed, said Wallace Stegner more soberly, "resistance to fact and logic." And yet these same reports told many nineteenth-century

---

* The miseries of the Dust Bowl years proved the fallacy of his notion that crops somehow manage to produce their own rain.

newspaper and magazine readers, all eager for tales of adventure and fortune, exactly what they wanted to hear. They were reports that won "Captain" Adams, as he liked to style himself, all too many allies, men "who were neither so foolish in their folly nor so witless in their rascality as he, but whose justification and platform was the same incorrigible insistence upon a West *that did not exist.*"

The West at the time was consumed by fiction. There were barroom rumors and penny-dreadful tales from heroic travelers—and tales more fanciful still from those who had no heroism about them. There were deathbed revelations related and embellished by survivors of massacres and accidents and unfortunate occurrences. There were fragments of doggerel and acres of bad poetry and mournful songs.

They told of scores of improbable and alluring things. Of a sacred mountain in Colorado that sported a pure white cross at its summit. Of a clutch of Aztec dwellings, complexes of adobe that showed where the last relatives of Montezuma had lived in North America. Of tiny craters that whistled and snorted and threw out blazing liquids, and huge gushes of boiling water that jetted into the sky at intervals.

It was all so wonderful, so improbable, so implausible—and yet not necessarily quite so impossible.

Hence the government decision, taken initially in 1867, to uncover and determine the truth about a West that did exist, as counter to the yarns about a West that might, could, or should be there. It was for that reason that what became known as the Four Great Surveys of the West were formally established and were then seriously publicly financed, properly staffed, and sent out for months and years in the field to find out once and for all what about the Western territories was fact and what was fiction.

Two of these surveys were administered by the Department of War. They had a kind of Masonic symmetry about them, the basis of one of them being a line that went from north to south, the basis of the other being a line going from east to west. Lieutenant George Wheeler, a surveyor and engineer, was assigned to lead what were called the United States Geographical Surveys West of the One Hundredth Meridian. And a Yale-educated civilian geologist, Clarence King, was in charge of an overlapping area defined by an imaginary line that was at an exact right angle to it: the United States Geological Exploration of the Fortieth Parallel.

The other two surveys—their creation a harbinger of the steady switch-over from the old military to the new civilian way of surveying—came under the authority of the Department of the Interior. A civilian geologist and medical doctor, Ferdinand Hayden, was given what sounds to have been an utterly unwieldy assignment: he would undertake what was to be called the United States Geological and Geographical Survey of the Territories. And most famous of all, there was John Wesley Powell, chosen to lead the United States Geographical and Geological Survey of the Rocky Mountain Region.

John Wesley Powell was a geologist and soldier, the first European American to explore the entire length of the Grand Canyon. He did this and all of his subsequent heroics with just one arm, his left, after having been hit with a minié ball at the Battle of Shiloh.

The Singular First Adventure of Kapurats

Powell was a big, blunt, cautious, skeptical, heavily bearded no-nonsense man, the son of an impoverished Englishman and licensed exhorter—a fancy term for a wandering preacher—who had migrated to New York and then to a farm in Illinois, in hope of better things. From the very beginnings the young man was marked out as curious: a dispassionate youth, a loner, a voracious reader, helped by a local farmer who seems to have learned the principles of natural philosophy from one of the New Harmony teachers who had fanned out across the Midwest in the 1830s.

John Wesley Powell, geologist and first explorer of the Grand Canyon, did all his surveying minus most of his right arm, which was shot off by a minié ball at the Battle of Shiloh. He became the second director of the US Geological Survey.

Wes Powell, as he was familiarly known, was also given to long-distance ramblings (four months tramping the wilds of Wisconsin in 1855, for example) and to solo canoe expeditions along the Ohio and the Mississippi. Charles Lyell, the Scotsman who twenty years earlier had written his famous *Principles of Geology*, had boated down these same rivers and had written two books of his American travels, with which Powell was familiar. The young man did much the same as Lyell, exploring, observing, and collecting—on the way amassing impressive numbers of rock specimens and fossils. So assiduous was he that he won an Illinois state prize for his glass cases of mollusks.

Publicly, he was little troubled by having his arm shot off. Wallace Stegner, writing in 1954, captured Powell's insouciance perfectly: "it affected Wes Powell's life about as much as a stone fallen into a swift stream affects the course of the river."

He saw action in nine further set-piece battles, from Charapini Hill to Vicksburg, before retiring from the army with the rank of major, which he would use for the rest of his civilian life. That was in 1865; later that same year, he took up the post of geology professor at a college in Bloomington, Illinois, and by Christmas was already dreaming of taking his students on a collecting expedition into the Rocky Mountains and farther west, mainly to find items to put in the glass cases of the university museum.

Two years later—the pivotal year, 1867—he decided that his planned adventure positively deserved some help from the government that he had served in wartime.

So he went to Washington, fighting his way through the crowds of other seekers of federal favor—all of them behaving like "rutting stags," as he put it— and eventually secured from his old commanders the promise of army rations for his party, even a military escort if he would agree to take his students collecting in the Sioux territories in the Dakota Badlands. But crucially, he ignored this second offer, for his old friend General William Tecumseh Sherman, of the famous march through Georgia, advised him otherwise. He should not go to the Dakotas at all. Instead of stirring up trouble among the Sioux of the northern plains, the understandably war-weary general told him, he should head south and probe the abundant mysteries of the Rockies of Colorado.

Sherman's advice changed Powell's life. It was a choice that set him on the path to win a reputation as perhaps the preeminent soldier-scientist-explorer of his day, for Powell and his men would in time head down into the Colorado River, and they would journey, for the first time any white men had done such a thing, from one end to the formidable other of that most immense natural wonder, the Grand Canyon.

The epic journey was formally named the Rocky Mountain Scientific Exploring Expedition and was technically not one of the Four Great Surveys—though its eventual success was the reason Powell was subsequently selected as leader of one of them. But success was by no means guaranteed, and in the planning of the expedition no sponsor was either especially enthusiastic or particularly generous. Aside from the promised army rations—food for twelve, to be drawn from Western army posts; cash to be given if rations were not available—Powell's only quasi-official support came from the Illinois State History Society and a local university.

The Smithsonian Institution gave him some scientific equipment; the Chicago Academy of Sciences, a little more. Otherwise, he was on his own—and such men as agreed to come with him went essentially as volunteers, for there was almost no money to pay them.

Not surprisingly, given such exigencies, it took him and his team some while—two full seasons, in fact—to put together the expedition and then to accomplish its intended mission. During the first season, Powell and his party, which would be steadily winnowed and trimmed to the six men who made it all the way, made their preparations and consolidated their friendships with the local Ute Indians. It was while the party was surveying and mapping the White and the Green Rivers, major tributaries of the Colorado, that the Ute gave him the name Kapurats. There was nothing poetic about it; it meant quite starkly Arm-off.

After a first winter spent in a frigid campsite and a good deal of throat-clearing—including another trip back to Washington by Powell to beg for a little more financial help and then a diversion to Chicago to have proper boats built—the party set off.

There is a particular historical irony about both the place and the time chosen for their departure. Powell's men slid their four initial boats* into the Green River on Monday, May 24, 1869. The location was the village of Green River, not far from the bridge, which had been built the year before, where the brand-new Union Pacific railroad tracks crossed the river.

The railroad provides the coincidence and the irony. For it had been just two weeks before, on Monday, May 10, and a mere two hundred miles to the west, at a place called Promontory Summit, in Utah, that the two

---

* The *No-Name*, the *Emma Dean*, the *Maid of the Canyon*, and *Kitty Clyde's Sister*. The first two would not survive the journey.

steam engines, the Union Pacific's Locomotive No. 119 and the Central Pacific's *Jupiter*, had met, cowcatcher to cowcatcher, on the lines that were to be joined by the famous golden spike. The transcontinental railroad had been completed, and trains were from this moment on crossing regularly between America's Eastern cities and those in the West. And in doing so, they would rattle across this very bridge, beneath which John Wesley Powell and his men were setting off into the unknown.

Trains with restaurant cars and pianos and sleeping berths and night porters and sundry other elements of comfort and luxury would pass overhead of their launching site, running between New York and California by way of Wyoming at thirty miles an hour, crossing the continent in only seven days. Powell's men pushed their craft into the waters immediately below the bridge and would not emerge from their own unprecedented epic for ninety-eight days of nearly perpetual terror, hunger, cold, and wonder. The bridge and its trundling traffic might provide evidence that the new world was catching up fast, but down below on the waterline it was clearly not catching up everywhere for everyone.

I camped once by the waterline of the Green River, down in the iron-red canyons that Powell's men encountered soon after they left, just as the stream begins to cut its teeth deep into the plateau. I found my site close to where the Yampa River joins the course of the mainstream, a place Powell named Echo Park. It lies in far western Colorado, close to the Utah line. To get there required a thirty-mile drive from the highway, then a further eight miles down a winding dirt track. The site looked like a tiny oasis, and indeed the few acres of sandy soil at the junction had once allowed the building of a tiny hardscrabble farmhouse, the ruins of which still lean and creak amid the sagebrush. The farm had stood in an amphitheater of cliffs seven hundred feet high, all as sheer and smooth as if sliced down with a cheese wire, all seemingly painted with an ocher gloss still called desert varnish. Eagles breezed past and soared on the thermals, their flight feathers seeming almost to graze the sunset-gleaming rock faces.

A shout made here is repeated four or five times (Powell counted twelve), its echoes repeating echoes until fading away and being replaced by the gurgling of the eddies, the sounds of each reflected back and forth from the canyon walls.

Steamboat Rock, a gigantic monolith of red and green and black-streaked Upper Carboniferous sandstone, stood sentinel beside the river, the waters washing around the base in a great lazy circle. I spread my tent on a sand spit across from it, lighting a fire with brushwood, and after pork and beans and beer, I sat gazing across at it until the sun dipped below the cliff edges. The huge rock turned purple and then black, and then became just an immense shiplike shadow that blacked out a full quadrant of the sky. There was no other company that night, and through the star-filled dark, there were no other sounds, except a dawn breeze shaking the cottonwoods and willows, and always the river lapping steadily and powerfully past.

Steamboat Rock, a massif of golden sandstone at the junction of the Yampa and Green Rivers, in what is now the Dinosaur National Park, straddling Utah and Colorado. The placid waters of the Green here are in sharp contrast to the terrific rapids both up- and downstream.

In his later memoir, Powell tells of climbing this rock—a fantastic feat even for a skilled climber with both of his arms. He tells of being stuck on a crag near the summit until his colleague George Bradley, a few feet above, did the decent thing and removed his long underwear, dangling it over the ledge and hauling Powell up with it. The story is true—except for a geographic miscue: it actually took place some hundreds of miles downstream, on a cliff somewhat less intimidatingly terrible than Steamboat Rock.

Upstream of Steamboat Rock, the expedition had already encountered some appallingly dangerous waters. The whirlpools and rapids of the Lodore Canyon—named for Robert Southey's famously dreadful poem about a waterfall in the English Lake District—claimed the *No-Name*, which carried all of their crucially important altitude-determining barometers.* Lodore and the Whirlpool Canyon below it had been "a chapter of disasters and toils," where the men learned very swiftly about being teeth-chatteringly cold and half starved, miserable at the loss of some food and the rotting of the rest. But Powell would not allow discouragement, and he urged the party ever onward,

---

* They were found in the wrecked stern of the boat a day later. All the clothes and food aboard had gone forever.

down from Echo Park toward the junction with the Colorado and into the true vertical magnificence and grandeur of what we now know as the Grand Canyon.

The junction came up unexpectedly, a full 538 miles into their voyage. They had run cleanly enough through some three hundred or so miles of mostly canyon country. The names they gave the defiles—in order: Desolation, Coal, Gray, Labyrinth, and Stillwater— suggested their mood on the day or their discovery of some locally unique feature (the lignite layers in Coal Canyon, for example, still very much on display). There was an Indian crossing point just after Gray Canyon, and a track leading westward; on the riverbank were some old rafts and other evidence—fire pits, debris—of mountain men and mule trains. Perhaps, the men surmised, some of these even dated back to the Spaniards' times. But there was also something chilling about the place: they had heard stories of railroad surveyors who had been killed a few miles west of here a dozen years before. The Powell party had shuddered at the realization, and hurried on downstream.

Shortly after the men emerged from the steady sweeping river race of the Stillwater Canyon, the topography changed swiftly. The high cliffs on both sides of the river eased. On the newly visible far horizon to the east, the men could now see mountains tipped with snow—most probably the Uncompahgres, the "hills that make the rivers red." The river widened, and a calm tide of cold water started suddenly to bleed in from their left.

This was the Grand River. It has now been renamed the Colorado, because hydrologists see it as an extension of the great stream, not a mere tributary (Powell's map had the Colorado River marked as flowing down only from where the Green and Grand combined). Names aside, the incoming stream is today just as Wes Powell had described it in 1869: cool, fresh, and swift-running.

I swam in it once, slipping into its coffee-colored waters from a boat just north of Moab, Utah, close beside what is now the Arches National Monument. The river ran through endless miles of treeless, grass-less iron-red desert: to be immersed in its limpid cool liquid was just the ticket.

The party reached the junction in mid-July. They camped here for a short while, to regain their breath, redraw their maps, sift the weevils from their flour, and contemplate the awful reality that probably no white man had ever been here before. The place was the meeting point for three dark canyons—two of them extending back upstream to the north and the east, and now the third, quite unexplored, heading down-stream to the south. It was here that Powell wrote the passage from which this section's epigraph is drawn. He ended it:

> *Wherever we look there is but a wilderness of rocks; deep gorges, where the rivers are lost below cliffs and towers and pinnacles; and ten thousand strangely carved forms in every direction; and beyond them, the mountains blending into the clouds.*

As refreshed as they ever might be they set off on July 21. Once they were around the bends, the stream became fast and furious. The river began to drop through wild successions of rapids that made the horrors of the Lodore and Desolation canyons seem like mere riffles on a millpond. This was serious, terrifying, deadly dangerous river work.

The sheer walls speared straight down to the water. There was no edge, no gradient, no beach, nowhere that would allow for a landing. The white fury of the stream, its roaring amplified and distorted by the echoes, allowed for no thought, no planning, no hope. For miles at a time, the men could rely on nothing but blind faith that strong wood and gravity would make it all right at the end, as their three remaining boats hurtled uncontrollably on their pell-mell way down the river. They called these first white waters Cataract Canyon, and they hated the place.

Briefly matters then improved, and greatly. The cliffs remained, vast and beautiful and colored like a rainbow, but they were not the hard limestones of before; they were made instead of a lower geological horizon, a soft and easily eroded sandstone. The river's long and lazy curves made wide sandy beaches where the men could rest, repair, and dry their clothes. They called this place the Glen Canyon: it was long—150 miles, at least—and compared with the miseries of the limestone passage, it had been a transit through the purest paradise.

If the Colorado, like so many great rivers, giveth at moments like this, then it most certainly also taketh away at others. A week later and around a bend, the waters began to churn white once again. The deadly hard limestones returned, and there were many more miles of the dire and dangerous ahead. This was the Marble Canyon—picture-postcard beautiful beyond belief, but with a beauty won at great cost. Few have ventured across some of its bigger rapids,* and boaters

---

* A Texan explorer named Clyde Eddy did manage to shoot this rapid in 1927, along with a black bear named Cataract, from the New York zoo; and Rags, an Airedale from the Salt Lake City dog pound.

and rafters still vanish without trace today in the treacherous boils of Marble Canyon, never to be found whole.

Here the men started to grow restless: the trip was taking too long and was bruising, cold, and dangerous. Major Powell, they complained, was interested only in science—in taking observations, in collecting, in writing. He seemed less concerned with getting the team down to calm waters, to safety, and to home.

There were mutinous rumblings, and they were scarcely mitigated when their leader realized— and promptly told his men—that it was going to get worse and take still longer. For with the cliffs fast rising around them and the outflow of a known river named by the old conquistador explorers the Colorado Chiquita bursting in on their left side, it was clear that the main part of the huge cleft, the hitherto untraveled Big Canyon, as it was then vaguely known—lay now directly in front of them. What had gone before were mere hors d'oeuvres. The main dish lay just ahead.

"We are three quarters of a mile in the depths of the earth," Powell would later write in his diary,

> . . . and the great river shrinks into insignificance, as it dashes its angry waves against the walls and cliffs that rise to the world above; they are but

*puny ripples, and we but pigmies, running up and
down the sands, or lost among the boulders.*

*We have an unknown distance yet to run; an
unknown river yet to explore. What falls there are,
we know not; what rocks beset the channel, we
know not; what walls rise over the river, we know
not. Ah well! We may conjecture many things.
The men talk as cheerfully as ever; jests are
bandied about more freely this morning; but to me
all cheer is somber, and the jests are ghastly.*

For now the journey, so deep into the ground, began
to fade into all-day darkness. No longer were the rocks
the soft, multicolored, and fossil-rich sandstones of
Glen Canyon, nor were they even the tough and mirror-
smooth marbles of the wilder rapids above. The river
had now dug itself so deep, had excavated its way into
what we now know is the ever-uplifting country rock,
that it was cutting itself down into the rocks of North
America's ancient basement.

It was now angle-grinding its way through mile-
thick swaths of lavas and tuffs and ignimbrites and
basalts and gabbros and andesites—rocks that hold no
fossils, but only sharp crystals, rocks that look tough,
uncompromising, lifeless, and generally dark-colored,
all of them buried from sight by the shade of the

mile-high cliffs beside them, which absorb or remove much of the light from the dreary scene. The men were in consequence ever more beset by claustrophobia and were perpetually tired, wet, and cold; and as all of their scientific instruments had long since been battered out of commission, and with the sun and the stars seen only fitfully (for the weather was poor, and it rained more than it is supposed to in what is now Arizona), any pretense that this was still a scientific expedition had been more or less abandoned.

There was a scattering of better moments. A creek that, in gratitude for relief, they later called the Bright Angel, produced for them a beach and clear drinking water. The unremitting granite abated, and there was a cave made of marble where they lit a fire and camped. High on the cliff walls, they once found the abandoned dwellings of the local Havasupai Indians, together with an ancient garden with squash still growing, which they took to supplement their own fast-diminishing supplies.

There is a Philip Larkin poem, "First Sight," which I have loved since childhood—and which tells of the astonishment that awaits a lamb, newborn in snow, at the unimagined grass that will soon appear in place of the cold's unremitting whiteness. The line that haunts—and that haunted me when I first went down into the darkness of Powell's canyon, when I was still a youngster—was that referring to *earth's immeasurable surprise*, the unsupposed wonder that awaited the creature in the days ahead.

It must have been the same for Powell and his men.

Here they were, cold and wet and hungry, a mile deep in the black bowels of the earth, the sky above so distant it was speckled with faint stars even at noon. How could they have possibly known that if they had only been able to climb some thousands of feet outward and upward, the blackness would be transmuted magically into gold and orange and purple, their dripping bastions of black lava sculpted into delicate cliffs and pinnacles of sunburned sandstone, endlessly repeating themselves in marvelous chaos, to produce one of the most incredible sights on the surface of the earth?

How could they have guessed—even considering that they had seen something of the nearby scenery, before the canyon swallowed them down—how magnificent the world was for hundreds of square miles above their watery prison? Up on the desert was a true wonder of the world: when Teddy Roosevelt inaugurated it as Grand Canyon National Park in 1908, he declared simply, and brooking no argument, "You cannot improve on it . . . what you can do is keep it for your children, your children's children, and all who come after you, *as the one great sight which every American should see.*"

None of this—none of *earth's immeasurable surprise*—could John Wesley Powell and his men ever fully imagine. And as they passed unknowingly beneath the cliffs of the South Rim, where in just a few years' time thousands would stand and gaze in awe, matters were exceedingly grim, worsening by the day. They soon ran into granite once again, and the rapids there were even more terrifying—at one time there was just no water to pass safely through, no beach to enable a portage. And when Powell climbed up a cliff to see if there was any escape, he managed to get himself trapped on a ridge, unable to move backward or forward, burdened by his disability. His men climbed up, exasperated, and got him down, but at no little cost to morale and comradely feelings. The incident was no doubt a contributory factor in the sudden decision by three of the men, all of them hired as hunters, to abandon the expedition.

This was "decidedly the darkest day of the trip." Soon after the men had shot the rapids below, after a night of discussion and imploring, the three men who decided to leave crossed the river to its northern side aboard the weakened and leaky *Emma Dean*, which all decided should then be abandoned, leaving the remaining party with just two craft.

The three men opted, bravely and generously, not to take any food with them—just two rifles and a shotgun, together with letters to be delivered (including one to Mrs. Powell) and a fob watch to be handed over in Denver to the sister of one of the men remaining. There were handshakes and weeping, and then the three climbed up a gulch and onto the top of the cliff. Their plan was to find one of the Mormon settlements, which they had been told were not more than seventy-five miles distant.

Powell wrote that he could still see the trio, gazing down from their vantage point on the cliff edge, as his two remaining boats with his five remaining men swept farther downriver, down through what they were later to call, understandably, the Separation Rapid. Once through it, Powell ordered his party to stop, to wait and maybe give the trio—whom they could still just see in silhouette on the high cliff horizon—another chance. There is some suggestion that rifles were fired, an invitation for the men to scramble back down. But they did not. And when they finally clambered over the ridge and vanished, they were never to be seen again. They never made the Mormon villages. It is assumed they were slaughtered, possibly a case of mistaken identity for a party of little-liked railroad surveyors, by Indians.

And the savage irony of this particular kismet is that the expedition they had abandoned at Separation Canyon was now just two days—*two days!*—from its successful completion.

There were more hellish rapids—the one at a place called Lava Falls, six miles below Separation Canyon, was reckoned the worst of the trip (but is now silted up at the head of Lake Mead, the immense inundation caused by the building of Boulder Dam). But once free of that, and when clear of another couple of stretches of extreme turbulence below, one of which very nearly saw the loss of one of the two remaining craft, finally on August 30, the two boats and their six passengers bobbed out into low cliffs, far horizons, and calm waters. They were set free from what Wes Powell had repeatedly called their prison. It was over.

And civilization reared its head once more, this time so much more benign and welcoming than that shudder-filled afternoon when they saw the old rafts and curious mule tracks back at the end of Gray Canyon. Down on the calms of this lower river, they met a white man, English speaking, friendly, welcoming. He was a Mormon named Joseph Asey, traveling together with his two sons and an Indian guide. They had been waiting by the river for weeks, Asey said. They had been instructed by Brigham Young, back up in Salt Lake City, to remain by the river as it left the lower canyons, and look there for wreckage floating by, perhaps even bodies.

Instead they found survivors. They found, alive, kicking, weary, and triumphant, six remarkable men: George Bradley; Jack Sumner; Billy Hawkins; the Scots youth Andrew Hall, who at nineteen was the baby of the expedition; and the veteran soldiers Walter Powell and his brother, the mastermind of the venture, the man the Ute Indians called Kapurats, John Wesley Powell, "Wes," the thirty-five-year-old son of an impoverished Methodist minister from Shropshire, now an American hero in the making.

With the meeting and the welcome and the weeks of exultation and congratulation, the private trip down the Big Canyon was now over. All was done, and all had succeeded entirely.

Now Major Powell would head back to the East, to safety, comfort, and congratulation. And there he would be asked, but this time officially, by his United States government, to go back to the canyon. This time, though, he would go as the chief component member of an official geological and geographic survey of the entire Rocky Mountain region, one of the Four Great

Surveys that would give the final detailed description of the still unknown quarters of the country.

And five years after leaving the waters behind, in 1876, John Wesley Powell first wrote out the name Grand Canyon. He appears to have concluded that Big Canyon simply didn't do justice to the almighty crevasse he had seen and journeyed through, so he replaced it with a pair of words soon to be on every American's lips and not much later on the lips of most sentient English-speaking beings on the planet, as one of the great natural wonders of the world.

The Men Who Gave Us Yellowstone

In the summer of 1867, Ferdinand Hayden, widely regarded as the preeminent field geologist of his time, was a restless man with a very full plate. The Civil War had briefly interrupted the fossil-collecting studies by which he had made his name, since he'd felt himself obliged to serve as an army surgeon on the Union side. But now that the fighting was done and Reconstruction was in full swing, he was in demand again by his government—first to survey and explore the prairies and badlands of Nebraska Territory, until it was promoted to statehood and left to its own devices, and then to do the same for the nation's remaining Western territories: Dakota, Montana, Colorado, New Mexico, Arizona, Utah, Idaho, and Washington.

It was a formidable task, one that—to judge from his ventures' vast archival legacy of maps, charts, reports, and books—Ferdinand Vanderveer Hayden mastered with the greatest acumen. Had this been his career's culminating achievement, he might have cemented something approaching a lasting reputation before vanishing slowly into the mists of geologic time as one more hammer-wielding wanderer among many. Indeed, his own road to eventual obscurity might have been taken rather faster. He was known as a difficult man—ruthless, impatient, prickly, and combative—and the fact that he died of syphilis at age fifty-nine, a fate then as now rather uncommon within the geological community, left eyebrows raised and reverence diminished. He was a man who amassed more enemies during his career than made good sense.

And yet he does remain today firmly annealed into the history books—and for a reason that can be summed up with a single word, recognizable around America and the world: Yellowstone.

Ferdinand Hayden was the first man on government business to survey and record the full majesty of what is now the government-run Yellowstone National Park. He came across its astonishing wonders through a cascade of unanticipated circumstances—and having done so, he then swiftly helped bring the place to the nation's attention. By his fierce advocacy in his writings and speeches, he also helped to unite the country around what he and others saw as the vital need to protect and preserve its dazzling treasures, as well as those of so much else of America's natural heritage.

Yellowstone is a hymn to active geology, created on a thunderously unprecedented scale. It is a uniquely concentrated tincture of seismic, thermal, and volcanic activity that—despite all the escaping steam, endless shudderings, and gouts of flame, mud, and smoke that could be seen and heard and felt and smelled for miles—had nevertheless managed to keep itself private, secret, hidden from the world for a very long time.

If geology was what made it so remarkable, geography is what kept it so little known. It is topographically as separate as a Vatican, walled off from the world. The peaks of the Absaroka Range stand to Yellowstone's east, together with the Shoshones and the Snowy

Mountains. The Wind River Range, the Gros Ventres, and the Teton Range lie to the south, and ranges that include the Centennials, the Crazy Mountains, the Big Belt, and the Pioneers all protect its acres far off to the west and the north.

A number of great rivers rise here among the peaks and lakes of what many like to call "the summit of the world." The Green River comes from Yellowstone and flows to the Colorado. The Snake River comes from near Yellowstone and flows into the Columbia. The Gallatin, the Madison, and the Yellowstone Rivers come from Yellowstone, joining to form the Missouri and thence flowing into the Mississippi. There is a plain within Yellowstone called Two Oceans Plateau, from which creeks trickle into streams that eventually pass into both the Atlantic and the Pacific. It must have seemed that the melting snows of Yellowstone's peaks eventually watered most of America.

Yet none of those passing on the Oregon Trail came close enough to see it. Nor did the hardy Mormon pioneers who were heading, not too many miles away, for the valley where they would build Salt Lake City. California-bound migrants also hurried on by, well to the south.

Even those who were heading for the Montana goldfields or had joined the railroad surveying parties turned away at the sight of the ranges of high and impenetrable-looking mountains. All of them seemed to balk at high passes that were still choked with snow even late in June. They swiveled themselves well away from whatever lay beyond. So all maps of the time marked Yellowstone as Unexplored or terra incognita, or else made no mention of it at all. It was probably, in 1869, the final important place in America to be so little known.

However, by 1870, Montana Territory, so long a wilderness, was becoming populated. Gold and silver had been discovered, and towns like Helena and Virginia City had been settled and were on their way to becoming substantial. Men who lived there and who by now had a little leisure, a measure of money, and a lot of curiosity soon began ferreting around the unknown corners of the territory, the better to know their homeland—and with luck to persuade the railroad barons back east to link it with a train service.

It was a group of such men who first offered a hint, a frisson of intelligence of the wonders to come. They were all swells: a local banker, a partner in a firm of Helena freight merchants, the president of a local hide-and-fur company, a lawyer, and a former tax collector. They had an army general as leader and surveyor, together with an escort for him from the Second

Cavalry and two African American cooks, named Nute and Johnny. They set out in the summer of 1870, heading up the Yellowstone River for whatever might lie in its upper reaches, beyond the deep and forbidding passes of its canyons. What they witnessed over the next six weeks quite beggared belief.

One of their party, a man named Nathaniel Pitt Langford from Minnesota, who had hitherto been best-known as a notoriously tough vigilante, brutally protecting the Montana gold fields in which he had invested, then wrote of his experiences. Two essays under his byline appeared the following year, 1871, in two early summer issues of the newborn *Scribner's Monthly* magazine.

They told of amazing things: of a geyser a hundred feet high erupting with such regularity that they named it Old Faithful, of impossibly deep canyons and enormous waterfalls, of boiling hot springs, of wildflower-filled meadowlands as green and fragrant as any in Switzerland, of towering snowcapped volcanoes, of immense pools of burping mud—such wonders in such numbers as had never been witnessed before by man or beast. The essays, moreover, were illustrated with woodcuts that made it all look stupendous, with peaks and waterfalls and declivities all constructed on the grandest scale imaginable.

In Washington, Ferdinand Hayden was just back from surveying Nebraska and was preparing himself to continue his push westward that summer into the Rocky Mountains in Nebraska's neighbor territory of Wyoming. He read Langford's pieces in *Scribner's* and was promptly captivated. The accompanying woodcut illustrations, however, were so wildly romantic and dramatic that Hayden, though intrigued, thought they looked quite absurdly contrived. Still, he could not so easily dismiss them, and after consulting with colleagues, who thought Langford's images equally fanciful, he made a fateful decision: that he had to go out and see for himself the formations they depicted.

So rather than mounting his next expedition of 1871 into southern Wyoming, he decided he would pitch his men up northward instead and use the exploring season to determine the full and unvarnished truth about this little-seen gathering of geological curiosities at Yellowstone. Moreover, and to make certain he would get the best out of what he saw, he would take along the very artist who had drawn the magnificently imagined *Scribner's* pictures.

The artist was Thomas Moran, who, along with Albert Bierstadt and Frederic Church, is now seen as one of the nation's greatest illustrators and colorists, a master of what would come to be called the

Rocky Mountain School of artists. What he drew and painted that year, and what Ferdinand Hayden found on his expedition, would have lasting consequences for America's perception of the glories of her countryside.

The new survey was a monster affair, its organization creating quite a stir within government, and persuading much pitching in. Washington agreed to finance the expedition to the tune of $40,000—fully eight times the allotment it had given to Nebraska—and furthermore Washington gave Hayden permission to draw such supplies as he thought necessary from any one of the convenient army forts nearby. The directors of two of the country's young railroads also agreed to carry the forty members of the team without charge to their base camp in Utah.

The eventual result of all this largesse was plain to see. Photographs from the time show an almost endless line of horses, mules, mule wagons, cooking vans, and ambulances passing eastward along the valley that would lead them into Yellowstone, with muscled cavalrymen protecting them all from any wayward behavior. Moreover, an army expedition of topographical engineers had been sent out separately and joined forces with the Hayden team for much of the trip: at times it must have looked as though Yellowstone was being invaded by a fair-sized punitive force.

The expedition, which would become best-known for the beauty of the imagery it produced, was initially intended to be heavily scientific. There were two geologists—Hayden one of them—two botanists, a zoologist, an entomologist (who doubled as an agricultural statistician), an ornithologist, and two topographers, and the maps they produced and the reports they wrote remain classics still, though gathering dust today in government archives and seldom consulted.

Yet it was the nonscientists who captured the Wow! factor of the place. William Henry Jackson took photographs like never before; Thomas Moran painted and sketched as if he were on fire. The results were stunning, and all America was enthralled.

For what sights they saw! What terrific geological violence was on display! It began almost as soon as they started southward up the Yellowstone River. There were indeed bubbling mud pools steaming up from the ground, just as Langford had written. There were limestone bands in some cliffs that had been squeezed by some cosmological vise into accordionlike folds; elsewhere they had been tipped almost vertical, bands of sandstone and quartzite alternating, and had quite weathered away, giving the appearance of a stairway reaching hundreds of feet into the air.

From one crag the explorers looked down on a gigantic mound, hissing and roiling with endless cascades of boiling water that bubbled from its summit, yet somehow looking frozen, with immense semicircular terraces of bright ceramiclike chemical deposits, seemingly stopped and stilled as they flowed down its flanks. At the top of the hill was a hot-water spring with water that, according to the official report, was "so perfectly transparent that one could look down into the beautiful ultramarine depth to the bottom of the basin . . . ornamented with coral-like forms . . . from pure white to a bright cream yellow." The water was hot—160 degrees—and the men cooked with it and drank it, and Jackson found his photographic plates would dry beside it in half the usual time, making the task of recording the thrilling sight all the easier.

There was more, much more, to come. Once the team had climbed high enough, they could see the backs of the ranges that formed the protective stockade that had kept Yellowstone secret for so long. And from the same vantage point, Hayden saw and wrote about the great lake that is the centerpiece, "a gem amid the high mountains, which are literally bristling with peaks." Down beside it the sight was even more marvelous, of "a vast sheet of quiet water, of a most delicate ultramarine hue, one of the most beautiful scenes I have ever beheld."

The grandeur of what they saw was soon being matched by what they heard—a roar, like thunder, which grew louder and louder as their horses loped steadily through the meadows and along the riverbanks. And then, the climax of the expedition—the river which had hitherto been heading westward without warning turned itself sharply south, and as it did so, it suddenly plunged over the lip of a vastly long canyon, and shot down once, and then a second time, hundreds of feet down toward the lake that glittered in the distance.

The roar of its waters was furious and quite filled the air; the mass of snow-white spray and spume and foam was overwhelming in its sublimity, terrifying the men and their horses, enthralling the photographers and, above all, the painters. Thomas Moran realized that the woodcut fantasies he had created for *Scribner's* the year before hardly did justice to the monumental sights before him. Poor Hayden, armed with only the words of a geologist, could hardly hope to do justice either, but he did his best:

The very nearly vertical walls, slightly sloping down to the water's edge on either side, so that from the summit the River appears like a thread of silver foaming over its rocky bottom; the variegated color of the sides, yellow, red, brown, white, all intermixed and shading into each other; the Gothic columns of every form standing out from the sides of the walls with greater variety and more striking colors than ever adorned a work of human art.

The artists had to linger here for days, easels and brushes and crayons and cameras and assistants stirring chemicals in the developing tents all going full tilt. The scientists were happily making their measurements, too, filling their notebooks with sketches and diagrams and pages of observations. But time was starting to press. It was late July before they reached the shores of the lake itself, and they began to notice a certain crispness in the morning air. They launched a small boat and mapped the lake, performed soundings down to some three hundred feet, cataloged the fish and bird life, the nematodes and the lilies. And then they pressed on, their horses clip-clopping now in a more westerly direction, back toward Idaho and to their base camp down by the railhead in Utah.

It was here that the vast powers of the magma chamber below them started to display themselves to the full. The mud chambers burped and bubbled all around, and so much steam was hissing out of ruptures in the ground that the scene looked positively industrial, like the factories along the Ohio River back in Pennsylvania. The fissures became ever more numerous and ever larger, until at the head of the valley they came across one geyser that was far larger than the rest, which from the center of a fragile disk of chemical-infused mud shot a fountain of boiling water and steam more than a hundred feet into the air—and, moreover, did so with just the regularity that Langford had noted the year before, which had prompted him so properly to name it Old Faithful.

It was the culminating triumph of their visit. The men then trekked away westward, being startled on more than one night by frequent small earthquakes, valedictory reminders of the violent processes that they suspected had created all the stupendous scenery behind them. There was ice on the streams each morning now, and the days were getting shorter; they had perforce now to hurry away down from the mile-high plateau, down to the railhead, and get themselves and their gear and their papers and plates and paintings back east. For they had to publish, to tell America of all the marvels that they had seen and heard and experienced.

In later years, Hayden liked to tell all who would listen that he alone was responsible for creating Yellowstone as a national park. But technically the distinction goes to a little-remembered Republican congressman named William Kelley, who late in the same year that Hayden had visited the region wrote, "Let Congress pass a bill reserving the Great Geyser Basin as a public park forever."

The plan was also supported by the very vocal members of the expedition of the year earlier, from which Langford had written his *Scribner's* essays. One can fairly say that a move to do something special to protect Yellowstone's seismic marvels had been gathering steam, as it were, even before Hayden went there.

Hayden's support proved crucial, however. His initial report, adding official grace notes to the more raucous timbre of popular journalism, was widely circulated to congressmen. Moran's brilliant watercolors and Jackson's stunning photographs were pored over, and evidently convinced some of the wavering members. The Department of the Interior, for whom Hayden worked, agreed that it would happily supervise the new entity. Congress duly passed the bill, both houses voting their support unanimously; and on March 1, 1872, President Ulysses Grant signed into law the Yellowstone Park Act.

A month later, Thomas Moran completed his massive work of art, the eight-foot-by-fourteen-foot *Grand Canyon of the Yellowstone*, and Congress duly bought it to hang in the Capitol—at the time, the grandest and, by the account of at least one art critic, the very best of all the paintings hung there. It sat for a while in the entrance hall of the Department of the Interior, symbolic of the National Park Service, which was founded in 1916 to administer and protect all of America's greatest treasures—of which Yellowstone was the first and, to many, the finest. Later, to win as wide an audience as possible, the department lent it to the Smithsonian, where it hangs now, seen over the years by millions.

Thomas Moran's immense oil painting The Grand Canyon of the Yellowstone *was commissioned by the US government on his return after he served as official artist of the 1871 survey expedition of the region. The painting, now in the Smithsonian, is generally credited with persuading President Ulysses S. Grant to declare Yellowstone a national park.*

And so far as geology and geography were concerned, there was a pleasing coda to the story. The very same Congress that had passed the Yellowstone Bill then went on in short order to appropriate a further $75,000 for Ferdinand Hayden and the work of his survey. The view was that, given the value that his activity had brought already to the country, Hayden's work could continue officially blessed for many more seasons to come. If ever a scientific endeavor needed further validation, the revelations of Hayden from Yellowstone had amply provided it.

Diamonds, Sex, and Race

It was a briskly cold day in late spring in southwestern Wyoming. Dark clouds were racing across the sky and setting down short, exhilarating squalls of snow. I had set off south from the old mining town of Rock Springs to drive down through the rock scrublands of Sweetwater County and across into the remote border country of northern Colorado. The plan was simple: I was to walk in the footsteps of a pair of swindlers who once committed the greatest diamond fraud in American history.

It was a spectacular episode of classic Wild West chicanery that began in early 1872 when a pair of strange-looking men paid an unanticipated visit to a bank in downtown San Francisco. When the subsequent cascade of bizarre occurrences ended, in the autumn of that same year at a lonely prairie railroad station on the Union Pacific line a thousand miles to the east, it made an immediate hero of the man who had cleverly uncovered the fraud—a thirty-year-old Yale-educated and Harvard-polished geologist and mountaineer named Clarence Rivers King, who would before long be appointed the first-ever director of the United States Geological Survey.

Not that King was by any account a man wanting in heroic qualities. He was exceptionally well born, mingled enthusiastically with the best of East Coast society, and was a clubbable, courageous, fearless explorer who had hordes of friends. Henry Adams* was one of the closest, and he wrote of King, after traveling with him on a climbing holiday in the Rockies, that he possessed "that combination of physical energy, social standing, mental scope and training, wit, geniality and science, that seemed superlatively American and irresistibly strong."

---

* Descendant of two American presidents—the second and the sixth—and author of the famous autobiography *The Education of Henry Adams.*

By the time of the events that made him famous beyond the world of science, King had already achieved much. At just twenty-two, he had joined as a volunteer geologist the great 1864 Survey of California, having crossed the country by horse and wagon with a party of pioneers, from Saint Joseph on the Missouri to the Sierra Nevada and thence by paddle steamer down the Sacramento River to the coast. He had seen all the famous sights—South Pass, Salt Lake City (where he saw, and wrote admiringly of, Brigham Young himself), and the Great Basin—just as all the homesteaders had done.

Not long after his arrival in California, the survey leaders recognized in King a youngster of talent and toughness and promptly hired him—without pay, at first—as a geologist proper. It was a decision that allowed him once again, as he wrote after all his sore weeks in the saddle, to be "one of those hammer-bearing sons of Thor."

He turned out to be a phenomenon, a whirlwind of a man. In the company of others who found it difficult to keep up with his energy and drive, he made sorties up into the High Sierra, astonishing and delighting his bosses by turn, making first ascents, naming peaks, and drawing maps of great accuracy that would last for generations. But he also lived the life of a swell, exuberant, happy, enthusiastic. It was around this time that he wrote in his diaries of his fascination with colored women, native women, slave women—all so much more attractive to him, he would say repeatedly, than the pale-skinned ladies who were the stuff of his colleagues' customary social fare. It was a liking that would play a part in his later years in a profound and fascinating way.

King returned to the East in 1866 after enduring three seasons of rigorous apprenticeship in California. Now that he was fully schooled in the art of fieldwork, despite his youth, he came up with an ambitious plan for surveying a cross section of the country along the route then being created for the transcontinental railway. In 1867, the year when the government in Washington so suddenly began to make a serious inventory of the country and created its Four Great Surveys to do so, lawmakers agreed with Clarence King that such a survey would be an excellent thing to undertake.

Five days after President Andrew Johnson signed the legislation, King was named geologist-in-charge of what would be the most ambitious of all the surveys, the Geological Exploration of the Fortieth Parallel. It was to be an army-directed exploration; it would take seven years to complete, and it would be, in the words of Henry Adams, "one of the classic scientific works of the century."

That so young a man, albeit one of the most daring of contemporary explorers, had won the top job was a matter of great astonishment in the capital. The secretary of war, Edwin Stanton, took King to one side: "Now Mr. King," he said, "the sooner you get out of Washington the better. You are entirely too young to be seen with this appointment in your pocket. There are four major-generals who want your place."

Armed with a government guarantee of $100,000 and at least three years of work, he did get out of town, promptly. Some of the finest geologists from around the world were eager to join, for even though King was merely twenty-five, he had an intelligence and a personal magnetism that teased out the best and the brightest from the scientific community, and having gathered his team together, he embarked on his assignment with a fury. The thirty-five-strong group reached California by way of Panama, crossed the passes of the Sierra Nevada, and were out in the field surveying in Nevada by the summer. They spent the following years painstakingly surveying and collecting along a swath of territory that ran along the ever-expanding railroad route for more than a thousand miles between Winnemucca and Cheyenne.

*Clarence King (left), the diminutive, phenom-
enally learned, and intrepid first director of the
US Geological Survey, had previously led the
six-year Survey of the Fortieth Parallel, which
mapped in detail the landscape with its flora
and fauna along a line between Sacramento and
Cheyenne.*

The discoveries King made during his years in the
field were legion; the maps he and his team drew were
innumerable; the volumes of his reports—massive,
leather-bound quarto-size monsters with lush and
lavish illustrations—have in recent years become the
stuff of collectors' passions. But the event that won him
his laurels occurred in 1872, five years into his official

work. He was approaching the end of his extended periods of fieldwork and at the time was situated with his exploring party well beyond the Rockies amid the flattening emptiness of the Great Plains. It was here that Clarence King first heard tell of what would come to be known as the Great Diamond Fraud. His success in solving it made him almost overnight one of the most famous geologist heroes America had ever known.

The scam first raised its head in February, when the aforesaid pair of ragged-looking men arrived at the offices of the Bank of California in downtown San Francisco carrying a bulging canvas bag. They were prospectors, they claimed; they were there to place the bag in the vaults. The cashier demanded to see what was inside. What he found were hundreds upon hundreds of uncut diamonds, rubies, emeralds, and sapphires, which if genuine would have unimaginable value. He called the bank's flamboyant founder, William Ralston, whose personal fortune, based on mining the great Comstock silver lode in Nevada, guaranteed at least his sympathetic interest. He pronounced the find at first blush quite extraordinary and demanded to know more.

The two visitors identified themselves as Philip Arnold and his cousin John Slack. They were prospectors from Kentucky and seemed frightened, at first saying little. But under pressure from the charming Ralston, they did let slip some details: they had been searching fruitlessly in a mountainous region quite some distance away from San Francisco for many months when, all of a sudden, they had happened upon a hill where many precious stones—rubies, garnets, and other colored stones mostly, but diamonds occasionally—were scattered.

The jewels were so abundant that a mere kick of a boot heel would reveal more of them among the dust and gravel on the mountainside. The bag they had brought to the bank contained no more than a tenth of what they had found; the rest they had left behind in secure hands. They told an enthralled Ralston they were confident that there was still an immense quantity left: they had simply carried out as much as their mules could bear.

The men refused point-blank to say where they had found the stones. They did reluctantly agree, however, to take two diamond experts—blindfolded—back to the field. Two weeks later, the team returned and declared that the find was even greater than imagined. Diamonds were to be found jutting out of the earth, were to be picked out of cracks in the rocks, were to be seen glittering in the dry beds of former streams.

They brought back with them a second bag, quite as large as the first, and deposited it alongside the first in Ralston's vault. The total assumed value of the gems was now a quarter of a million dollars, no mean sum in 1872.

Ralston moved quickly. He formed a company, naming it with suitable swagger the San Francisco and New York Mining and Commercial Company, and persuaded twenty-five of the city's business elite to chip in $80,000 each, a capitalization of $2 million. The word got out: madness erupted. A diamond frenzy spread like a forest fire through the San Francisco financial aristocracy; and by the summer's end, twenty-five other firms had been founded, with a total capitalization of a quarter of a billion dollars.

Ralston then brought onto his own board the great and the good and the not so good, including a lawyer who had run unsuccessfully against Lincoln, a member of the US Senate, and the magnificently named adventurer Asbury Harpending, who later wrote a book about the affair, most of it more wildly exaggerated than was necessary.

The banker also sent two of his lawyers armed with sample stones across to New York, to the Manhattan offices of Charles Tiffany, the greatest jeweler of his times. What, pray, was his considered opinion? Back came the word: the stones were all genuine, all precious gems of enormous value. The small sample he had seen was worth $150,000, at the very least. If they had more, then the value was probably in the millions.

Yet even Tiffany's word was not an absolute guarantee. To secure once and for all the faith of any future investors, Ralston hired the country's foremost mining engineer to perform all the due diligence that a highly risky mining operation demanded and pronounce upon its worth, for many others had had their fingers badly burned by placing risky bets on dubious geology. The engineer's name was Henry Janin. He had examined more than six hundred mines in his career so far—and he had never once been wrong.

Janin promptly traveled out to the mystery location, along with Arnold and Slack, in a train with the shades pulled over the windows. The journey took a day and a half. He then duly accepted the customary blindfold as he left the railroad depot and rode with it on for two blazing days on the back of a mule. He eventually, and by all accounts miserably, reached the site.

The diamond fields were on the northern side of an unusual cone-shaped mountain, a peak that rose quite memorably out of the scrubby landscape. After spending the better part of twenty-four hours camped there, having walked across the prodigiously jewel-strewn hillsides, he pronounced the find entirely genuine. He offered his professional view that the company and its backers were set to make a fortune. And the one thousand shares he had just accepted (along with $2,500 cash) for giving this opinion would be likely to make him also a very rich man.

Once they heard the news, San Francisco's private investors agreed, and hundreds of them began clamoring for stock. The firm turned down offers of as much as $200,000 for claims on the promised land. No less a figure than Lord Rothschild, five thousand miles away in London, succumbed to the fever as well and demanded that his American agent buy the company outright—a move that Ralston resisted, although he did invite the Rothschild man onto his board. So certain was the San Francisco and New York Mining and Commercial Company of its future that Ralston opened a gleaming new office in a marquee building in the city center and hired nearly two dozen clerks to run its increasingly complicated business affairs.

And then Philip Arnold and John Slack, not men for the boardroom lunch and the briefcase, suddenly overcome by nostalgia for their artless world of pick and shovel, mule and map, decided to cash out. They left behind confidential instructions on how to reach the mystery diamond fields, negotiated a price of $300,000 each for their interest in the new company, took all the proffered cash, and for good measure added a demand that they be paid on an ongoing basis a percentage of any future profits.

This last was purest cheek. For there would be no profits, nor would there be a company.

For it turned out that Clarence King, the government-commissioned geologist-extraordinaire, had heard the story and had smelled a rat. The supposed diamond find sounded to him like a blatant absurdity—for the simplest of geological reasons: diamonds can occur in one place, sapphires in another, and rubies in a third, but never, in King's experience, were they all to be found in *one* place. It was not a geological impossibility, but it was very, very unlikely.

If this wasn't enough, then the suspected location of the supposed find seemed fishy. Arizona was widely suspected—yet it seemed unreasonable to King that it would take a thirty-six-hour railway journey and a further two-day trek by mule to reach anywhere in Arizona. To find out just where it was likely that the men had gone, it seemed essential that those who were skeptical of the prospectors' claims now interview the man who had last accompanied them to the diamond field, the so-called mining expert, Henry

Janin. Clarence King bearded him in his den, a celebrated San Francisco restaurant. King demanded to know two simple things: what had the weather been like during Janin's trip, and when he was on their mules, in what direction did he imagine he had been traveling?

Janin said he'd determined, from the arguments he overheard between Arnold and Slack and from the slight amount of sunlight he could discern through his blindfold, that they had traveled generally southward.

And as for the weather—it had been hot, and he had suffered mightily in the saddle. The journey had been a kind of purgatory: his thirst and weariness had almost overwhelmed him.

This proved to be the Sherlock Holmes moment. For Clarence King knew first that a thirty-six-hour railroad journey on the lines that then existed would take someone a thousand miles *east* of San Francisco, well beyond Salt Lake City. At the time that Janin had passed through, Salt Lake and most of Utah and Nevada had been enveloped in a drenching rainstorm. The weather had become dry only once they had passed farther east still, and into the rain shadow of the Rockies, in Wyoming. For Janin to have begun his journey with the muleteers in weather that was warm and dry, he would have to have been on the eastern side of the mountain range, somewhere near the town of Cheyenne, and to have proceeded south from there.

So King jumped onto a train himself and headed out toward Cheyenne, to the very same countryside that he was already in the process of mapping. Indeed, the "cone-shaped mountain" that Janin had mentioned sounded faintly familiar to him: it was entirely possible his survey teams had already reconnoitered it, had maybe even mapped it. So King had some reason for optimism. He traveled east with his old school friend James Gardiner, now chief topographer of his survey: the pair arrived thirty-six hours later at the lonely Wyoming station of Rawlings Springs, just a few miles from the Green River bridge where John Wesley Powell had set off for his expedition to become the first down the Grand Canyon.

And clues were accumulating. From asking myriad innocent questions of the attendants on the train it was clear they were, as one might say, on the right track. The two men were beginning to suspect just where the diamond field may have been, within a radius of about fifteen miles. These suspicions were bolstered by what they were told at the station—that a small party of men on horses had been seen thereabouts just a few weeks before, and they had headed off in hot weather, in a southerly direction. Janin's recollection of his sightless journeying seemed to have been correct.

But it was October now, and cold. The small expedition—King and Gardiner and four other helpers, all on horseback—set out armed with camping gear, sieves, and long shovels, in case they got snowed in. They wore several flannel shirts and long wool mufflers. They noted that balls of ice clung to their horses' legs each time they crossed a frozen river, and that this ice clicked like castanets. After six days of cheerless riding, they reached the Colorado boundary. They were now at seven thousand feet, and though there was still a bitter wind, the snow had either melted or been blown away from the hillsides. There were tracks of horses, apparently made quite recently. And there was a mountain ahead of them, shaped like a low cone.

Here they came across a crude wooden sign nailed to a tree. It was a claim to the water rights to a nearby stream, signed by the man whom they had met in the San Francisco restaurant, Henry Janin. This was, just as King had calculated, the likely place.

Within moments, as they walked along a ridge below the cone-shaped hill, they began to find gems. A ruby was the first—then several dozen more after an hour or so of searching. That night they discovered just three diamonds. The next morning King got forty-two more stones, but barely any diamonds. There were also amethysts, spinels, and garnets, an assemblage of minerals that was as improbable as the sapphire-ruby-diamond combination that had already alerted King's skeptical mind. And then he found a diamond perched on a small finger of rock—and this awakened every remaining suspicious fiber in his body: for how could it have stayed there, perched so precariously amid the Colorado gales and snowstorms, over the hundreds of years since it had been made?

There is an apocryphal story that a German mining engineer who had accompanied the team sealed their suspicions by crying out when he found on the ground a diamond that had already been cut. "Look, Mr. King, this diamond field not only produces diamonds, but it cuts them also."

But there was no German on the team, nor was a cut diamond ever found. The final confirmation came more prosaically, in an anthill. For the men noticed that at the base of nearly every anthill on the ridge there were a number of tiny holes, eight inches deep or so, seemingly made with a stick pushed into the soft ground. And at the bottom of each hole lay a precious stone, put there by whoever had drilled the hole with his stick.

The entire field was a fraud, a swindle, a hoax. The field had been *salted*, to use the miners' term, had been cleverly peppered with cheap stones in the hope that someone would invest and help make those who did the salting very wealthy men indeed.

As Philip Arnold and John Stack, now hastening away to somewhere with $600,000 in their pockets, had instantly become. It turned out they had bought their stones some months before in London and Amsterdam. They had made a $25,000 investment in a few hundred offcuts and rejects from unknowing stone cutters and gem dealers in Europe. They had then placed these stones, in all their varieties, in holes in the bases of ant-hills in a windy desert on the Colorado and Wyoming border and had waited as they slowly emerged onto the surface and into the sunlight.

Charles Tiffany, who had famously valued the waste diamonds as "a rajah's ransom" managed some-how to reburnish his tarnished reputation. Asbury Harpending fought a libel suit to try to regain his. Henry Janin—who had already gotten wind of impending trouble, and sold his thousand shares for $40,000 before the discovery of the fraud—remained friends for life with Clarence King. William Ralston paid his investors back and put the stock certificates under glass as a warning to be more prudent in the future—but three years later his entire financial empire collapsed, there was a run on his bank, and the day afterward his body was found floating in San Francisco Bay.

Philip Arnold was arrested in Kentucky and charged. But his state refused to extradite him, and even displayed some small sense of pride that a doughty Southern son had managed so successfully to tweak the tails of some Yankee profiteers. In the end he cut a deal, paying back half the money in exchange for the charges being dropped. He used what was left to set up a bank of his own, only to be shot in the shoulder by a jealous rival. He died painfully of pneumonia caused by complications from the wound. His cousin, John Slack, was never found.

As a consequence of his detective work, his high intelligence, and his apparently unimpeachable code of morals, Clarence King was to enjoy fame and good fortune for almost all the rest of his days. Everyone applauded. One newspaper said it all: "Fortunately for the good name of San Francisco and the State there was one cool-headed man of scientific education who esteemed it his duty to investigate the matter in the only right way, and who proceeded about his task with a degree of spirit and strong common sense as striking as his success."

The successful exposure of the diamond fraud coincided with the effective end of the Great Survey of which King had been in charge. The Four Surveys were all done now (the fourth, led by the non-geologist soldier George Wheeler, was in the main a mapping expedition). It remained only for the usual squabblings and rivalries attendant on such government-run matters to conclude in 1879 with the formation of the US Geological Survey, which exists still to this day. Clarence King was appointed its first director; John Wesley Powell was its second—the appointments honoring two of those who had made so much of their Western experiences. Ferdinand Hayden, the great explorer of Yellowstone, had to be content with a professorship instead.

It took me a long cold afternoon to drive clear across Sweetwater County. The Rockies were in the distance on my right. Low mountains shelved downward to my left. Ahead of me the countryside was beautifully desolate, with sandstone buttes and mesas, small winding creeks and little waterfalls and runnels, an uncared-for expanse of scrub and tumbleweed, with occasional gatherings of rusting oil tanks beside farms of quietly dipping nodding donkeys, each of them sucking the last remnant drops of precious oil from long-forgotten wells. The road was good, paved and well repaired with oil money, no doubt, through the entire mileage of the county. Then, WELCOME TO COLORADO said a yellow sign, peppered with bullet holes—whereupon the road became dirt the moment we crossed.

The Bureau of Land Management looks after this remote, unwanted wilderness and does its best to keep away people who have no business there. I had explained to its office in Rock Springs that I was curious about the history and wanted to walk the land where Clarence King and Henry Janin and Arnold and Slack had once walked. A kindly man there eventually relented and gave me a map. I used it for navigation over the last few miles, for which the GPS aboard the car had no information at all, identifying it as simply an immense empty expanse, as if it might be the sea.

But it was, of course, a place littered with prominent and long-ago-mapped features—all of them the result of the King expedition of a century and a half before. And then Diamond Peak, the immediately recognizable low and cone-shaped hill on the flanks of which the gems had been planted, was lying squarely ahead of me, its bulk slightly to the right.

The wind, which had been stilled ever since I left Rock Springs, now picked up again, and there were more flurries of snow. The top of Diamond Peak was pure white, with ice crystals glinting in the low spring sun. I spent an hour or so there, fossicking around in the ground, kicking the anthills, turning over a promising selection of small boulders, dredging channels in the dust with my boot heels. It was all fun and good exercise, and I had the intense pleasure of knowing I was at the time perhaps the most isolated man in America, scores of miles away from any other human being, in the absolute middle of nowhere.

It would be pleasing to be able to report that after spending a while on this lonesome search, I saw a sudden pink glint in the dirt. It would be agreeable to say I then fell to my knees and within seconds picked up the gleam and discovered a ruby, a gemstone that had been bought in a job lot in Hatton Garden just a few miles from where I had grown up in London, and had then been lodged in the Colorado dirt here by a Kentucky con man and adventurer whose nefarious salting-doings had been uncovered by a son of Rhode Island, a Yalie, who had once ridden a horse clear across America. It would indeed have been pleasing to report this. But it never happened.

For fossick and kick and dredge as I might, everything on the slopes of Diamond Peak these days is dust and gravel, tumbleweed, scrub, and bird feathers, with the occasional organic traces of rabbit and prairie dog. I looked around, considered the view, enjoyed the sublime isolation—and then the snow began to fall more steadily, the wind began to howl in a way peculiar to the Great Plains, and the view of the Rockies in the west faded from sight. I got back into the car, bumped along ten miles of dirt, and then found solid Wyoming pavement once again, and was back on the interstate highway by nightfall, without a stone of any worth to show for my troubles.

There is a curious small coda to Clarence King's story, a happenstance that serves to render this heroic figure more distinctly human than might be suggested by his great cascade of achievements. It all goes back to his fondness for dark-skinned and native women.

In a book review for the *Atlantic Monthly* in 1875, King outlined the basis for his predilections for such ladies:

> *Whoever has strolled at dusk where palm groves lean to the shore and watched the Indian women sauntering in the cool of evening with a gait in which a ripple of grace undulates—whoever has seen their soft, dark eyes, and read the expression of tenderness and pathos which is habitual on their*

faces, can but feel that here simple nature has done
all she can for a woman.

It was all Rousseau, Gauguin, the *Bounty* muti-
neers, and Robert Louis Stevenson, with a touch of
Arthur Munby, all wrapped up in one. And from his
writings it seems that it was sincerely felt—a feeling
deep within King's heart that women "in the primitive
state" were the very best, and that "Paradise, for me, is
still a garden and a primeval woman."

By contrast, white women held few charms. When
his friend John Hay once tried to arrange a meeting
with an especially attractive and eligibly patrician
lady in Washington, King recoiled: "To see her walk
across a room, you would think someone had tilted up
a coffin on end, and propelled the corpse spasmodically
forward."

Well aware of the social dangers involved, King—by
now living in New York—tried his level best to find
the "natural woman" of his dreams. Night after night,
after leaving the celebrated clubs of which he was a
member, and after his virtuoso performances as the
greatest storyteller in the city of the day, he liked to
leave behind the haut monde and make his entry, with
robust enthusiasm, into the much more exciting demimonde, the black demimonde in particular.

He liked to spend his late nights trawling the insalubrious corners of the cities in which he lived and stayed. He would travel on official business overseas. During his travels to London in particular, he behaved rather like Dorian Gray, with his visits to "the dreadful places" near Blue Gate Fields—though in King's case not, it seems clear, solely or even principally for the purpose of sex. He went to mingle with people who, he claimed, acted more on instinct than on intellect, to indulge in behavior that took him back somewhat to his happier and more liberated times, when he was in the field and had been very much a man of action, when he had spent his days as a climber, a mountaineer, a hunter.

It was in New York in either late 1887 or early 1888 that King met a nursemaid named Ada Copeland. The fair-skinned and blue-eyed Clarence King, a Yale-educated geologist and former senior government official from a good and old family in Newport, met and fell in love that winter with a young black woman from the banks of the Chattahoochee River in Georgia. She had been born into a slave family. He was forty-six; she was twenty-eight. He was white, and she was black.

But he had no wish for embarrassment or social ostracism. He could not afford to be embroiled in the scandal that would be certain to erupt if it ever became known that he was courting a woman who in those less sophisticated times would have been called a Negress. And so he came up with a plan—an ill-thought-out, heat-of-the-moment plan that, because it involved an ever-more-complicated web of deceit, was doomed from the start for eventual failure.

As soon as the couple met, he embarked on a lie. He decided to tell Miss Copeland three things—and by extension, four. He would not tell her his real name, instead announcing, without thinking, that his name was James Todd. Second, he made no mention of Newport but said instead that he came from Baltimore. And third, as if the first two lies did not bring complication enough, he said he used to be an itinerant steelworker but now had employment as a porter on the Pullman express trains.

This third falsehood was the single most important of all. The job of Pullman porter had long been a reserved occupation—and Miss Copeland was to infer and understand from what she was told that the man who had tipped his hat so effusively to her was, despite outward appearances, a black man. Not perhaps as fully black as she, but black in law and by custom nonetheless.

Under Victorian social rules, if he had only a drop of Negro blood in his veins, his genetic makeup could legitimately be claimed to be that of a black man. He could, in other words, assert without too much fear of contradiction that he was a man of the same basic color and caste as the woman he would later marry.

Though a white man from Rhode Island, Clarence King had a yielding respect and affection for women of color. He eventually met and married Ada Copeland, from a Georgia slave family, though renaming himself "James Todd" and insisting he was a pale-skinned black Pullman porter. The pair had five children together; one son, Wallace, is pictured here with his mother.

And when James Todd did marry Ada Copeland, in September 1888, at a small ceremony conducted by a Methodist minister in a private house in Manhattan, one feature of the occasion suggested his subsequent fate. None of his friends or relatives attended the wedding. James Todd was alone, and he would in a sense remain alone for almost the rest of his days—because Clarence King had told no one, and indeed could tell no one, of his decision or of the life he intended to live with the woman who would in short order bear his five children.

Because James Todd was a chimera, he could have no relatives or old friends of substance. And because Clarence King had plenty of both but could admit his other life to no one, a strange separation slowly got under way. King embarked on his final thirteen years as two distinct and very separate people, of which until the very end his new wife knew only one.

The complications of such an arrangement were legion. His world was now awash with deception, avoidance, excuses, feigned illness, absence. (At one geological congress, he met a young geologist whose name was James Todd, which must have given him momentary pause, at least.)

And there were the costs—of among other things, maintaining two households. As Clarence King, he lived alone in a residential hotel on Manhattan's Eleventh Street. As James Todd, he lived with his fast-enlarging Todd family in a small house in Bedford-Stuyvesant, across the East River in Brooklyn. To this man from a family that was socially grand but financially threadbare, such an arrangement was likely to prove corrosive, probably ruinous. To sustain himself, King borrowed money from an unsuspecting John Hay; the first of six loans would be nearly $200,000 in today's money; in the end, he owed the equivalent of millions.

From time to time, there were episodes of domestic panic. To help pay the bills, King tried mining in Mexico and cattle ranching in the West. All of his ventures failed, and in his later years, he was reduced to acting as a mining consultant and a professional expert witness in geology-related court cases—a mournful fall from grace for so bright a figure as he had once been. John Hay felt for his friend: "I fear he will die without doing anything, except to be a great scientist, a delightful writer, and the sweetest creature the Lord ever made." But Hay was never told the truth, or the reason.

Racial passing—the profoundly difficult effort that had to be made, day in and day out, without ever allowing a mistake, in order to convince others that you belonged to a different race than you appeared to—imposed a formidable burden on King. As he approached the mid-1890s, he began to suffer both physically and mentally from its strictures. He had long been a dandy, given to wearing dinner dress in field camps and exotic velvet suits in New York. But now his friends began to notice that his appearance was shabby, his beard more often than not unkempt, his clothing frequently ragged and soiled. And he flew into rages. One afternoon he was arrested after an altercation in, of all places, the lion house in Manhattan's Central Park Zoo, and then he so offended the magistrate that he was committed for two months to an insane asylum, his long-suffering friends paying the bills and offering attentive support. His medical report diagnosed him as suffering from acute melancholia.

On his release, he became ever more furtive, ever more terrified that his friends were about to find out what he thought of as his awful secret. He moved the family out to Flushing, Queens, farther from Manhattan, hoping to make it less likely that the worlds of the Kings and the Todds would ever collide.

Here, pinioned in so distant and economical a part of town, the family did start to enjoy some modest prosperity: in January 1900 there was a New Year ball mentioned in the local paper and celebrated by "Mrs.

Ada Todd"—suggesting that the former nursemaid now enjoyed a newfound status and believed she could count on a more stable future in the new century for herself and the four children who survived. Yet because she did not know the truth, she never realized the irony of the newspaper notice, which announced the theme of the celebration ball as *a masquerade.*

This remains the most puzzling aspect of all. For almost all of their marriage, and while helping to raise their children, King clung tenaciously, even in the privacy of his home, to the complex cascade of fictions. Maybe he intended to keep Ada from ever knowing, but within the year of the ball and her momentary flush of social success, she did find out. She did so because King, torn apart by strain, fell ill—terminally ill, as it turned out—with tuberculosis. He moved out of New York, alone, and fled to the warmth of the American Southwest to convalesce and maybe to become cured.

But once he finally accepted that he would not survive, he wrote to Ada and told her some of the truth. He did not tell her all. He disclosed little of his background. But he did inform her, at long and perplexing last, that his first name was not James but Clarence, and that thanks to his surname she was not in fact properly Mrs. Todd but should be called from now on Mrs. King.

Not long after this stunning revelation, Clarence King died, on Christmas Eve, 1901. His doctor in Phoenix, where he spent his final days, was the one person who knew King had died a married man. The doctor accordingly sent a telegram to Ada with the news. And when he came to inscribe the death certificate for her, he—with an infinite kindness—filled in the line that asked for a description of the deceased, and which offered Black as one choice, with a single typewritten word—*American.*

It seems possible he did this to fulfill a wish the hapless man may have expressed on his deathbed. Some years before, King had written an essay, in which he stated that he imagined an ideal America in which "the composite elements of American populations are melted down into one race alloy, when there are no more Irish or Germans, Negroes and English, but only *Americans.*" Perhaps the doctor knew that. Perhaps, in some way, by marrying into a black family and siring five children who were of mixed race—"melted down into one race alloy"—King played his microcosmically small part in helping his country achieve that. To the highly complex and multilayered business of the uniting of the states, Clarence Rivers King contributed by his vital role in the great geological surveys and perhaps also by crossing the lines of race and class.

The Great Surveys had been officially concluded in 1872. Clarence King died in 1901. The frontier, which had been so much a part of pioneer America's experience, was effectively closed and ended as a phenomenon in the 1890s. So it is reasonable to say that by the last quarter of the century, the states of America, at least in terms of knowledge of their surface and subsurface components, were fully known and linked with one another so entirely and so intimately as to be, now in fact just as in name, United.

Yet other components still needed to be clicked into place. For an American in Maine to feel true kinship with a brother American in Arizona, for a New Yorker to be able to feel at one with an Oregonian or Alabamian, for a Kansas farmer to be able to send his wheat to market in Florida, or for a Massachusetts mill owner to sell his shoes or textiles to a store in California or North Dakota, people and the things they made needed to be able to move with speed and ease from one corner of the nation to another. To achieve this, to overlay this kind of ability on the bedrock of all the knowledge that had been so painstakingly assembled, it was now vitally necessary to create the ways and means for Americans to enjoy real and true mobility, so they and their goods could reach out to all corners, to all the nooks and crannies of the nation.

These means—which had to be in conception far more sophisticated than the crude simplicities of the wagon trains and cavalry squadrons and survey parties—all now had to be conjured out of the national imagination. Transporting devices had to be planned, invented, tested, and then made and manufactured. And just as important, so also the ways across which these means would then travel—the waterways, roadways, railways, and skyways—all had to be identified, built, improved, made more useful, made permanent, made safe.

The waterways would be the first. They were already there—the nation's rivers. They were obvious, they were free, and for years they would be preeminent. For a long while, they would remain so until, in a nation growing as fast and furiously as the new America, even the immense network of rivers became inadequate to the task. They turned out to be neither numerous, extensive, nor convenient enough. Then Americans had to build new rivers for themselves, which they did with all the energy of a new race of pharaohs, devoted to the task of uniting America by water.

The London Times *is one of the greatest powers in the world—in fact I don't know anything which has more power—except perhaps the Mississippi.*

—ABRAHAM LINCOLN, QUOTED IN WILLIAM HOWARD RUSSELL, *My Diary North and South*, 1862

Low bridge, everybody down
Low bridge for we're coming to a town
And you'll always know your neighbor
And you'll always know your pal
If you've ever navigated on the Erie Canal

—THOMAS S. ALLEN,
Erie Canal, 1905

. . . the ties which bind us together would be strengthened and multiplied by these ship-canals, creating another Mississippi from Saint Louis, and Kansas, and Saint Paul, to New York and Boston. It has been well said, that the myriad-fibered cordage of commercial relations, slight in any individual instance, but indissoluble in their multitudinous combination, produces such unity of purpose, unity of interest, intelligence, sentiment, and national pride, and social feeling, and that homogeneousness of population which unites peoples and maintains nationalities. Such will grow up with a power which no sectional feeling can break between East and West, when connected together by these canals.

—NATIONAL SHIP-CANAL CONVENTION,
CHICAGO 1863, FINAL COMMUNIQUÉ

PART III

When the American Story Traveled by Water

1803–1900

Journeys to the Fall Line

The very earliest Europeans to fetch up on the American shores—the brave and sea-weary boatloads of Norwegians, Spaniards, Florentines, Basques, Portuguese, Frenchmen, and Britons—generally took a good long time, once they had landed, to catch their collective breath.

The basic business of settlement kept them frantic and fearful. The building of houses and stockades, the planting and growing of crops, and the establishment of protocols for hunting unfamiliar animals and for dealing with the often not unreasonably hostile local inhabitants were all very complicated, time-consuming, delicate, and dangerous. So it is perhaps understandable that they didn't make it a first priority to engage in any serious exploration of the countryside that lay beyond the coasts.

This began to change in about 1520, when a group of sailors from Toledo undertook an expedition into the lowland coastal forests of what is now North Carolina. Then in 1535 Jacques Cartier did much the same, exploring the interior of what is now Quebec. Half a century on, Henry Hudson was to be found deep into the state of New York. These explorers, and a host of others similarly bent on extending their coastal settlement inland, may on occasion have ventured inland on foot. Generally, though, they traveled into the interiors by the most logical passageways provided by nature: they went by water, along America's rivers.

For almost all of them, the pattern of their adventures was similar, no matter the latitude, the weather, the kind of foliage or wildlife, or the varied inclinations of the locals. The Europeans would fashion their canoes, outfit them with supplies, bid tearful farewells to those left behind in the stockades, and then spend many subsequent days or weeks cautiously paddling upstream, pausing every so often to write notes, paint pictures, or gingerly explore the flora or fauna in this creek or that.

Then, in almost all cases, they would be forced to stop. There would suddenly be waterfalls, rapids, boulders, twists, and turns in the stream that made each promptly non-navigable. The would-be navigators had encountered a topographical phenomenon that is common to most rivers but which seems particularly dramatic in North America, and which proved in each case of singular importance to the eventual human geography of the land they were exploring. Villages and then towns and in many cases cities were thrown up where they had been compelled to halt their boats, and all of them remain today. The men who in essence founded these settlements, canoe-borne venturers all, enjoy varying degrees of fame or notoriety today for their geographically enforced achievement.

One of their number won his fame for a rather different reason. He was a big, bluff, bearded, heroic, troublemaking, and (so far as his writings were concerned) highly inventive man, Captain John Smith. Geographers know him somewhat for coining the term *New England* for the territory he later explored and mapped. John Smith is far better known, however, for a singular episode of high romance that occurred during his time as the de facto governor of Virginia.

He had already been amply prepared for adventure in the Americas, case-hardened by some vivid experiences in Europe, where he had been captured and forced into slavery, had beheaded three Ottoman Turks, and had briefly earned a living as a pirate. On the ship that brought him out to Virginia in 1607, he had been so troublesome that his captain sentenced him to death. He was on the verge of being executed when sealed orders were opened showing that all along the merchants who had sent him planned for him to be one of the leaders of the Jamestown colony.

Early the following year, after the first true English settlement in the New World had begun, Smith then set sail around the vast expanse of the Chesapeake Bay. He encountered bands of local Indians and was captured by one of the Virginian chiefs, Powhatan. He was about to be ceremonially clubbed to death when the chief's beautiful teenage daughter, at the time known as Matoaka, intervened—by laying her own head upon Smith's and preventing her father from smashing down with his war club.

Matoaka was, of course, Pocahontas. The young girl's spontaneous intervention saved John Smith's life—and then led her on a dizzying series of adventures that have since become the stuff of legend and have proved ceaselessly fascinating to generations of novelists, historians, and filmmakers. The young woman was in due course captured by the Britons. She converted to Christianity, took the name Rebecca, forswore her native lifestyle, married an Englishman named John Rolfe, bore a child, moved to London, met King James, settled in Brentford, shopped and went to the theater, and after a year, set off back to Virginia—only to fall ill while sailing down the Thames estuary, die, and then be buried in Gravesend. Outside Saint George's Church there remains a life-size statue of Pocahontas, the noblest of all the noble savages for whom she has ever since been ambassador.

Pocahontas and John Smith did meet once again during her sojourn in London. But despite breathless suggestions to the contrary, no serious historian believes they were ever lovers. Perhaps more important, her marriage to the tobacco-farming Mr. Rolfe and the birth in 1615 of her son, Thomas, ensured that her genes have since been spread liberally through a vast swath of contemporary American society: from the Jeffersons to the Reagans, a piece of Pocahontas seems to be just about everywhere.

Smith's real significance to history and to the story of America's eventual knitting together has little directly to do with the saga of Pocahontas. Rather it comes from three separate expeditions he undertook—the first before his near-death moment with the Indians, the second two after his release. In all of them, he tried to venture short distances into the American interior along the country's eastward-flowing rivers.

The first try was in the summer of 1607 soon after he arrived from London, when he sailed his boat up the James River—a last-ditch attempt on behalf of the London Company back home to locate gold, to find the so-called Lost Colony of Roanoke, and, of course, to discover the fabled waterway to China with which all explorers of the day were obsessed. He found none of these things. But what he did find held an importance he did not at first realize.

As he sailed up the James, he came to a point where he could see rising from the plains some distance to the west the pale blue dusting of a long range of mountains. And just a short time after seeing these hills, he found that his progress along the stream was firmly blocked, because the river ceased flowing quietly and serenely and was instead roiled by a series of waterfalls, rapids, and shallows. The city of Richmond now stands almost precisely where he was forced to stop and turn his boats around.

John Smith repeated this experience on no fewer than three other rivers. Irritated by his lack of westward progress, he sailed back down into Chesapeake Bay and worked his way clockwise around it, venturing into one after another of the great swarm of streams that bled into its vastness. When he sailed up the Potomac River, he was stopped by rapids just a little above where Washington, DC, now stands. When he went up the Rappahannock, he met falls at what is now

Fredericksburg; and along the Susquehanna, he came to shallows and falls at a place where there is now a Maryland town named Conowingo, an Indian word for "at the rapids."

Much the same thing had happened to Jacques Cartier, sixty years before. Dangerous waters had stopped him, quite unexpectedly, in his tracks. He had been traveling westward up the Saint Lawrence River and found his way interrupted by what are known today as the Lachine rapids.* He turned around. A small way station was later built for such explorers as decided to portage around the rapids; the camp became consolidated into a village, and in time what is now the city of Montréal arose, partnered by topographic accident with Richmond, Washington, Fredericksburg, and a score of other places.

A pattern was emerging. Crude maps readily showed it. There was a ragged line to be drawn across this corner of the American colonies, linking a variety of places where the behavior of the rivers suddenly and dramatically changed. The phenomenon repeated itself time and again, and the puzzlement and annoyance it first caused would eventually prompt the beginning

---

* So named because it was assumed that Cartier was on his way to, or had indeed found, China.

of a phase of engineering that would alter the face of America for all time.

The phenomenon, now well known to geographers wherever it may occur, is the appearance of what is known as the fall line. In the eastern part of America, it is a line marking the place where the hard rocks of the Appalachian Mountains give way to the plains formed of the sediments that were eroded from them. Hard rocks make for steep slopes down which rivers tumble in narrow, fast-running streams, impossible to navigate. But when the river reaches the fall line, there is a sudden moment of transition, a final set of cliffs over which to tumble, a last moment of white water and spray—and then the rivers broaden, they still, they run smooth and silent and deep, and as they near the sea, they start to become estuaries, all entirely friendly to boats and boatmen.

When upstream explorers—Smith, Cartier, and Hudson first, then a host of others—reached the fall line, it became a customary necessity for them to stop, to change their cargoes onto different and more suitable conveyances, to stay overnight or over a week, to pick up guides or stores or victuals. To provide such services to them all, merchants gathered, and this coalescence of people, larger and larger each traveling season, gradually became the makings of a town.

The cities of the eastern American fall line are well known today—Baltimore, Washington, Richmond, Fredericksburg, Philadelphia—even though the part that the very similar accidents of geology and river behavior played in their origins may have been long forgotten. Hints of the origins can still be spotted, though. Invariably there will be a bridge in each city, spanning the river that brought the first boatmen here hundreds of years before. Cast your eyes over the parapets, and the stream below will quite probably be running through the bridge fast and furious. This is where the first sailors stopped, and the speed and temper of the river are the reason why.

The same kind of demarcation line—made by subtly different geologies but resulting in similar topographies—also affected and afflicted most of the other English colonists who had arrived in more distant places. Once their settlements were up and running in the New England possessions, they began to explore and push up the rivers, just as John Smith had done in the Chesapeake.

Men long forgotten did the exploring. Edward Winslow went up the Connecticut River in 1632, was stopped by fall-line rapids, and settled at Hartford. Walter Neale went up the Piscataqua. Simon Willard sailed up the Merrimac. Bangor in Maine turned out to be the city waiting to be built on the fall line of the Penobscot; Augusta was that city at the rapids on the Kennebec; the town of Fall River would be erected near the rapids of the Quequechan.

And perhaps the most economically and historically important of them all—there was the Hudson River and the point where it, too, crossed the line. This was where it became suddenly inconvenient for Henry Hudson, an Englishman then exploring for the Dutch, to pass any farther northward by sail. Here he stopped—and by his simple brief presence inaugurated the beginnings of the city that would become the New York State capital, Albany. As it happens, his discovery inaugurated a whole lot more besides.

So on all of these rivers, beside all of these halts, there were now settlements to create. And upstream of them all there now were mountains to cross. How best to do this—especially when it was realized just what was to be found on their other side—was to become a dominant business of the America of the late seventeenth century.

It was a business that would be dominated by the awe-inspiring immensity of the great rivers that would then be discovered—rivers that made the eastern streams puny by comparison. And it was also a time when men would start to become obsessed by a need to link all these newly found rivers together in a great vortex of travel, trade, and settlement. They would do this in large measure by the construction and use of what to America would be a wholly new invention: the canal.

The Streams beyond the Hills

Surprisingly few of America's big rivers empty directly into the sea. A lot of small ones do, particularly all of those that drain down from the Appalachians into the Atlantic. But they are somewhat modest rivers; some of the rivers waiting to be found in the America of the sixteenth century were true monsters, and it so happens that most of them keep their distance from the ocean.

In that simple geographic sense, the making of the North American continent has been achieved rather differently from elsewhere. Most of the iconic European rivers—the Rhône and the Rhine especially, as well as the Thames, Shannon, Danube, and Seine—do empty into oceans. The Amazon does too, and the Plate and the Orinoco. The Niger, the Gambia, the Congo, the Nile, and Kipling's "great grey-green greasy Limpopo," all

ease out into the Atlantic Ocean, the Mediterranean
Sea, or the Indian Ocean. The Yangtze and the Yellow
Rivers spend themselves in the East China Sea, and
all of South Asia is watered by rivers, from the Indus
in the west to the Ganges, the Brahmaputra, and the
Irrawaddy in the east, which disgorge themselves into
the Arabian Sea or the Bay of Bengal. Even Russian
rivers—the Lena, the Ob, the Yenisei, the Volga, and
the Black Dragon (which is shared with China), end
in oceans. And Canada's—the Saint Lawrence, the
Fraser—do too.

But the same is palpably not true for the United
States.

Of all the great, face-of-the-nation, visible-from-
space, known-by-all rivers that dominate the American
landscape—the Ohio, the Tennessee, the Mississippi,
the Missouri, the Colorado, the Arkansas, the Snake,
the Platte, and the Columbia—just three of them open
directly into salt water. Most of the shorter East Coast
rivers do, but most of the continent's big boys do not.
They simply flow into other rivers. They are huge
tributaries, or tributaries of tributaries, that feed into
the supergiants that then do the work of carrying their
waters on and into the oceans.

With so few American rivers reaching the sea, the
others simply did not exist in the minds of any passing

mariners. Those could not be seen and so were left undiscovered for much longer than their size and subsequent importance might suggest. Instead the Europeans first discovered the three rivers that did drain directly into the ocean—the Colorado,* the Columbia, and the Mississippi.

The Mississippi was the first of the three to be seen. A sinuous line approximating its position had already been marked on Spanish maps made as early as 1513; it was most probably first properly viewed six years later, in 1519. There was the certain and definitive encounter with the river twenty years later still, when Hernando de Soto, plundering and destroying his way across the southeastern quadrant of the country, stood on a limestone bluff just south of where Memphis lies today and, astonished, saw the great brown river unwinding slowly hundreds of feet below. These men, the first Europeans definitely to see America's defining river, assembled crude log barges and managed to cross the stream, which at this point was "almost halfe a league

---

*   The claim that the Colorado reaches the sea is these days somewhat fanciful. So much water is now extracted by irrigation canals and as drinking water supplies for Los Angeles and San Diego that the river reaches the Gulf of California in Mexico more often as just a smear of damp sand. It is hard to remember that until the first dams were built in 1908, there was a moderately healthy paddle-wheeler service to Yuma and beyond.

broad," and as an English translation of one the conquistadores' diaries had it,

> . . . if a man stood still on the other side, it could
> not be discerned whether he were a man or no.
> The River was of great depth, and of strong
> current; the water was alwaies muddie; there came
> down the River continually many trees and
> timber.

The Colorado, fifteen hundred miles away in the
Far West, was next. Its delta was spotted from the sea
in 1536 by a Spaniard exploring the Gulf of California.
Just as with the Mississippi, its first major exploration
was made from overland, when Francisco Coronado
heard rumors from local Indians of a large river across
the desert: he accordingly went, saw the lower parts of
the Grand Canyon, but, for some inexplicable reason,
left distinctly unimpressed. Perhaps rivalry played a
part: perhaps he suspected that de Soto was stealing his
thunder by managing to cross the much more impressive Mississippi back east. It has never been made clear
just how much the two great conquering expeditions
knew of each other: de Soto certainly heard rumors of
Coronado's journeys in the desert, but of the reverse,
little is clear.

The third of the big sea-reaching rivers, the Columbia, which gets to the Pacific Ocean in today's Oregon, was not to be found for fully two more centuries, and when it was discovered and sailed into, the explorer was not an outsider at all, but a homegrown American.

The Royal Navy's George Vancouver sailed past its entrance in the spring of 1792, as did an American fur-trading sailor named Robert Gray, aboard a merchant-man named the *Columbia*. But then, while at anchor close to where Seattle now stands, these two men exchanged notes, Gray insisting he had seen muddy river waters dirtying the seas off Cape Disappointment. The world-traveling Captain Vancouver sagely expressed his doubts. But something then prompted Gray to turn back, to investigate the suspect waters, and eventually to find the chain of sandbars that marked and protected what he surmised was a river entrance. He lowered a small boat and in short order sailed past the sandbars and in through a safe passage, into the estuary of the most important river on the American West Coast.

He named the river for his ship; only a small tributary stream was named the Gray. George Vancouver, who then used Gray's charts to make a journey up the river a short while later, is by contrast memorialized almost everywhere, the presence of two cities of

Vancouver, one in Canada and the other nearby in Washington, being a rather overgenerous memorial to one man's failure to find the most important geographic feature in the region.

In the late seventeenth century back in the East, English explorers were boldly crossing the river fall lines (by portaging: they had not yet begun to construct canals) and were also starting to clamber up and over the Appalachian Mountains. They were becoming very well aware of the presence, some hundreds of miles away to their west, of the Mississippi River.

More important, so were the French. While the English began somewhat timidly making their way toward this river from their colonial outposts in the East, the French headed swiftly toward it from their clusters of forts and settlements around the Great Lakes. In doing so, they were determined not merely to find this river but to own it.

It was Louis Jolliet and a Jesuit priest named Jacques Marquette who succeeded in locating it and formally seizing it. They first crossed from Lake Huron into Lake Michigan, then paddled up the Fox River past where Green Bay stands today, made a portage of a mere *two miles* through a tangle of swampy grassland before reaching another, south-flowing stream—today's Wisconsin River. A simple seventy miles of canoeing with the current down this pretty little river, and then, voilà!—the rolling gray immensity of the country's mightiest waterway. It was June 17, 1673. The river they had first heard of from Iroquois lake traders back up in the frigid north was now here in its vast reality; and from the strength and direction of the stream, there could be no doubt that voyaging down it would take them eventually all the way to the Gulf of Mexico.

They sailed their tiny craft through 450 miles of rips and boils and roaring river waters. They slid past the confluences with other giant rivers—the Des Moines, the Missouri, the Ohio, and the Arkansas. And then suddenly, a frisson of nervousness: they became troubled at the increasing likelihood of meeting armed scouts of the fiercely protective Spaniards, who they knew controlled the river's lower reaches and of whom local Indians now began to bring reports.

So they laid their claim: to the north of the junction with the Arkansas, all of this huge river and its watershed now belonged to the majesty and dignity of New France. Until the Louisiana Purchase was signed in 1804, this claim was widely accepted—with one major exception: the status of one of the Mississippi's largest tributaries to the east, the Ohio.

This the French keenly wanted, too, but for decades after the Jolliet and Marquette expedition, they found themselves running into stiff and unyielding resistance from the British. And for a quite understandable reason: the Ohio River's watershed drained lands on the far side of the Appalachians, lands that Britons had now energetically started to settle and farm.

The simmering dispute over the ownership of the Ohio reached its boiling point 1749, when a magnificently named New French naval officer, Pierre-Joseph de Céloron de Blainville, sailed down the Ohio and claimed it in a manner that was traditional in Europe, but which to this new world was famously inventive. Beneath prominent trees that stood at the confluences of half a dozen of the tributaries of the Ohio, Céloron secreted specially made lead plates, formally announcing the annexation of the surrounding territory for his monarch.

Each of these plates was hammered from solid metal, shaped like a rough tablet about a foot long and seven inches wide. And on each was engraved a portentous message of seizure in the name of the royal governor of Canada, the equally exalted-sounding Roland-Michel Barrin, Marquis de La Galissonière. A blank space was left in the middle to insert the name of each tributary being claimed: few imperial pronouncements can have been so sonorous:

> L'an 1749 du regne de Louis XV Roy de France, nous Céloron, commandant d'un detachement envoie par Monsieur le Mis. de La Galissonière, commandant general de la Nouvelle France, pour retablir la tranquillité dans quelques villages sauvages de ces cantons, avons enterré cette plaque au confluent de l'Ohio et de Tchadakoin ce 29 Juillet, près de la rivière Oyo, autrement Belle Rivière, pour monument du renouvellement de possession que nous avons pris de la ditte rivière Oyo, et de toutes celle qui y tombent, et de toutes les terres des deux côtes jusque aux sources des dittes rivières ainsi qu'en ont jouy ou du jouir les précédents rois de France, et qu'ils s'y sont maintenus par les armes et par les traittes, spécialement par ceux de Riswick d'Utrecht et d'Aix la Chapelle.

As translated about 1877 by historian Orsamus Holmes Marshall:

> *In the year 1749, of the reign of Louis the 15th, King of France, we Celeron, commander of a*

detachment sent by Monsieur the Marquis de la Galissoniere, Governor General of New France, to reëstablish tranquillity in some Indian villages of these cantons, have buried this Plate of Lead at the confluence of the Ohio and the Chatauqua, this 29th day of July, near the river Ohio, otherwise Belle Riviere [Beautiful River], as a monument of the renewal of the possession we have taken of the said river Ohio and of all those which empty into it, and of all the lands on both sides as far as the sources of the said rivers, as enjoyed or ought to have been enjoyed by the kings of France preceding and as they have there maintained themselves by arms and by treaties, especially those of Ryswick, Utrecht and Aix la Chapelle.

Fine sounding this may have been to Gallic ears, but to the English colonists on the far side of the Appalachians, this annexation by buried plates was as preposterous as it was bizarre—a gesture akin, as one historian put it, impresciently, "to planting a flag on the moon." And it was one of the many factors that helped to spawn a period of intense skirmishing between locally based French and British militias. It was part of the fighting that was to become known in

North America as the French and Indian War, or "the war that made America," as some have called it.

In the rest of the world, this was part of the very much greater Seven Years' War, a costly episode that by the eventual British victory—formalized by the Treaty of Paris of February 1763—altered borders and the shapes of nations and resulted in land exchanges stretching from India to the Caribbean. Most important of all, the trouncing of the French in North America led to the establishment of Canada as a British possession.

In the context of this particular story, the war led also to formal cession in 1763 of the Ohio River valley to the British. The initial idea was that these newly acquired lands west of the Appalachians were to be left as a reserve for Indians; in fact they became de facto British territory, because so many land-hungry British settlers were trekking over the hills and decorating the region with homesteads. The settlements of these colonists had the effect of pushing the frontier of French-owned lands several hundred miles farther to the west, across to the west bank of the Mississippi. And to make certain that matters in these newly demarcated territories remained stable and secure, the British then built a series of forts, the first of them constructed at the confluence of the Monongahela and Allegheny

Rivers, and where the Ohio River starts: today's city of Pittsburgh.

The British victory helped win prominence for a twenty-six-year-old scion of the Virginia tobacco-growing gentry, George Washington. His own victory while fighting the French in the Ohio Valley would help in his appointment forty years later as the first president of the United States. But it also played a more immediate role in his visionary decision to use the country's rivers—and to build artificial rivers, as well—as a means of joining the newly discovered vastness of America into one.

It all started because a grateful King George II back in England had promised his loyal American lieutenant some twenty thousand acres of prime Ohio Valley land as reward for his battlefield services. The newly won British control of the Ohio watershed then gave Washington and his friend and colleague in Virginia, Thomas Jefferson, the gleam of an idea: in order to encourage the settlement of this valley by Americans, should they now perhaps attempt to forge a physical link between, on the one hand, those settlers who lived in the valleys of the Potomac and James Rivers on the eastern side of the Appalachians and, on the other hand, those in the newly conquered lands on the western side?

A physical link could mean creating a network of roads, of course. But it occurred to George Washington that it would be far more effective and economical to construct a brand-new river of sorts, to carve it clear across the intervening hills, linking valley to valley by water. It had been done before, in Europe, where it had proved a considerable commercial success: the valleys could be connected by *a shipping canal.*

The land Washington had been given—which he later expanded by buying still more, eventually owning sixty thousand acres—lay at the confluence of the Ohio and the Kanawha Rivers. The Kanawha's source on the west side of the Appalachians and the sources of both the Potomac and the James Rivers on the east side of the same mountain chain were a scant thirty-three miles apart. So it seemed eminently reasonable at least to try to create canals linking these particular streams.

And so in the late 1780s, once the Revolutionary War was over, the first practical steps were taken, plans were drawn up, and constructions were commenced that might in time create a series of passageways. The East of the country could in theory now be united with its Middle West across fully a thousand miles, entirely by water. George Washington would be the first to attempt to do so.

The Pivot and the Feather

Rivers—whether broad, deep, and influenced by tides; or cold, narrow, and interrupted by rapids and waterfalls—had always been central to the development of the young George Washington. And in his maturity there was one expedition, taken entirely by river, that changed everything.

Washington had been born beside a creek that flowed into the Potomac, and he spent much time fishing and swimming in the inlets of Chesapeake Bay. He rode out across the Allegheny foothills to explore the upper reaches of the James River and the Shenandoah, and he cut his teeth as a surveyor in their wilderness valleys, learning the craft, winning the license that would propel him to his first job, at seventeen, as the official colonial surveyor of Culpeper County.

During the wars against the French, he became all too familiar with the winding progress of the Ohio River, and he had been granted land at the confluence of the Ohio with the Kanawha. His heroic crossing of the Delaware River on Christmas night, 1776, the first crucial move in the War of Independence, remains one of the most famous riverine expeditions mounted in American history.

But it was a very much longer river expedition, undertaken after the war was over, in September 1784, that cemented Washington's lifelong belief that waterways were vital to America's expansion and consolidation. By now he had retired from the military, had stunned the world by resigning his military commission, and returned to Virginia to continue planting tobacco.

But he was restless. His fond memories of the victorious Ohio campaign against the French thirty years before, and his new awareness that settlers were now streaming across the mountains to make homes for themselves in these new-won western lands, prompted him to mount an expedition to the western frontier, to see for himself a little of frontier life. His plan was to venture from the bluff overlooking the Potomac where stood his family home—Mount Vernon—and make his way by horse along the river valleys to the small British fort-city of Pittsburgh.

He began by traveling due west along the Potomac, passing rivers named the Monocacy and the Conococheague, the Cacapon and the Antietam, and his old friend the Shenandoah. As he pressed on and up into the hills where the forests closed in, the streams became smaller and faster, rivers became creeks, creeks became rills. In the far west of Maryland most of them—the Savage and the Bear and the Muddy and the fancifully named Rhine—ceased being in any way navigable, even by small canoe, and became mere springs, trickles, and sources.

There then came a defining moment of local geography: one low ridge of mountains lay ahead, and on its far side Washington and his team discovered the beginnings of a downward slope where all the trickles and creeks and rills poured no longer eastward but instead *to the west*. The ridge marked a true tipping point. They were crossing the Eastern Divide; they had left the basin of the Atlantic Ocean behind them and had entered the drainage of the Mississippi River. They had entered what was then, spiritually if perhaps not technically, the West.

The Youghiogheny River, which the men found flowing north with the grain of the mountain ranges, is the only river that rises in Maryland—very much an Eastern state and one of the original thirteen—and empties its waters into the faraway and very Western entity of the Mississippi. They hoisted their canoes down into it; they bumped down it to still water, and found it then duly connected with the Monongahela, which the group then took downstream to its confluence with the Allegheny, at the fort where Pittsburgh now stands. The junction of these two by now quite large streams forms the Ohio, and downstream from the small settlement stretched its own broad reach of water, spearing southwestward to lands that were still quite empty but evidently full of promise and filling with newcomers.

The Youghiogheny, the only river to rise in far western Maryland and send its waters to the Mississippi, eventually gave George Washington a navigable route to the Ohio Valley. Downstream from here it joins the Monongahela, which eventually merges with the Allegheny at what is now the city of Pittsburgh.

There were settlers at the confluence—hundreds only, but those among them skilled as boatbuilders were making a bustling business. Washington took his expedition onward down the Ohio and could hardly avoid noticing how it was becoming a highway, almost crowded, with newly built Pittsburgh boats running in tandem with him, all carrying young men and their families downstream. On the prow of each craft, a man would invariably be standing, gazing this way and that, peering intently at the passing shores. Once in a while he might cry out at the sight of a favorable-looking inlet or a flattening meadow-to-be or a stand of interesting-looking trees, and he would order his boat to turn in to land and would moor it and leap ashore to see what promise the land might hold.

Some of these families might then stay put. The passersby could see places on the riverbanks with scatterings of survey flags or crude signs marking a claim to ownership. Farms, hamlets, and small villages were springing up all along the river, and every so often, tracks were curving up the banks and vanishing into the woods, marking the sites where men were cutting roads into the hills, making farms even farther away. The place was a bustle of activity. America was being built here, farm by humble farm, along the banks of the Ohio. George Washington was witness to a moment of creation.

After two hundred miles of steady sailing, he reached the spread of lands that the former English king had given him, at the point where the Kanawha

River comes into the Ohio from the left. This seemed as appropriate a place as any, and Washington ordered the party to leave the broad, slow Ohio and turn up into the Kanawha. They sailed back through the hills as far as possible, noting where it narrowed, speeded up, and became a creek, and finally with an inevitable short portage, he and his men crossed back over the divide, high on the Alleghenies. Once on the hills' downward slopes, the men reached the white-water headstreams of the James River, and once it became navigable, they floated back down eastward through Virginia to the fall line at Richmond and thence across the alluvial flats and back into Chesapeake Bay, up to the Potomac, and finally to Mount Vernon and home.

If not especially long by later pioneering standards, this had been a remarkable journey, and Washington returned from it with a singular conviction.

The settlers he had met in the Ohio Valley, he reasoned, were in a peculiarly vulnerable position. They were still few in number, and they were sur-rounded—by the French to their west, by the Spaniards to their south, and by the English loyalists and royalists up north in Canada. If these Ohio settlers did not feel intimately conjoined to their fellow Americans across the hills back east, they might well throw in their lot with some non-Americans who might take better care of them. They might well trade first and foremost with the Spaniards, sending their farm produce down the Mississippi to New Orleans. It would be an altogether easier and less expensive means of making a living than trying to sell goods that needed to be hauled and hefted back to the East.

So Washington issued a dire warning to the governor of Virginia, in a letter he wrote on October 10, 1784. The Western settlers, he said (and underlining the point that he spoke *from his own observations*), "stand as it were upon a pivot. The touch of a feather would turn them any way."

The implication was obvious. Were this feather touch ever to turn them to Spain or France and against the original colonies, the newborn America for which Washington had fought so bitterly would be doomed. All that would remain would be a small coastal country clinging to the edge of a vast continent ruled by Europeans of uncertain—meaning non-British— reliability. Washington found this prospect utterly unacceptable.

He knew of one certain solution, however. He knew that the one true cement that would indissolubly bind the new settlers to the men and markets and institutions of the United States would be money—by the practice of trade and commerce.

The settlers—energetic, adventurous men and women, full of spunk and ideas and courage, who were made of the stuff that would make the ever-enlarging nation truly great—could worship whichever gods they chose, could adopt whatever local politics they wished, and could choose customs and callings quite different from those they left behind. But Washington confidently asserted that if they could perform all their important business dealings not with foreigners but with the United States, if they could feel themselves bound to their forebears by money, then they could and would remain contentedly members of the Union. The United States would, as a consequence of the new-forged economic bond, remain powerfully and perhaps permanently united.

Yet how exactly could these newcomers be bound physically with the people back east? The answer was in theory inescapable: physical links needed to be built across the hills that now separated them.

There were some roads, true—a small scattering of niggardly and rutted tracks that wound through mountain passes—but they were ill suited for more than the occasional horsemen, let alone wagon trains of trade goods. But rivers, by contrast, were there already, several broad waterways along which cargo-laden boats could move with ease and speed. All that was needed was that they be connected with those on the west, and George Washington and his colleagues knew from the experiences of traders back in Europe that the building of canals would solve this problem.

So he backed two schemes, first becoming president of the Patowmack Company, set up in 1785 and charged with building a canal in the direction of Cumberland, then naming himself a sponsor a few months later of the James River Company, intended to bring a water-way up to link with the Kanawha and a more southerly section of the Ohio Valley. If either succeeded in actually building a canal, Washington would be doubly enriched—by helping to create major pieces of unifying infrastructure for the new country and by bringing trade and prosperity to the lands he already owned.

But neither project went anywhere. It mattered little that Thomas Jefferson was an early believer in one of the schemes. "Nature has then declared in favour of the Potowmack," he wrote to George Washington, noting the river's seeming closeness to the Ohio, "and . . . it behoves us then to open our doors to it." Nor did it matter that the auguries had been so good. An engineer named William Weston came over from England to supervise and lend his advice. A clever and eccentric builder of steam-powered riverboats, whom Washington had met on his 1784 expedition, had also pledged to help with the more technical aspects of the construction. But the expense of the thing was the problem, and soon money was flowing faster from the Patowmack Company coffers than waters from a pound lock sluice gate.

Confidence in the commercial sense of the project sputtered out after only a few miles, after the building of five sets of locks around the Great Falls, which lay just a few miles north of the starting point in Georgetown. The canal never got anywhere close to the Ohio, and it took seventeen years to build to where it was termi-nated in western Maryland.

To be sure, Washington's canal did eventually allow boatmen to travel all the way from the hill town of Cumberland down to the Tidewater—a journey that might take three days going downstream, two weeks for men who had to pole their boats all the way up through the mountains. The cargoes carried on the little keel-boats were typical for any passageway between a devel-oped and a developing society—raw materials, flour, whiskey, tobacco, and iron ore on the downstream leg, manufactured goods like guns, clothing, and hardware on the journey back.

But the tolls that could be charged for the boats scarcely covered the interest on the company debt. When in 1828 the firm was taken over by the

Chesapeake & Ohio Canal Company,* it had debts of around $200,000, immense at the time. By this time, George Washington was long dead; he had died two years before even the Grand Falls locks were opened and essentially saw nothing of his vision and never knew of the thwarting of his dream.

Much the same fate befell the canal that was supposed to link the James and Kanawha Rivers, and again, it failed in spite of Washington's keen involvement. It ran out of money; wars (such as that with the British in 1812) interrupted its construction; technical problems plagued it; and when it was abandoned in 1851, it had reached only the mountain town of Buchanan, 150 miles west of the starting point in Richmond, with no plans to take it over the summits and down into the Ohio Valley.

In the matter of canal making, Washington was more of a dreamer than a builder. True, it was he who first came up with the idea that waterways could one day knit the land together. His plans for doing so foundered early on, however, and men of more practical bent were left to design and dig waterways along routes very

---

* The C&O eventually built its own canal, 185 miles long, which went all the way up to Cumberland, with seventy-four locks, eleven aqueducts, and a total rise of six hundred feet. But it was obsolete by the time it was completed: the Baltimore & Ohio Railroad got to Cumberland first.

different from those Washington preferred. He had the vision and set the tone, but his particular dreams were not to be realized, and the country would in the end be connected by waterways that ran westward from starting points a very long way from his home country in Virginia. And so far as the ultimate development of the nation's geography was concerned, that was perhaps just as well.

The First Big Dig

In one of those happy synchronicities of history, by the time the first real shipping canals were being excavated in America—an explosion of construction that started in the 1790s—the technology had just been perfected in Europe. In England, France, and Germany, engineers now knew how best to drill the tunnels and create the flights of locks and the aqueducts needed to make navigable artificial waterways.

So it is no wonder that American engineers made strenuous efforts to learn these new techniques. Before even starting to make a waterway, they needed to know how best to survey the land and how to ensure that the bottom of the ditch they planned to build across it was kept level across great distances. Special equipment that was quite unknown in America at that time was required to achieve this—in particular leveling telescopes, special theodolites, and most particularly an elevation-measuring device known as the Y (or wye) level, developed by William Troughton in London.

The engineers had to learn how to dig enormous trenches, removing millions of tons of earth, and how to employ explosives safely to deal with heavy nuisances like tree trunks and embedded rocks. They needed to know how to puddle sand and clay together on a ditch bottom, making it watertight, and how to concoct the proper formula for a cement that would set and remain strong while totally submerged and so prevent water from leaking away into the canals' sides.

They also needed to know how to design and build proper pound locks—to *impound* water between gates to raise or lower a craft passing up or down the waterway. They would have to fashion wooden lock gates strong enough to hold the immense tonnages of water yet light enough to be opened and closed by hand by passing boatmen or by lockkeepers and their wives. They needed to know how to make and operate the special valves and sluices and reservoirs that would move the waters into and through the gates, allowing watercraft to pass safely along, up and down the hills through which the canals would be cut.

They needed, in short, to learn everything in order to begin the frenzy of what came to be known as America's Canal Era. Britain had become the new center of canal construction, so the Americans went off to towns like Birmingham and Manchester and Gloucester and London to watch and listen and to learn from the great engineers, men like Thomas Telford, John Rennie, Benjamin Latrobe,* John Smeaton, and William Weston, men who were weaving Britain's canals into the complex and clever system that still exists to this day.

Among the most celebrated visitors to Britain were the Loammi Baldwins, a father-and-son team of American canal engineers who played vital roles in the churn of invention and practical imagination that seemed suddenly to grip postrevolutionary America.

Some like to call Loammi Baldwin Sr. the father of American civil engineering. (This was a time of many fathers, in every field from astronomy to zoology, as well as engineering to canal making; to make matters even

---

* Following his success with canal construction in England, Latrobe, a Yorkshire-born classical architect, came to America, where he was appointed one of the lead designers of the United States Capitol building—both before and after its partial burning by British troops in the war of 1812.

more confusing, some also like to give Loammi Jr. the same title.) The elder Baldwin had an extraordinarily varied career. He was the son of a Massachusetts carpenter and, thanks to his early but unexplained interest in hydraulics, was first employed by a local firm that made water pumps. He was destined for greater things, however, and in his later youth became greatly inspired by hearing the weekly Harvard lectures given by John Winthrop, a mathematician and astronomer and one of the first American intellectuals to be taken seriously by the academics of Europe.

Baldwin would regularly walk the ten miles from his home north of the city to hear John Winthrop's talks, often in the company of a polymath scholar friend, the British inventor Benjamin Thompson, who became famous in his own right.* The Revolutionary War interrupted matters, however. Baldwin accepted a commission as a major on the rebel side, and his regiment was one that crossed the Delaware with George Washington and fought in most of the famous early battles—Lexington and Trenton most notably.

---

* As Count Rumford, while living in Bavaria, where he acquired the title, Thompson became fascinated by the phenomenon of heat and the science of thermodynamics. He won lasting repute for inventing a more efficient fireplace, a coffee percolator, a highly nutritious soup of pearl barley and sour beer, thermal underwear, and the dessert known today as baked Alaska.

On his return home to civilian life, Baldwin briefly became a county sheriff but soon reverted to his interest in hydraulics, physics, and mathematics, all essential in canal building. Massachusetts, in common with most of the early states, had seen the wisdom of waterborne trade, and men like the Baldwins were summoned to help build structures that would change the fate and the future of the American economies and the shape and size of American cities for years to come.

Few of these canals looked as pretty as those that had been handcrafted by canalsmiths and masons and carpenters back in Oxfordshire or outside Bath or deep in the valleys of Wales, and they were certainly not in the same league as the Canal du Midi in southern France, breathtakingly lovely still. The American canals were industrial monsters built for the demands of big cities and big business. Most of those that remain look that way still. The canal landscape of today's United States is dominated by rusting iron gates, cement walls, and greasy hydraulic gearings rather than by limestone cottages, wooden lock gates, and baskets of gillyflowers. But in most cases, the canals that were finished worked exceptionally well; they each made local history, and they all changed national geography by doing so.

There is endless bickering about which canal was America's first. Much depends on definitions, and proponents press the claims of dozens of early river-improvement schemes and bypass canals. The South Hadley Canal, in central Massachusetts, built in 1792, is a fine example. It was short, only a couple of miles long, and was designed to bypass the rapids on the fall line of the Connecticut River. But it employed something quite majestic to do so: an enormous water-filled iron bath on wheels, a caisson, in which ships floating serenely inside could be winched on iron rails up the slope by a pair of gigantic chains, the whole mighty ensemble powered by waterwheels turned by waterfalls. The South Hadley Inclined Plane has now been lost to time and rust, but a seal memorializes the great engine in an engraving, and the fact that it was built—by an otherwise long-forgotten engineer named Benjamin Prescott—without any template or precedent is a reminder of just how good Americans were quickly becoming as makers of great and complicated pieces of machinery.

The Dismal Swamp Canal, on the border between Virginia and North Carolina, is an early claimant too. It was basically a dredged passage through the marshes, and it enjoyed brief fame because of a hotel built on its banks directly on top of the state line. Young eighteenth-century swells would hold duels here, one man standing in Virginia, the other on the far side of the border, making their crime legally ambiguous—particularly important if one of them died. And gamblers could scurry across the hotel lounge into North Carolina whenever any Virginia marshals arrived to break up their game.

The Santee Canal, begun in 1793, connected the South Carolina port of Charleston with the newly made state capital of Columbia, well inland. Designed by a Swede strangely named Senf, built by slaves who worked under terrible conditions of summer heat and insects and snakes, it briefly saw cargoes of indigo, rice, and cotton carried from the plantations through eight locks that took it up and over a series of rises totaling about thirty-five feet and thence down to the sea. But it was poorly made; it was shallow, narrow, and could take boats that carried only a few tons of goods. It swiftly fell into disuse.

The Santee's fate mirrors that of all too many of the early canals, whether they were waterfall-bypass constructions or full-dress navigational waterways. For despite all the arguments over to which are due the laurels, there is one reality about all of the major American canals that were planned before the Loammi Baldwins and their successors entered the business. Each one of them ultimately failed, and most of them did so very quickly. Even those that were imagined, planned, and begun by great dreamers and unifiers like George Washington himself turned out to be pipe dreams. They were pioneering constructions of enormous scope and ambition, with profound symbolic importance, but they never turned a dollar of profit, and they joined the country together only in theory, not yet in practice.

It was shortly after the first sod for the James River Canal was cut that men like Baldwin started to take over from the dreamers and the speculators. The specter of failure that had dogged the pioneer builders for so long finally started to abate. With the involvement of professionals, the true worth of American waterways began swiftly to be realized, and a network of sorts began to take shape.

Baldwin cut his teeth first on a project near Boston, a navigation canal that would be a model—technically but not financially, for it didn't do too well either—for most of the other great triumphs of canal engineering that would follow. This first was the Middlesex Canal, which connected the Merrimack River to the port of Boston.

In applying to be in charge of its design and construction, Baldwin was somewhat hesitant. "It might be said," he admitted to the directors of the newly formed Middlesex Company, that he had "no Experience at all. It is true I have studied the theory for many years, and have been at Considerable pains to get possessed of the principles of Canaling, but I have never seen one foot of Canal which has been completed in a proper manner."

Nonetheless, he was hired, and he quickly got help. The directors agreed to pay $2,000 to William Weston, who had given technical advice to George Washington; he came up from Pennsylvania to lend his British-won experience to Baldwin. And the combination seemed to work: after eight years of construction, the canal was opened in 1803. It was a formidable achievement on many levels. It was the longest canal in the country at the time—and it immediately helped the city of Boston to develop its muscles as a full-blown commercial city and, for a while at least, a great American port.

The construction, which had been prompted by a fall-line chain of rapids and waterfalls along the Merrimack that made the river essentially non-navigable, coincided with and then accelerated the planned development of the city of Lowell as one of early America's great mill centers. Long rows of cloth factories, roaring with water-powered loom noise, were promptly thrown up alongside the waterways where the Merrimack and the Concord Rivers join. Bales of Southern cotton were then brought in, and millions of yards of finished textiles were then sent out—and all of them along the Middlesex Canal.

The twenty locks, seven aqueducts, and well-cemented walls (and the newfangled technical marvel of a floating towpath) of this beautifully built and scrupulously maintained waterway allowed supply boats to get down from Lowell to Boston in only eighteen hours, the goods then being sent on for export or transit to the rest of America. Freight rates dropped almost overnight: the rate from Lowell to Boston fell from $8 a ton to $4, the up-country rate from $13 to $5. The city grew in a fury of construction, and even as a shadow of its former self, it remains at a respectable hundred thousand souls today.

Unintended consequences were legion. Organized labor came to Lowell in the mid-1830s. Irish navvies came in to dig yet more waterways within and around the city, and thousands of young Irishwomen were brought across the ocean to join refugees from the hardscrabble New England farms who came to Lowell to work in the mills on the endlessly thwacking looms. They became the long-remembered Lowell Mill Girls. These women eventually decided to protest their initially appalling working conditions. They struck blows for the rights of the woman and against the ills of wage slavery. With strikes, slowdowns, and work-to-rule actions, they won reforms that eventually elevated Lowell to national renown as a seat of industrial and social experiments that have affected all of America ever since.

This is not the place to discuss in detail these changes to the nation's labor arrangements, but suffice it to say that the reforms made within the immense brick mills of Lowell were born of a simple fact: that this particular industrial town was located where it was, was planned to make what it made, and was designed to do what it did in large part because there was now a well-run artificial waterway to take its goods speedily and inexpensively to market. The Middlesex Canal was not merely "an example of early American engineering at its finest," as one history suggests. It was an indication of a new trend: canals were beginning to alter the social fabric of the nation.

Albert Gallatin, President Jefferson's Swiss-born (and notably French-accented) treasury secretary, remarked that the Middlesex Canal was "the greatest work of the kind which has been completed in the United States." Gallatin—among whose many legacies is New York University—made this declaration in a formal report that outlined how important he and his government now believed canals were. Together with good roads, he said, they were central to the making of the nation. They should wherever necessary be financed by governments. They already were changing the geography of the country. And they were now well on their way to achieving a far more profound effect on American society than even the unfulfilled dreams of George Washington had ever supposed.

Moreover, though commercial success may still have eluded it, the Middlesex was a technical triumph like no other. Those who had worked on its construction learned much that was applicable to later projects, most of them far greater and grander than this. In particular, the success of this waterway was to lead to the construction, with work starting just twenty-three years later, of what is both symbolically and practically the most important canal ever built in the nation. The Middlesex was a pioneer canal, just 27 miles in length, and it changed the face of manufacturing on America's East Coast. The Erie Canal was to be 363 miles long, and it would change the face of America.

The Wedded Waters of New York

It was a notorious geography question, well known to teenage schoolchildren of my generation back in the Britain of the 1950s. "Discuss the significance," it usually went, "of the Hudson-Mohawk Gap."

Probably few children today—even Americans—would even try to answer the question. But the Hudson-Mohawk Gap is a thing of undeniable importance: It is a geographic feature, as plain as the nose on your face, composed of a pair of joined valleys that scythe through the hills of the Eastern United States. It offered an obvious route between the Atlantic Ocean and the lakes and rivers of the American Midwest.

At one end of the gap lies the city of New York. At the time that this saga begins, in the early nineteenth century, it is only a modest-size port, bustling but in no sense a metropolis of world-class power. At the gap's other end are the twin cataracts of Niagara Falls, together with the long westward chain of the Great Lakes and, south of their southern shorelines, the endless rural sprawl of the American interior.

If there was any natural feature in the American East that might link these two, this gap, along the valleys of the Hudson and Mohawk Rivers, was it. It might fairly be claimed that New York City owes its very existence as a world-class city to the presence and human exploitation of the Hudson-Mohawk Gap.

For it is all so obvious. Spearing due northward from the sea and the glacial moraine that is Long Island, the Hudson, a broad river of formidable power, flows deep and serene more than 140 tidal miles down from Troy and Albany to the sea. After reaching Troy, an upriver traveler on the Hudson sees it shrink suddenly and visibly, turning into a narrow and much faster-flowing stream that tumbles ice-cold out of a tiny lake called the Tear of the Clouds, high in the Adirondacks. The important local reach of the upper Hudson is not its source, however, but the place where its sudden shrinkage occurs. For this marks the point on the river not only where the Atlantic tides cease to have any effect, but also where another enormous stream joins the Hudson, as a tributary, on the river's west bank.

This is the Mohawk River, not quite as wide as the Hudson but large nonetheless. There is a change in direction here, too. For while the Hudson flows north to south, the Mohawk runs almost precisely west to east, passing through a wide valley that separates two massifs, the Adirondack Mountains to the north and the Catskills to the south. This valley extends some 150 miles toward (but never quite reaching) Lake Erie. Its basin is huge and topographically prominent; thanks to ancient glaciers, it extends a long way beyond the source of today's river, in essence reaching all the way along the southern shore of Lake Erie to the escarpment that causes Niagara Falls.

Joined together, these two river valleys—the north-south valley of the lower Hudson from the ocean to Albany and the east-west valley of the Mohawk up to its source near Constableville, New York—make up the famous gap.

The Mohawk River was first seen by European explorers in the sixteenth century. It had been a popular route west, beautiful to look at but so isolated as to be somewhat feared. At the time of the first declared interest in building a canal, it was still sparsely settled. Its valleys were dark with thick forests; there were huge waterfalls and a thousand tributary streams. Along the banks, a few intrepid settlers had set down tiny log cabins and cleared an acre or so of land. There were a few isolated outposts of perpetually frightened soldiers, and fleeting in and out of the shadows behind them were Indians in their thousands, who had been known to collect scalps from white men who were too insolent or aggressive in their settlement and acquisition of land. Even as late as the 1830s, when Alexis de Tocqueville famously passed by, nature here was still, in his words, "vigorous and savage."

The Mohawk Valley may have been something of a mean-spirited notch through the mountains, but with the Hudson it was the *only* major gap cutting through the Appalachians. And since it headed relentlessly westward, perhaps it could be adapted and protected for human use.

It was here that the dream was born. To the new settlers and builders of early-nineteenth-century America, suddenly the logic became inescapable: it could be commercially prudent and politically useful, perhaps in time even profitable, to build a canal along the valley, a large shipping canal that would bring and take goods and people all the way west to Lake Erie. The Hudson would thereby be linked to the Great Lakes. New York City would thereby have a ready-made route directly to America's heartland.

The eventual Erie Canal, named for its up-country destination, was a creature with many parents. Most of the names of those who had the earliest visions are lost to all but the keenest of local historians. There was, for example, the marvelously alliterative Irish-born duo of Cadwallader Colden (a surveyor) and Christopher Colles (a sewage-pipe inventor), who, though unknown to each other, wrote early on of their conviction that one day the Mohawk would provide the necessary link.

On the higher side of the social scale, there was Gouverneur Morris, a blue-blooded politician, high-class dreamer, perpetual optimist, and tireless orator in support of the plan; on the lower side, one Elkanah Watson, an indentured servant turned heroic social climber. Watson persuaded George Washington to bring him tea in bed while he was a guest at Mount Vernon. He made a dangerous trip all the way to the Mohawk headwaters in 1788 and on his return declared without fear of dissent that "a canal communication will be opened, sooner or later, between the Great Lakes and the Hudson."

Watson went on to state in no uncertain terms just who he thought should pay for it. The people of "the state of New-York have it within their power, by a grand stroke of policy, to divert the future trade of Lake Ontario and the great lakes above, from Alexandria and Quebec, to Albany and New-York." In other words: act quickly and decisively, lest Canada and Britain steal from Americans the future that was theirs by right.

Of all the early backers of the Erie Canal, one of the more intriguing and perhaps even the most influential was a flour merchant named Jesse Hawley, from the beautifully named northern New York town of Canandaigua, formerly the chief settlement of the Seneca, in whose language it meant "chosen spot." In 1805, Hawley was grumbling out loud about his inability to send his flour cheaply and efficiently from the mills around his hometown down to the bakers in New York City.

The roads, he complained, were execrable, worse than impossible. In summer these old Indian trails, barely improved, were chokingly dusty or, in the low-lying parts, mosquito-filled swamps. The use of logs laid transversely over marshy patches to keep them open made for a monumentally uncomfortable ride: not for nothing were such trails known as corduroy roads. And if summertime was not purgatory enough, the roads were routinely blocked by deep winter snows or in spring by marooned carts stuck wheel-deep in pits of glutinous black mud.

The only alternative for a trader like Hawley was the river—except that the tolls then being charged by the Western Inland Lock Navigation Company, which had built a small number of bypass locks around unpassable rapids on the Mohawk, were by common agreement outrageous. There was no doubt, Hawley declared to anyone who would listen, that this highly agricultural part of New York State could survive and prosper only if a canal was to be built—a proper, full-scale, cargo-dedicated Middlesex-like canal that would slice down the valley of the Mohawk to the point where it joined the Hudson. Only then could the wheat and barley of upstate New York get to market at reasonable cost and allow the farmers to make a decent living.

Jesse Hawley, a New York wheat farmer and debtors' prison inmate, wrote fourteen columns under the pen name "Hercules" in the weekly Genesee Messenger, arguing for the construction of what became the Erie Canal. Once freed he was among the first honored when the man-made waterway was eventually completed.

In the early years of the nineteenth century in this part of the world, such sentiments were hardly unusual. They could fairly be said to be the talk of the valley. But there was something most unusual about the place from which Jesse Hawley eventually circulated these grumblings in a series of well-remembered essays. From 1807 onward he was in the Canandaigua Debtors' Prison, serving a two-year sentence for racking up

undischarged debts—most of them to the very navigation company to whom he had been paying the allegedly exorbitant fees. He had originally skipped town,
to Pittsburgh, to try to escape his creditors; but in the
end, he manned up and returned to face the music, only
to get himself packed off to jail. It was during this long
enforced sojourn that Hawley, under the nom de plume
Hercules, wrote fourteen essays for his local paper, the
Genesee Messenger, all of them urging the construction of a brand-new waterway.

These essays were by no means the angered ventings
of an embittered miscreant. They were all concerned
with the future, as he saw it, of the region. They were
well considered and elegantly written, and they mixed
eloquence and prescience in equal measure. Were a
canal to be built, he wrote, "the trade of almost all the
lakes in North America would center at New-York for
their common mart. This port . . . would shortly after
be left without competition in trade except by that of
New-Orleans. In a century its island would be covered
with the buildings and population of its city."*

---

* Hawley's uncommon predictive ability extended well beyond his homeland. "A marine canal, the most noble work of the kind . . . would be cut
across the Isthmus of Darien," he wrote in his thirteenth essay, published
shortly before his release in 1808. He had foreseen the Panama Canal more
than a century before its opening.

They were also quite specific, offering in great detail recommendations for the route the canal might follow. He forecast the numbers of locks that would be needed and the rate of ascent of the suggested route. Lake Erie is 565 feet above sea level; the Hudson at the mouth of the Mohawk, just 5. The resulting 560-foot climb, he reckoned, would have to be accomplished with thirty-six locks over the 363-mile route.

Hawley also predicted with some accuracy what revenues might be expected, and he answered unerringly such questions as where the canal's water might come from, how much the construction would cost, and how similar the design should be to the great European canals, of which he knew a great deal. Finally, Hercules argued vocally and very well that the state—New York—should finance the project. It was just the kind of grand venture, he declared, that only a government could afford and that good government was bound by duty, moral force, and the pragmatic considerations of commerce to undertake. He concluded with a fine rhetorical flourish:

> By the falls of Niagara [the Creator] has given a
> head to the waters of Lake Erie sufficient to flow
> into the Atlantic by the channels of the Mohawk
> and the Hudson, as well as by the Saint Lawrence.
> He has only left the finishing stroke to be applied

by the hand of art, and it is complete! Who can
reasonably complain?

Many people were reading and listening, none more avidly than DeWitt Clinton, mayor of New York City, who later became one of the most revered and memorialized governors of New York State. Memorialized in name, perhaps—fifteen American cities are named for him, and six counties (two in Illinois, one named DeWitt County to avoid confusion)—but in truth, these days he is all too little remembered. And yet he should be: for as the historian Daniel Walker Howe reminded readers of his *Oxford History of the United States,* "The infrastructure he worked to create would transform American life, enhancing economic opportunity, political participation, and intellectual awareness." And the jewel in the crown of that unifying infrastructure was the Erie Canal.

Clinton persuaded the New York Senate to put up the $7 million seed money for construction. Naturally, with a sum like this taken from the treasury, there would be critics and naysayers: Clinton's Folly, the project was called. Clinton's Ditch. But their objections were brushed off: on Independence Day 1817—ten years after Jesse Hawley's first essays were published— the first sod was cut, in the town of Rome, New York. The place had been chosen well, for it was almost exactly halfway between the lake and the Hudson. The workers who flocked in to build the structure were to move outward, west and east.

The sunrise ceremony was an impressive spectacle, with high panjandrums gathered from around the state, with marching bands and immense breakfast tables and gaily colored streamers and flags, and with speeches, speeches galore, the tone and tenor of the occasion well matched to the task to which the state had set itself. The day's excitement related more to the impact of the canal on the world outside than to its effect on the inner workings of the country that stretched away to the west. It was as though everyone already knew and accepted how profoundly the country would be affected: it was time to nudge the world beyond into believing even more. One of the dignitaries forecast that unborn millions would use this "great highway" to "hold a useful and profitable intercourse with all the maritime nations of the earth."

After the boom of a starting gun sounding from the roof of the local arsenal, a local judge loosed his team of oxen, and at Canal Commissioner Clinton's order, he let them pull a symbolic plowshare forward to dig out the first few feet of trench. The construction was properly begun.

Yet it strains belief that the canal was ever built at all. The participants didn't seem to know what they were doing. The men who laid out the routes were not surveyors, but judges. One of the principal builders was an arithmetic teacher. Though many of the bosses had read about the building of the Middlesex Canal, almost none in the early Erie had any experience in the field, and few could even imagine knowing how to work a theodolite, pay out a survey chain, or construct a pound lock gate.

And yet with eight years of heroic endeavor, they did indeed build it. With thousands of red stakes, they marked out the sixty-foot-wide path through virgin forests miles from any other habitation. They then set about clearing a track through the forest, braving clouds of mosquitoes. They decided it was far too time-consuming to fell trees with axes, so they pulled them down with ropes fixed to their topmost branches, bending them over until their trunks broke with great echoing snaps. They removed tree stumps using infernal homemade devices built of chains and gigantic iron wheels, which brutally ripped the roots out of the ground. They then got down to digging the ditch—forty feet wide and just four feet deep (the boats that would use the early canal were to be pulled by mules, with no screw-driven craft that might scour the canal bottom). The earth from the dig was piled up on the north side of the ditch and tamped down to form the towpath, from which the animals would haul the canal narrowboats.

It was grueling, backbreaking, miserable work. But it was work—and the fact that New York State provided the early funds (President Jefferson had turned down flat a request for federal funding)* meant that workers were kept in full employment, despite the various financial crises that were roiling the country and the region at the time. Thousands of Irishmen came across the ocean to take the jobs, open to anyone who was willing and able. Canal laborers were paid "fifty cents a day and found," meaning that free food, drink, and crude quarters were provided in addition to the wages, from which they generally cleared about $12 a month.

They learned to build by building, and efficiency gradually improved. The first section, to Utica, was finished by 1819—though the fact that it took two years for a paltry fifteen miles suggested that it might take three decades to complete the whole thing. But it didn't. Matters accelerated rapidly as more and more men arrived—though fully a thousand died from various fevers, malaria most probably, as the route edged westward across the Montezuma Marshes, close to where Syracuse now stands.

---

* In a rare misstep, Jefferson declared consideration of the canal to be "little short of madness," and asked supporters to return in a hundred years, when it might be a more suitable time.

The city of Syracuse played a vital role in the building of the canal. There had for some years been a prosperous local salt-extraction industry, and the Syracuse salt barons, together with the flour merchants of Genesee and Canandaigua, had long argued for the building of the canal. They got their way early in the process, for Syracuse was linked to the waterway no more than two years after Utica. But there was a sting: as soon as boats laden with the sacks of salt crystals began passing along the waterway, a canny New York revenue authority saw to it that a tax, at the rate of twelve and a half cents per bushel, was levied on it.

So an elaborate, classically styled weigh station was built at Syracuse village; it lifted the barges bodily from the water, and as they hung briefly in the air, dripping, the revenue men calculated the tonnage of salt aboard. Though it never happened elsewhere, the tax seems never once to have been resented—it was seen as a necessary means of raising money for the construction of something that would ultimately be of benefit to all. But today people in Syracuse, now a giant city of half a million, sometimes remind visitors that, in their view at least, the Erie was "the canal that salt built."

Work was completed in 1825 with the final flights of locks built over the Niagara escarpment, hoisting the waterway up the last eighty feet of limestone, beyond which Lake Erie's waters brimmed. This was particularly trying work, most of it performed by hand, with picks, shovels, and muscles. It was decades too early for the "mountain howitzers," the great steam-powered excavators that would be later used at Panama. Instead there was much employment of highly unstable black powder explosive, with consequent loss of life and limb, as the last few miles were completed to the terminal point.

The citizens of Buffalo, then a smallish lakeside town, embarked on a brief campaign, led by a local judge named Wilkeson, to clear their own eponymous riverway and so tempt the canal engineers to route the Erie Canal to a terminus nearby. Energetic lobbying, together with the clearance of the creek, evidently worked, for the engineers did eventually end their labors there, and the fact that more than a million people now still brave one of the country's cruelest climates (with roof-topping lake-effect snowfalls drowning the city each winter) to live in and around Buffalo is testimony to the wisdom of Judge Wilkeson and the city fathers of 1825 in doing all the persuading, as well as dredging and prettifying the banks of Buffalo Creek.

The completion celebrations, a nonstop bacchanal the like of which had perhaps never before been seen in America, then occupied the better part of twelve crisp autumn days, as well as involving every single one of the 363 miles of the canal. The party began on October 26, 1825, just after breakfast.

There was a convoy of narrowboats drawn up in Buffalo; it was allowed to begin its lap of honor along the waterway only after a long line of artillery pieces, arranged in sequence within earshot of one another all the way down to Sandy Hook, New Jersey, had each fired a pair of near-deafening shots in a salvo that rippled the length of the canal to the open sea, which was followed by a second that rippled all the way back again. The entire gun-after-gun-after-gun process took two full hours to play out, and the narrowboats waited patiently in Buffalo all the while.

But then, once the blue smoke from the final cannon mouth had drifted away and the echoing around the Lake Erie shores had stilled, the orations began. The first was from Jesse Hawley, he of the Canandaigua Debtors' Prison and the Hercules letters, and he made a gracious version of an I-told-you-so speech. There were others, most of them less brief and less memorable. Then the tow horses were prodded into motion, the boats in the flotilla eased away from the bankside, and they began to glide slowly in line ahead, eastward to the sea.

On the deck of one of the boats, Governor Clinton presided over two ornate American-made oak barrels, both of them filled with water from Lake Erie. They had a symbolic purpose that was clear from the start: they were to participate in what all knew was to be a wedding ceremony.

The floating procession took a full week to reach the halfway point at Albany, where the boats passed through the canal's eighty-third and final lock,* and then entered the broad reach of the Hudson. The horses were then unhitched from the narrowboats, and a flotilla of well-polished and exuberantly decorated steamers then took the unpowered canal craft in tow. The entire fleet and its cargo of dignitaries, visionaries, politicians, and hangers-on—with the two barrels of precious lake water still intact—sailed majestically down the tidal stream. There were endless thunderous volleys of gunfire and fireworks displays from every community onshore, until after a day and a half of sailing and partying, the small navy reached New York City and the waters of the Atlantic Ocean.

It was November 4, a Friday. The wedding ceremony was scheduled for noon. It was a crisp, cool, gleaming morning. The official party, barrels and boxes and bottles in hand, boarded steamboats that were suited—as canal narrowboats were not—to the roll of the open

---

* Today there are just fifty-seven locks, thirty-six of which are numbered. Efficiency and rerouting put paid to the remainder.

ocean. And that rolling duly began, as most on board promptly noticed uneasily, just as the craft headed out past Governors Island and through the Narrows—over which the Verrazano-Narrows Bridge arches today— and into the wide expanse of Lower Bay.

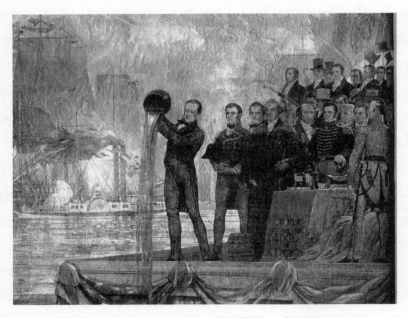

New York State governor DeWitt Clinton pours a bucketful of newly transported Lake Erie water into the Atlantic Ocean off New York City, celebrating the opening of the long-planned Erie Canal in 1825. Shortly thereafter a keg of ocean water was taken to be poured into Lake Erie, to make matters even.

At the bay's southern end, two low spits of land interrupt an otherwise perfectly flat marine horizon. On the left, as seen from the steamships, was Breezy Point,

New York; and on the right, Sandy Hook, New Jersey. A line drawn between them would mark the limit of New York Harbor and the beginning of the open ocean. The boats hove to. Governor Clinton then picked up one of the kegs of water, and with squadrons of artists drawing furiously, he withdrew the bung, tipped the barrel, and poured a long gush of Lake Erie water down into the salt waters of the rolling sea. (As if this were not enough, one of the governor's friends had brought with him thirteen additional bottles, which—he claimed— held waters collected from the Thames, Seine, Rhine, Danube, Nile, Gambia, Ganges, Indus, Orinoco, Plate, Amazon, Columbia, and Mississippi. These were all tipped into the Atlantic, too, though the precise reason for doing so went unexplained.)

This, at last, was the Wedding of the Waters, an event that had been designed and choreographed to its final seconds and was designed to be one of the iconic events in the history of a republic that was not yet half a century old.

No speech could do the moment justice, though many orators made the attempt. William Leete Stone Sr., a then popular journalist who wrote a massively long memoir of the day on behalf of the state, perhaps summed it up most succinctly. He took care to do so two weeks later, once the burghers of Buffalo had had their mirror moment, when they poured a barrel of ocean water into the waters of their lake.

All Europeans, he declared, had already begun to admire, and all America could now never forget, that those who planned and constructed the Erie Canal had "built the longest canal in the world in the least time, with the least experience, for the least money, and to the greatest public benefit."

It was a warm early summer Sunday morning, and I was staying in a hotel, on the third floor of what had once been a textile mill in the village of Little Falls, in the Mohawk River Valley of central New York. The building first had been a flour mill; for most of its existence, though, it made army uniform cloth, notably for soldiers fighting the Mexican-American War. Even though stone buildings can often be cold, this one happened to be uncomfortably hot, and the night before, I had grumbled. The innkeeper had found a key and opened the window for me—letting the roar of the Mohawk River swell up from below and lull me to dreamless sleep.

The chasm, with its sheer dolerite cliffs, had long before squeezed the river into a fury of falls and rapids here. When the first upriver settlers arrived, they had been compelled to portage their canoes and flatboats up into calmer waters; in later years they paid fees to the builders of a small bypass weir and eased their boats around the white water.

But since 1820, the Erie Canal had sliced like a knife through Little Falls. A channel of perfectly flat and perfectly still water replaced the torrent of the Mohawk, and for years watercraft of considerable size could ease through town without difficulty. In midcentury the entire waterway had been enlarged, widened, deepened, and strengthened to meet the enormous new demands of commerce and bring millions of tons of goods downstream from the factories of the Midwest to the markets of New York—until the twentieth century, when the railways and the highways began to eat into the profits, as they would everywhere else in the world.

Though commerce had long since died away, and pleasure boats have a virtual monopoly instead, in Little Falls the widened and straightened Erie Canal is still there, in full working order, running halfway up the hillside on the far side of the river that I could hear churning over the rapids down in the gorge below. So I saddled up my bicycle and rode over the Mohawk River bridge and up onto the canal towpath that ran beside it.

At first the waterway was quite empty. The only sound was a faint hum coming from a tiny power station that took water from the canal down through its penstocks. Maybe once upon a time the station had provided the power for the factories here that made bicycles, tissue paper, bookcases, milking machines, and felt, or maybe for the plants that turned out endless rolls of seamless cheese bandage or churns and milk coolers for the local cheese-making industry that once dominated this corner of Herkimer County. Today it merely feeds into the grid, with just the faint hum and a white sigh of water passing down its spillway.

And then around a great rock bluff were the closed upstream gates of lock number 17. This is the lock with the largest rise of any on the system—forty-five feet, according to some guides; forty-one, according to others. On this Sunday morning, I expected the lock to be quiet, but the keeper—a new immigrant from Belorussia, a young man who was clearly having fun operating the machinery—was waving excitedly to a small boat lying in the stream below the downriver gates. He was gesturing to the crew to back up some yards, as he was about to open the gate's sluices to empty the pound and let them into the lock.

Once the boat had backed safely away, he then pressed a button, and a white froth of boiling canal water erupted just below the gate, turning a previously quiet stream into a furious maelstrom. The skipper of the little boat looked briefly alarmed as his craft began to buck and dip in the torrent; he gripped his tiller firmly, white-knuckled, and did his best to keep his boat pointing head-on to the raging water. From my vantage point, I watched as the surface of the pound waters dropped, foot by foot, leaving the walls of the lock slimy with algae and weed.

It took ten minutes to empty the lock totally. Then the man from Minsk, businesslike and whistling cheerfully, pressed other buttons and sounded a horn. The immense steel lower gate, a counterbalanced portcullis-like affair, was slowly winched up and out of the water, dripping sheets of water as it did so. The entranceway was now fully open, the lock closed at its upper end by the two miter gates I had seen from my bike, holding back a wall of water forty-five feet high, a daunting thought.

Once the chamber waters had quieted, the small boat inched inward, settling itself halfway along. The skipper and his shipmate each took hold of a pair of dripping rope lines suspended from bollards on the berm above, which would hold their boat steady. The lockkeeper closed the lower gate and, with a quick warning and thumbs-up to the boatman, opened the upper sluices, which are not on the lock gates but are set into the walls of the chamber itself, and started filling the lock again.

There were ten minutes more of boiling upwellings, with the boatman smiling happily as he was steadily hoisted up into the warm Sunday sunshine as the chamber was filled. A final press of a final button, well-greased gears turned and well-oiled wires spun, and the miter gates swung open—and with a cheery wave, the skipper let go the lines, started his engine, and headed back into midchannel.

Before a minute had passed, he was rounding the bluff and sailing on toward lock number 18. I walked beside him as he was chugging slowly along, warming up his engine. He said he was going next to the small town of Ilion, where the Remington Company had once had a plant that made rifles and until 2007 still owned a museum that showed its complete range of typewriters, the more constructive side of its business.

He had started his voyage back at Albany. Thus far he had traveled 87 miles along the canal, and in locks like this—though none so tall—had been hoisted up a total of 410 feet above sea level. He now had 276 sailing miles to go, and though there were a few downhill miles to negotiate around Oneida Lake and the old Syracuse salt works that had helped finance the canal, in total he had another 155 feet to climb before he and his craft would be deposited on the wide inland ocean of Lake Erie. Where he might go after that—Chicago, perhaps, or Detroit, or even Duluth, because he was a Bob Dylan fan—he had no idea. But he would be up there in the high heartland of America, which world was then his oyster.

I asked him what he thought of the canal. He tried to answer, saying something along the lines that he blessed its existence, but at the very moment he opened his mouth, two fast freight trains went roaring by on the CSX lines that lay on the far side of the Mohawk, and they quite drowned out his voice. Because long-distance trains had put paid to the commercial future of most American canals, as they had elsewhere around the world, it seemed properly symbolic that his reply was being so obliterated. But then, when the expresses' roaring had died away, he reminded me that the Lake Shore Limited, no less, had crashed spectacularly here in 1940, killing many people.* "Canals much safer," he shouted, as he slid gently around the bend.

---

* The New York Central tracks had been built on a very tight curve through the same defile that is occupied by the Mohawk River and the Erie Canal. On the night of April 19, 1940, the passenger express to Chicago was running twenty minutes late out of Albany, and its driver approached the curve at the ill-advised speed of nearly sixty miles per hour. The engine came off the track and plowed into a rock wall, causing terrific damage and thirty-one deaths. By ironic chance, a senior New York Central official was on the footplate, and while the driver and firemen were killed, he lived on quite uninjured, passing away at the age of 102.

I put my bicycle over my shoulder, walked down the steps beside the lower gate, and cycled on down the towpath through early-morning mist. All soon fell quiet again, and before long it was just the empty river gliding by, the navigation path picked out by small buoys, and on both sides fields, cattle, and deep forests, little different from when the canal had been started two centuries before.

A few miles along the path, I came to an old mansion, the Herkimer Home. It had been built in the 1760s by a German migrant, Nicholas Herkimer, who had farmed here and sold provisions to those preparing for the trials of the portage. He had sided with the revolutionaries in the War of Independence and had been fatally injured, dying a true American hero. An obelisk stands as his memorial in a graveyard across from the main house.

It was perfectly silent. There were no cars, no visitors, no one apparently even in residence in the gatehouse—though maybe the occupants were asleep, I supposed, as this was early on a Sunday morning. There was just the ancient brick house set down in a clearing in the forest, a few old stone walls, green meadows sloping down on the one side to the water and rising on the other three sides up to the tree line, and beyond them the woods, endless, dark, and deep.

The Wedding of the Waters and the building of this canal had quite changed America, everybody said. And yet from down here on the waterside, and as I cycled slowly back along the towpath on that peaceful summer morning, it didn't seem to have been changed at all.

The Linkman Cometh

There is a rougher, ruder side to the American canal story, too. The cutting that was made along the Mohawk River Gap may have transformed itself in later years into something of an idyll, but the Chicago Sanitary Canal is, as its name suggests, a thing of raw utility only, an industrial, no-nonsense kind of waterway, and as we shall see, one that has unwittingly brought with it consequences today of more sinister moment than even its most ardent opponents can have imagined. Back when it was first conceived, it was a monster of a construction; its eventual existence, along with the wild nexus of railroads, did much to promote Chicago to its status as America's second city.

The purpose behind its making was simple enough. The mid-nineteenth-century trade goods of the Great Lakes hinterland could pass through the Saint

Lawrence River (if the Lachine Rapids allowed) to the ports of the northern Atlantic; they could be dispatched along the few hundreds of miles of the Erie Canal to the wharves of New York; or they could pass down south to the Mississippi River and then be shipped by way of New Orleans out into the Gulf of Mexico.

Of the three routes, the last made the most sense, most particularly for the Midwestern goods, those materials that were grown or mined or made around Lakes Michigan and Superior. But there was a problem: there was no physical connection between the Great Lakes and the tributaries of the Mississippi. Indeed, there was a geographically infuriating obstacle in the way, a low plateau of wide expanse, just a few dozen feet high but some miles wide, which separated the growing lakeside city of Chicago from the rivers of the West.

Two men are ultimately responsible for successfully breaching this low hill, and both are now all but forgotten for having done so. One is William Gooding, a self-taught engineer who cut his teeth on the Erie Canal, helping dig a bypass around Niagara Falls. The other, more nobly born (his family was loosely connected to the Jeffersons) was Isham Randolph—a Virginia forester who was first employed as an axman felling trees for a small country railroad. Their twin achievements, made serially, half a century apart, are little memorialized; the place where the infamously annoying hill once rose is now ingloriously obscured by what is claimed to be the world's largest sewage treatment plant.

It was Jolliet, the French Canadian who had explored the upper Mississippi in 1673, who first remarked on how easy it should be to cut a small canal across the hill (though back then it was more hilltop marsh than hill), which he and his fellow explorers had crossed on their venturesome way from Lake Michigan westward. By doing so, Jolliet later wrote, traders could link the lake (and thus the Atlantic) with the Mississippi (and thus the Gulf of Mexico).

Jolliet already knew a little of artificial waterways. Though he was a native-born Canadian, he was well aware that a giant canal was at the time being built in France: the Canal du Midi, a 150-mile structure linking the Mediterranean to the Atlantic, had been begun six years before, though it would be another eight years before it was completed and opened to traffic. It had been delayed for years and had seemed quite daunting and complex. But here in America, said Jolliet, a similar linking of two oceans demanded only the building of a waterway no more than *half a league* (about a mile and a half) in length. It would, in other words, be almost trivial, a cinch.

Yet despite the project's fine good sense and apparent simplicity, it was to be a long, long time before Jolliet's dream was realized. Neither the British nor the French colonial administration pursued it during their respective paramountcies. Only when America was independent and Illinois became a state in 1818 did any serious discussion begin. By then the local Indian tribes—the Miami—had been either routed or forced to sign treaties beneficial to the settlers. One had given away absolute settlement rights to the site where Chicago now lies; the other ceded the strip of land that could one day be used for building a canal. It remained only to do the surveying, draw up the engineering plans, and find the money.

This all took time. For one thing, engineers and surveyors soon discovered that Jolliet's Panglossian half-a-league forecast was utterly wrong; a workable waterway would have to be scores of miles long. Nevertheless, the economics of the idea made sense, and so on Independence Day 1836, with a company established— the Illinois & Michigan Canal Company—and with William Gooding hired over from the Erie Canal and with a goodly sum of money raised from the sale of three hundred thousand acres of Illinois farmland as a gift-in-aid by the federal government, ground was broken and building began.

The implications for Chicago became obvious from the moment the first pickax was thrown. Connect the lake to the Mississippi, and trade would commence as never before. The city consequently began to attract immigrants in huge numbers: in the dozen years it took to finish the canal, the population rose from zero to twenty thousand. Six years later it had tripled. It reached a million people by 1890—and all due to William Gooding, several thousand Irishmen and other workers brought over from New York State and around the world, and twelve years of backbreaking labor.

It took so long because the canal was much bigger than the one Jolliet had suggested. It was a fully formed waterway, sixty feet wide at the top, thirty-six feet wide at the bottom, and six feet deep. It was ninety-six miles long and had seventeen locks, four aqueducts, a giant pumping station with feeder streams from nearby lakes and rivers, and towpaths for the mules that would haul barges. It allowed vessels to pass without interruption from the lake to the new city of La Salle, on the Illinois River. From there it was a mere deepwater hop down the Illinois to its junction with the great stream beyond.

When the steamer *General Thornton* arrived in the middle of Chicago on the afternoon of April 19, 1848, bringing a load of sugar from New Orleans, a cascade of hitherto unimaginable events occurred: Chicago got its first telegraph wire, the Board of Trade opened its doors, the first steam-powered grain elevator started

working on the lakeside docks, the first railroad connection was started—and suddenly the city seemed
poised to become a vital fulcrum for commerce and
business, conveniently halfway between East and West.
Journeys that had taken fur traders three weeks and
nineteenth-century farmers ten days could now be
accomplished in less than a full day's sailing. A torrent
of trade goods flooded in: lumber, wheat, corn, stone,
salt, and—perhaps in the long term most crucial—
livestock bound for slaughter, the packinghouses, and
the dining tables of the nation.

People suddenly found travel into the American interior delightfully uncomplicated. Pioneering, the Gold
Rush, the peopling of the West—it all became so much
simpler when the first half of the journey was so easily
waterborne. Voyagers had only to embark on a series of
watercraft that would bear them onward by way of the
Erie Canal, the Great Lakes, the newly built connecting
links of the Illinois and Michigan Canal, the Mississippi,
the Missouri, the Platte . . . In short, with the arrival of
the *General Thornton*, everything about the nature and
being of Midwestern America began to undergo deep and
permanent change, connected and civilized at a stroke.

It would be agreeable to report that Isham Randolph,
the aristocratic Virginia axman, swept onto this vibrant
scene half a century later like a knight on a white horse,

improving and expanding an already great waterway and turning it into something truly magnificent. But the reality is somewhat uglier, even though the structure that remains today, what chart-makers and boatmen call the Illinois Waterway, does have an undeniable nobility to it.

The problem with late-nineteenth-century Chicago was that it had fallen victim to its own success and swift expansion in one particular way: it had far too much sewage. It had no way of getting rid of it. People were dying of the diseases that it spawned.

From time to time, huge rainstorms acted to flush this mess, and others equally terrible, out into Lake Michigan, from which the city drew its drinking water. Epidemics of dreadful waterborne illnesses were feared—though these fears were never actually realized. The hot, crowded city, its waterways often smelly and sickly, unfairly won a reputation akin to Calcutta's, as a pestilential place killing itself slowly. Huge amounts of money were then spent to reverse the flow of the Chicago River and deepen the cut through the low hill in the west, to try to flush the wastes out and away, well away from the lake. But ever greater rainstorms— especially one of fantastic ferocity in 1885—undid all the good work, and it was finally decided that a truly enormous new waterway, with the unlovely name of the Chicago Sanitary and Ship Canal, should be cut instead. This waterway should provide a new, better, and bigger passageway for ships, but it should also allow the reversal of the flow of the Chicago River and take all of the growing city's sewage out west, away from the lake, and send it hurtling down into the Mississippi, out of Chicago's sight, out of the city's mind.

Isham Randolph had come to know Chicago through working on its new railway system. He was not the natural choice to create the new canal, but it turned out that he had the vision and ambition to oversee the greatest earthmoving undertaking that had thus far been conducted in American history. His creation is only twenty-eight miles long—a dwarf in length compared with either the Erie Canal or the Illinois and Michigan Canal—but it is a giant in all other respects and still stands intact. It is more than two hundred feet wide and twenty-four feet deep. It required a new generation of machines to excavate it and a new generation of engineers to perfect its making.

Throngs of tourists came to see the men and their iron leviathans at work. They held their children up to see and hear the dynamite blasts. They came day by day to watch the limestone walls appear, running straight and true for miles, and to see great new canyons form in front of them, like one of the huge chasms out west that excited travelers were reporting having seen. This was becoming Chicago's Grand Canyon, and it was made by man. And when Randolph and his crew broke the final wall on a bitterly cold January day in 1900, it seemed to a writer on the *Chicago Tribune* an almost apocalyptic moment:

> *"It is open! It is open!" went up from scores of throats as the water at last . . . had been made to start down the toboggan slide into the canal. The fall . . . was 24½ feet. . . . "It is the Niagara of Chicago," Mr. Eckhart said as he stood, watching the waters, together with the ice and boulders, sweep down the chute and drive far into the wide canal, whose surface already was beginning to take on a rich mahogany brown. . . .*
>
> *Engineer Randolph stood at one end of the structure and waved his hat triumphantly. The flooding waters sent heavy spray over the feet of the men on the pier, and threatened to carry the group, the pier and all the rest into the canal beneath. Like schoolboys on vacation, the drainage officials waved their arms and shouted.*

Admiral George Dewey, hero of Manila, came up to open the canal in May 1900, so proud was Chicago of what it had made, so hopeful were its merchants that it would encourage even more trade—and also so relieved were its people to be able to dispose of their waste and send it off toward the cotton fields and poor country towns down in the Deep South.

Yet it has proved something of a disappointment, in truth. The railways built to occupy the same valley have taken the lion's share of the goods traffic; barges carry grain, coal, and iron ore worth about $1.5 billon from the lakes down into the Mississippi Valley each year, a respectable though not sensational amount. Over the years, the canal has been plagued by troubles of one kind and another, including court battles of Dickensian length and tedium, usually involving disputes over the amount of water being taken from Lake Michigan each year.

For decades the canal was an overlooked part of the city's infrastructure, a gray and listless swath of fetid water hidden away behind its cement walls, lying at the back end of the brickyards and treatment plants of the city's western suburbs. In recent years, it has reemerged in the public consciousness for the worst of reasons: for though it has not proved to be much of a passageway for the promised tonnages of freight, it does now appear to be a potential entranceway for a most unwelcome visitor, making Isham Randolph's proud venture the site of a most inglorious new war.

A non-native species of fish, the highly aggressive Asian carp, known for its ability to jump out of the water at high speed, terrifying humans who get in its way, as well as for its ruthless competition with native fish, was introduced into the lower Mississippi in the 1970s by commercial fishermen wanting to use it to clean their breeding ponds.

The spectacularly athletic Asian carp, which can leap up to ten feet out of river water, has long been at home in the Mississippi and its tributaries, but is not wanted in the Great Lakes, by either the American or the Canadian fishing industry, both of which it would decimate. Electric barriers across the Chicago Shipping Canal strive to keep it from entering.

It became astonishingly successful, and over the next three decades migrated in vast numbers upriver, traveling into the Illinois and Des Plaines Rivers and, much like Jolliet and Marquette three centuries before, arriving at the low hills that stood guard above the lowlands of Lake Michigan. But now, since the canals had breached those hills, there was a passageway into the lakes, with only the locks on the waterway acting as barriers. Were the carp to get across the barriers, they would upset the environmental applecart like no other introduced species before.

All efforts have since been made to keep the fish out, and the 112-year-old Chicago Sanitary Canal is now the last line of defense. Two powerful submerged electric-fence barriers have been installed, a high-voltage stockade to keep the fish away. But already, in 2011, one or two suspicious-looking creatures, bearing carpish genetic signatures, have been found in Lake Michigan, and it is feared that the barrier has not entirely held and that the lake may soon be fully fouled by a pest that would cause untold problems for the Great Lakes and beyond. It is a problem that no one foresaw, not even Louis Jolliet, who long ago imagined that a mere half-league ditch might unite the lakes and the gulf. It brings up the sobering thought that enforced unity can be a mixed blessing.

That Ol' Man River

I was once given an assignment to visit every town in America named Paradise. There were at the time sixteen of them, scattered randomly across the country from Pennsylvania to Nevada. Why, the editor wanted to know, would settlers want to call their new town by such a name? And if the towns had each seemed like Paradise back when they were founded, did they still warrant the naming? Was each Paradise a paradise nowadays?

It took two delightful months of wandering to find out that only one of the towns truly deserved the name: I have long kept its exact location under my hat, other than offering that it is hidden somewhere deep in the wheat fields of the Midwest. While there, I was invited to stay with the town's best-liked couple, who were improbably named John and Mary Angel. As if staying with the Angels in Paradise was not sufficient, Mary Angel one day collected cherries from a tree in her garden and baked me a pie. It all seemed a dream from which I never wanted to wake.

But most memorable of all was my discovery of the melancholy little community that had once been named Paradise, Arkansas, a dilapidated hamlet standing on a bluff above the right bank, the western bank, of the great American river, the Mississippi.

Paradise, Arkansas, is a town that no longer officially exists. It used to, and its name stands foursquare in the center of Drew County, on the Arkansas page of the *National Atlas of the United States*. But that was published back in the Nixon era, and much has changed in the years since. On the AAA road map of Arkansas today, most of Drew County is seen as being of so little consequence that the mapmakers put their compass rose right in its center. There are no places listed at all in this part of the county, least of all Paradise. When I called the Drew County clerk in Monticello—there are a lot of Jefferson-era place names in these parts, memorials to the Louisiana Purchase, which, among other things, made Arkansas a part of America—she said she didn't know of any Paradise, though she did know of a place called Possum Valley.*

I insisted that there once was such a place as Paradise here. Maybe it was dead today, but I had been there for my 1984 essay, perhaps when the town was already

---

* Drew County is that kind of a place: the Historical Society lists among its most distinguished residents the first man to observe and prove that a squirrel can run down a tree faster than a bolt of lightning.

just about done dying. It was hardly any sort of a community—no more than a gaggle of broken-down shacks perched on a low rise under a stand of live oaks. A few impoverished-looking men lolled about in the drowsy heat, fanning themselves.

When I sat with them, they told me that at the turn of the century, the hamlet had been a fairly prosperous site of a cotton gin and that bales would be sent by horse and cart down to the bayou nearby and loaded onto steamboats for shipment to New Orleans. But then the abler men drifted off to the Northern cities—to Detroit, mostly—and in more recent years, the Georgia-Pacific Company had come along and bought up the land for pine-tree plantations and pulp making, and Paradise started to die. In a year or so more, the men agreed, it would be quite gone, erased from the maps forever.

But then one of the older men stood up and took me to the edge of the live-oak copse. "You'll be wondering why it was called Paradise," he said. "Well, look down east a ways and try to imagine."

I squinted into the sun, peering through the gray haze shimmering with restless thermals. Below was the bayou, a complex network of swamps and curved lakes and relic streams that glinted like hammered steel in the high sun. And then in the far distance, soundless and immense, was the Mississippi, wide and brown, a huge painted swath turning lazily down through the valley.

"Can you imagine," asked the old man, "just how hot it must be down there today?"

It was certainly as hot as Hades up here on the hill.

He continued. "Can you imagine those men from long ago, paddling a raft for miles across the river, coming down from Memphis, or from closer places like Tunica or Clarksdale, and then hauling themselves and their supplies across the swamps, with all that mud and the mosquitoes and the steamy heat like today's?

"But then they'd see this hill. A low hill, nothing more, but a place that catches the western wind a little, and has a grove on top where there is some shade. The men would climb up and take a rest under the oak trees. And then the sun would set over the hills, and a slight breeze would spring up to cool them down. Can't you imagine them lighting cheroots and saying to themselves, 'Now this is surely heaven! After what we've been through, this is *paradise*. Let's set ourselves down and stay awhile'?"

That was how the town was founded, he said. Perhaps, perhaps not. I am sure there is truth in his story, even though the Drew County Historical Society knows nothing and the Monticello archivist has never heard the tale. Yet Paradise, Arkansas, is still on the *National Atlas* map. It has a set of coordinates: 33° 32' North, 91° 47' West. And I went there, many years ago. All that is quite enough.

But the other significant thing for me was just as the old-timer had said. Despite all its evident shortcomings and poverty and nowhereness, this oak-lined bluff must have once seemed a heavenly place to those who for days before had been mired and swamped in the great miles-wide tract of torrid unpleasantness that is the middle course of the Mississippi River.

It has always been my experience—quite different from the writings of Mark Twain, who made this river into a kind of magic carpet—that the Mississippi is a thing shorn of all romance and is instead more generally a displeasing mess of foul water, mud, slime, and desolation.

Charles Dickens saw "nothing pleasant in its aspect." Mrs. Trollope wrote that she "never beheld a scene so utterly desolate." And I suspect most others who see the Mississippi, at least in these reaches, feel about it much the same. It is *mighty.* It is important. Some say it is perhaps the most important river in the world. But pretty it is not.

If driving, you become aware of its presence miles before you reach it. The landscape falls away. There are swamps on either side, dense hedgerows and copses, miles of small lakes of curious shape. There are the distant chimneys of power plants and factories, sited at the river's edge because, presumably, the owners are eager to take advantage of some neighbor-state's tax benefit. And then finally there is a bridge, usually an iron behemoth of a thing. Often bridges come in threes: there is the one being driven over, another beside it built later on to accommodate the unanticipated extra traffic, and also perhaps a rusting railway crossing nearby, built in much earlier times, and maybe also the empty stone pillars of one that long before had been half demolished or destroyed.

The road bridge then heaves itself up and over yet more swamp, more trees, then yards of soggy ooze, then an inlet and an islet and a stranded barge or two, then a fuel depot and another islet and yet more brown sludge, before finally the river itself appears below, *eau de nil* in hue and filled with floating mysteries and perhaps a slow-moving barge or two, its waters rumbling past beneath the span—until, a mile or half a mile on, there is more ooze and low forest, and the process repeats, a mirror of what went before, and then there is a bullet-riddled "Welcome" sign, telling you how good it is to be in Arkansas or Louisiana or Missouri and reminding you that Mississippi, Kentucky, Tennessee, and Illinois are all in the rearview mirror now, that East has become West, and the West will be everything for the next two thousand miles or so.

Despite Saint Louis's best efforts to make its own stretch of riverfront pretty and rich with symbolism by having a master craftsman like Eero Saarinen put up his mighty steel-clad Gateway Arch on the right bank, the crossing of the Mississippi in America's midregion is seldom a pleasant experience. Though the waterway may be of enormous national significance, both real and symbolic, I imagine the transit of it to be only a little more pleasant today than it was when those nineteenth-century settlers flung themselves ashore on a cool Arkansas bluff and declared that they had at last found paradise. That they did so was not because Arkansas was heavenlike but because what had gone before on America's greatest river had been so very disagreeable.

The story of America's unification has, by and large, been a saga played out along an east-west axis known by the well-worn phrase "from sea to shining sea." But the country's most obvious physical feature, the great gray-green greasy Mississippi, does not run that way at all. It lies at right angles, almost precisely north to south,* a fact that renders it at first blush less a unifier than a dividing line, at least symbolically, a boundary delineating where the American East becomes the American West. But a new-made nation demands to be unified across all of its dimensions, Minnesota and Louisiana, North Dakota and Texas being quite as worthy of connection as are Maine and California, Virginia and Oregon.

The river itself spills out of a small lake amid the birch forests of northern Minnesota. Henry Schoolcraft, the geologist-explorer who first found its source in 1820, liked to give pronounceable pseudo-Indian names to features he found, and he gave this lake the name Itasca, which is not even remotely Indian, but a cornily truncated version of the Latin words for "truth" and "head." Though one of the streams that feed the lake is presumably the Mississippi's real source, it has become convenient to present a spillway at the lake's northern tip as the accepted beginning

---

* Over its 2,300 miles, the Mississippi deviates just a little over five degrees in longitude between source and sea, making it almost die-cut straight. Only the Nile is straighter, with a deviation of a mere four degrees over almost twice the Mississippi's length.

of the river, and there is a marker, a stone walkway, and a large car park to denote the official start of the nation's greatest stream.

From this point, the stripling continues northward for a few miles, runs cold and narrow and waving with fronds of wild rice that the local Indians are permitted to grow, then doubles back and proceeds southward, steadily widening and deepening as it goes. It is artificially dammed and nudged by its state and federal caretakers into staying between its banks, and for five hundred miles it thrusts itself fast and rockily southward. Then as it leaves the northern states, it matures and steadies itself, and after seventeen hundred more muddy miles of slow curves and recurves, it finally reaches the sea in the Gulf of Mexico.

Thus it does link the nation, top to bottom. It has its origin close to one American frontier, and its terminus is close to another. It passes from almost Canada to almost Mexico. In doing so, it progresses from the nearly arctic to the subtropical. It flows through an unimaginably vast valley filled with numberless tributaries. It leaves endless trails of oxbow lakes as relics of where it once had been. In places the state boundaries that the river once marked are now quite out of date, their own twists and turns out of sync with the river's new twists and turns. Finally it eases itself down into a long and muddy delta that is scored with dependent streams known as distributaries, and though it wiggles about a good deal, it does so basically moving in a series of fairly straight-trending lines, from north to southeast. Finally, at a place known as the Heads of Passes (a point to and from which all the lower-river distances are measured),* it divides into three long navigable tubeways of sand and mud that splay like a chicken's foot and passes out into the warm salt waters of the open sea.

The basin of the Mississippi encompasses a good two thirds of the contiguous forty-eight states, thirty-one of which—together with two Canadian provinces—contribute waters to its flow. The very existence of this million-and-a-quarter-square-mile wedge-shaped watershed, which enfolds places as different from each other as Montana and Maryland, New Mexico and

---

* There are two numbering systems on the Mississippi, and both increase the numbers going upstream, not down with the flow of the waters. From Head of Passes to the town of Cairo, Illinois, locations are given in RMs, river miles, counted up from the south. But at Cairo, which is at RM953.8, the system begins all over again, with Cairo now at zero and points like Saint Louis, Saint Paul, and Lake Itasca being given RM numbers, too (Saint Louis is at RM195). Other American rivers, like the Ohio, are counted *down* from their sources; this makes for much bewilderment among first-time navigators. In all cases, the mile numbers are painted white on buoys or lights that mark the edge of the navigation channels.

Kentucky, Wisconsin and Oklahoma, Idaho and Iowa, has created, if unintentionally, a distinct kind of oneness in the middle of America.

A rancher beside the Yellowstone River, a canoeist on the upper reaches of the Monongahela, an ice fisherman jigging for walleye in northern Minnesota, a plantation owner with a lawn sweeping down to the Red River in Louisiana—all of these and millions more can take some kind of conforming comfort in knowing that the Mississippi connects them all. As with the midrib of a leaf or the shaft of a feather, the river provides for all some kind of half-imagined structure, offering a kind of geographic strength, a degree of certainty, stability. The river is, after all, always there, an ultimate destination for their own fresh waters, for their goods, their supplies, and their commerce, as certain and comforting and immortal as the still more distant sea.

Some relish the romance of such connections. The great Depression-era filmmaker and poet Pare Lorentz was one. His 1938 short government-financed documentary *The River*, which called attention to the damage that man was doing to the stream, won awards in Venice, and his script was nominated for a Pulitzer prize. The opening minutes are accompanied by a catalog of an enduring loveliness:

Down the Yellowstone, the Milk, the White and
  Cheyenne;
The Cannonball, the Musselshell, the James and
  the Sioux;
Down the Judith, the Grand, the Osage and the
  Platte,
The Skunk, the Salt, the Black and Minnesota;
Down the Rock, the Illinois, and the Kankakee,
The Allegheny, the Monongahela, Kanawha, and
  Muskingum;
Down the Miami, the Wabash, the Licking and
  the Green,
The Cumberland, the Kentucky, and the
  Tennessee;
Down the Ouachita, the Witchita, the Red, and
  Yazoo.

Down the Missouri, three thousand miles from
  the Rockies;
Down the Ohio, a thousand miles from the
  Alleghenies;
Down the Arkansas, fifteen hundred miles from
  the Great Divide;
Down the Red, a thousand miles from Texas;
Down the great Valley, twenty-five hundred miles
  from Minnesota,

Carrying every rivulet and brook, creek and rill,
Carrying all the rivers that run down two-thirds
    the continent—
The Mississippi runs to the Gulf.

For most of the two hundred years that American mankind has lived beside the river and its feeder streams, the human connection has been based almost wholly upon commerce and trade. In this sense, the Mississippi has much more in common with the Yangtze, in China, than it has with the other two great world rivers, the Nile and the Amazon. Both of these latter flow through and water the soils of many countries and are important less for internal navigation and mercantile matters than they are for their benefits of irrigation and nutrition. The Yangtze and the Mississippi, however, each exercise a monumental influence over essentially only one nation.

Each plays an immense role in the cultural and social geography of its national home. The east-running Yangtze divides China more or less amiably into two peoples. Those to the river's north are generally thought of as being taller, paler skinned, wheat growing, noodle eating, Putonghua-speaking, aesthetically aware guardians of the ancient Chinese ways. Those to the south, by contrast, are seen as smaller, swarthier, rice eating, coastal dialect speaking, arithmetically able, and commercially adroit peoples who are more concerned with business than with culture. The river is a national treasure, regarded with pride and awe by all.

In this sense, Ol' Man River is much the same—running in a ninety-degree different direction, true, but crudely also parsing the country into two—with industry and academia and antique culture on the eastern side, pioneering and pastureland and cattle culture on the western side. And here, too, aboriginal peoples once bestrode every inch of the river's banks—and for no tribe did the river itself serve as an entity that fully divided them. So up in the north the Ojibwa and the Santee, the Illinois and the Iowa were to be found on both sides of the Mississippi. In the middle of the country, the unknown "mound people" who built the city of Cahokia, whose structures Lewis and Clark encountered when they first set off up the Missouri, constructed their earthen marvels with a careless disdain for which side of the river they were on. The Quapaw and the Chickasaw people—the former settled on the river's west, the latter on the eastern side—fought occasionally as proxies for the French and their enemies, though because the river is so wide, it was harder for those on one side to reach the other, so the battles were minor and not at the Indians' own choosing. Farther south still, the Tunica, the Natchez, the Choctaw, and the delta-living Chitimacha people all saw the Mississippi as a source of water and fish and good living for all, and only infrequently did they fight one another across it or over it or because of it.

No such humiliation as the presence of a foreign gunboat was ever visited on the Mississippi, at least not in the two centuries that have elapsed since it became fully a part of American territory. (To imagine a flotilla of Chinese warships cruising by absolute right between Hannibal and Vicksburg today, their captains dispensing instant justice in the riverside towns they passed, is perhaps to know briefly how it must have felt to the Chinese in the early years of the twentieth century, to know imperialism from the other side.) No foreign warships; but foreign cargo ships aplenty, because the commerce on the Mississippi is similar to that on the Yangtze, only far, far greater. The Mississippi is a working river, an industrial river, like few others.

Today there are fewer passengers boats, true—the car and the plane take most Americans places. And there are a handful of resurrected stern-wheelers, cliché ships with calliopes and Dixieland jazz ensembles and Mark Twain cocktail bars; but they serve a small tourist market and have no business with those who travel out of need. Most rivercraft today are enormous and almost unpeopled barges and tows, some as much as six acres in extent, heading ponderously north and south with vast tonnages of materials of one kind and another in bulk. The southbound vessels, in particular, disgorge immense volumes of wheat, corn, soybeans, coal, and lumber, and they load it onto cargo ships drawn up at the docks in New Orleans and Baton Rouge, which then take it onward to customers around the world.

The Mississippi, indeed, sends so much of America's produce down and onto the world's oceans that it seems more often than not that the river is where the country is being turned inside out. Back in Depression times, this was happening quite literally. Pare Lorentz and others chronicled a great American tragedy as billions of tons of topsoil were torn from Iowa and Louisiana and Missouri by steel plows and leached from Southern cotton plantations to be gushed out as liquid mud into the sea and away from America.

It was as if the country was having its very heart torn out, and during those years, the years of the Dust Bowl, the forced migration of the hopeless, and John Steinbeck's *Grapes of Wrath* and Lorentz's *River*, it must have seemed as though the nation's greatest river was indeed bent on destroying the very countryside that had given it birth. Campaigning for the river's rights began in earnest once the New Deal had been struck and the war was done. Nowadays the Mississippi is not merely loved and revered and admired by all; it is now also quite well cared for, protected and preserved, and has in large measure reverted to its original role and purpose—it transports, it irrigates, it nourishes and sustains, and it links the peoples of its valley together. For many years it did not, but now, generally speaking, it does.

The federal government is substantially involved, even though the national mood has lately become hostile to immense taxpayer-funded bureaucracies. Even the source of the river has a connection with the taxpayer: the stepping-stones that mark its official exit point from Lake Itasca were put in place by youths of the Civilian Conservation Corps, a body established during the Depression as a way of providing young men with work.

Lower down the river, the US Army Corps of Engineers is in pole position, and because it is a direct descendant of the Army Corps of Topographical Engineers, which performed much of the country's early surveying and mapmaking, it is a body that considers its role fully deserved. The corps has been responsible since the 1820s for keeping the river open to navigation and preventing it from flooding, roles that, in times of peril and poor weather, can be mutually exclusive. The navigational side is not normally so challenging. The corps's mandate requires it to keep a nine-foot-deep channel open between Minneapolis and Vicksburg, a twelve-foot channel down to Baton Rouge, and then for large oceangoing vessels, a channel forty feet deep down to the gulf.

Flood control is something else. Huge levees on both sides of the river try to keep the mightily ponderous torrent in check, even when the combined snowmelt from the Rockies tears down the Missouri and threatens flooding on an epic scale. The levees were seldom easy to build, not least because of a persistent want of material strong enough to put in them. Engineers have tried concrete, mats of grass, articulated cement mattresses, curtains of woven wattles. And of course, rocks in great abundance. Except there are no rocks in the Mississippi Valley. For tens of thousands of years, the valley has been essentially a grassy wilderness of alluvial mud and clay, with not a rock bluff or a cliff of anything truly hard in sight. So it can fairly be said that every piece of rock larger than a football that now lies in the levees or lines the banks and even the floor of the river was brought and dumped there by barge by the men and machines of the United States Army Corps of Engineers.

This same body of men that built Cape Canaveral, the Los Alamos atomic bomb labs, the Washington Monument, the Panama Canal, and the Pentagon now tries to control this river. The corps owns and runs the two dozen locks and dams that lie along the Upper Mississippi. At some of the dams, it generates electricity, the US Army competing with the major utility companies. And it has built and still runs the mighty diversions and spillways and pumping stations that try, at times desperately, to keep this gigantic and unimaginably powerful body of water from overreaching itself and bringing catastrophe to millions who live close by.

Time after time, the corpsmen have been severely tested, most recently in 2005 by Hurricane Katrina, which resulted in the overtopping and failure of many miles of levees, and by the great Mississippi flood of 2011, which for the first time in thirty-five years resulted in the opening of gates along the enormous Morganza Spillway, dumping trillions of gallons of floodwater into a river known as the Atchafalaya—a name that sends a shiver down the spines of most who live and make their living in New Orleans and Baton Rouge.

It does so because of one of the more pressing realities of hydrology. The Mississippi is a river that like most has a mind of its own, and according to today's hydrologists, it has lately wearied of flowing into the gulf by way of New Orleans. It seems to want instead to move its mouth scores of miles to the west and emerge into the sea along the course of what is now simply a deep and short distributary stream, the Atchafalaya. Such a natural shift in the watercourse* would spell ruin for the two great cities of Louisiana, by quite simply taking their river away. Everything—every dockside crane, wharf, loading bay, waterfront hotel, and business center—would then have to be shifted, most probably to a currently little-known port, Morgan City (whose motto, presciently, is "Right in the middle of everything"). On the advice of the body that runs the river, the Mississippi River Commission, troops from the Corps of Engineers constructed the gargantuan four-thousand-foot-long Morganza Spillway in 1954 and what is known as the Old River Control Structure in 1963 to make sure that no such thing is ever allowed to happen.

---

* Nature is not alone. Human greed is also responsible, the main culprit being an otherwise heroic figure, Henry Shreve (the founder of Shreveport, Louisiana), who is better known for designing the grande-luxe sternwheeler river steamers and for clearing a legendary 150-mile-long logjam on a Mississippi tributary. But in 1831, to allow his steamers faster passage upriver, he cut a deep channel through a hugely inconvenient oxbow bend, unwittingly changing the flow of the Mississippi and making it significantly more likely to pass down into the Atchafalaya.

The purpose of these enormous confections of iron and concrete and cranes and gates is to prevent avulsion, the natural process whereby a river suddenly abandons one channel and finds another better suited for its purpose. The Atchafalaya, which these days offers a deeper and more direct route to the sea, must seem awfully tempting to the Mississippi waters, and most hydrologists believe that one day, despite humankind's best and costliest efforts, they will take it.

In his 1883 *Life on the Mississippi*, Mark Twain warned of man's puniness when pitted against the overwhelming avulsive power of the river, and he cautioned against hubris:

> One who knows the Mississippi will promptly aver
> . . . that ten thousand River Commissions . . .
> cannot tame that lawless stream, cannot curb it or
> confine it, cannot say to it Go here or Go there,
> and make it obey; cannot save a shore which it has
> sentenced; cannot bar its path with an obstruction
> which it will not tear down, dance over, and laugh
> at. But a discreet man will not put these things
> into spoken words; for the West Point engineers
> have not their superiors anywhere; they know all
> that can be known about their abstruse science;
> and so, since they conceive that they can fetter and

handcuff that river and boss him, it is but wisdom
for the scientific man to keep still, lie low, and wait
till they do it.

The notion of keeping still, lying low, and waiting may well have been suited to the wise men of science, but to the Americans of the later nineteenth century, bustle and hustle and ambition and yearning were becoming the more familiar watchwords. To be sure, the Erie and the Chicago Sanitary, the Mississippi and the Middlesex and the Santee all helped to knit the states into one, but the men and the craft that made use of these hydraulic pathways did so in a necessarily ponderous fashion. By now impatience was beginning to become an American virtue. A growing sense of fierce urgency was starting to shoulder out of the way the old ideas of tranquil floating progress. Before long came times when the country, while still bent on uniting itself ever more fully, became convinced by technology that it should now try to do so at a far greater speed. The moccasin and the covered wagon and the canal boat were not fast enough. Now it was necessary to harness raw power and bring engines of one kind and another to bear in bringing the nation together.

And engines—whether steam engines, the external combustion engines propelling riverboats and railway

trains, or internal combustion engines moving automobiles or trucks or buses, or electric motors or jet
engines—all required fuel to be burned to produce raw
energy, usually in the form of heat. Ultimately it was
various kinds of heat, harnessed with care and cleverness, that allowed America and her now perpetually
hurried Americans to become physically united. The
harnessing of these various kinds of heat allowed them
to be propelled up and down and around and across the
country at a newly useful speed, and now in contraptions that were always and essentially powered by fire.

On April 27, 2005, the Federal Railroad Administration (FRA) . . . published the Final Rule on the Use of Locomotive Horns at Highway-Rail Grade Crossings. Effective June 24, 2005, the Final Rule requires that locomotive horns be sounded at all public grade crossings 15–20 seconds before entering a crossing, but not more than one-quarter mile in advance.

The pattern for blowing the horn remains two long, one short, and one long sounding to be repeated as necessary until the locomotive clears the crossing. Locomotive engineers will retain the authority to vary this pattern as necessary for crossings in close proximity and will be allowed to sound the horn in emergency situations.

—UNION PACIFIC RAILROAD INTERPRETATION OF THE
NEWEST RULING ON THE USE OF TRAIN HORNS

Afoot and light-hearted I take to the open road,
Healthy, free, the world before me,
The long brown path before me leading wherever I choose.
Henceforth I ask not good-fortune, I myself am good-
    fortune,
Henceforth I whimper no more, postpone no more, need
    nothing,
Done with indoor complaints, libraries, querulous
    criticisms,
Strong and content I travel the open road.

—WALT WHITMAN, "SONG OF THE OPEN ROAD,"
Leaves of Grass, 1856

Today in El Paso all the planes are asleep on the runway.
—MATTHEW ZAPRUDER, FROM "APRIL SNOW," 2010

PART IV

When the American Story Was Fanned by Fire

1811–1956

May the Roads Rise Up

Long before there were engines to run along them, there existed thousands of miles of American roads. The country was covered by a tracery of highways, many of them laid out and built by soldiers and most of them then used by postal dispatch riders. British colonial forces had built the first of them, facing considerable challenges in making them because of factors unimagined in the tiny kingdom back across the sea: the sheer vastness of the countryside, the often extreme weather, and a topography that was pitiless and challenging. The forests were dark and mysterious, the rivers wide and fast, the land filled with exotic and often unpleasantly dangerous animals (the Britons seldom saw bears or mountain lions back home) and an indigenous people who were less than eager to have their traditional grounds crisscrossed by a network of alien pathways.

That is not to say the builders balked at the challenge. The King's Highway, initiated by the colonial governors in 1660 on orders from Charles II following his return from exile, was a road designed to run for 1,300 miles from Boston to Charleston and connect ten of the original thirteen colonies. The first section, between Boston and New York, was opened in 1670 to riders carrying postal packets: a letter carried between the cities cost the sender nine pence, the same as a letter back to London. It took two weeks, door to door.

There were many other such roads nearby. The Fall Line Road linked all of the Eastern river towns that had been built where waterfalls interrupted upstream water travel. The Old Federal Road ran between Augusta, Georgia, and Mobile, Alabama; the Federal Horse Path ambled down toward New Orleans; the Great Valley Road swooped scimitarlike down the valleys of the Alleghenies from Philadelphia to North Carolina; the Mohawk Trail ran from Albany to Lake Erie. And later, out west, there were the Oregon Trail, the Mormon Trail, and the California Trail, which had been blazed by the pioneers lured out by all of the talk about gold and silver in the hills and limitless fine fields to plow and plant.

Most of the Eastern routes were old Indian trails. Certainly the Boston Upper Road was, having for years been known as the Pequot Trail and said to have been so heavily traveled for so long that it had eroded into a shallow ditch. Down in the South, the Natchez Trace, which the National Park Service maintains today as a living museum of Southern roadways, is also very obviously ditchlike. But it was cut by sturdier feet, arranged in fours, not pairs. Buffaloes migrated north each summer to the salt licks of western Tennessee and, sated, would return home before the cold settled on the land. Their constant to-and-fro eroded a twenty-foot depression four hundred miles long, which local Indians then arrogated to themselves, until they in turn were followed and displaced by French and Spanish colonists, later by British soldiers and settlers, and by tourists today.

The story of most early American roads is much the same: first came the Indians, then the soldiers, then the mails, then settlers, then commerce, and finally something approaching permanence. By the middle of the eighteenth century, a network of such trails, all of them dirt, few of them maintained, and nearly all of them notoriously difficult to travel along, knitted the entire eastern half of the nation loosely together. They allowed for the slow, halting, and unsure passage of men and horses, Conestoga wagons, and hay-heaving buffalo carts.

The earliest of these roads were included in an atlas published by Christopher Colles in 1789. *A Survey of the Roads of the United States of America* was a bound series of strip maps, showing only the roadways, their junctions, and points of interest along the way. It illustrated the charmingly random nature of the country's road-building efforts thus far. More important, the map foreshadowed the inevitable: that during Thomas Jefferson's presidency, twenty years later, some semblance of order would finally be brought to the nation's highway-building programs and that the federal government would become properly interested in designing, building, and preserving a road network for the future. Once Lewis and Clark had come home and reported their findings, Jefferson decided to create, if not a network of roads, then at least one single fine highway that would cross the entire country.

The town of Cumberland in Maryland—connected to the sea both by a rudimentary canal and by roadways used by troops in George Washington's day—was chosen as the ideal jumping-off place. Once Congress had given formal approval, construction began in the summer of 1811 on the first section of the Cumberland Road, financed in large part by federal land sales. By 1825, when the first six hundred miles of what clearly was going to become a coast-to-coast highway had rolled out across the nation, Jefferson's project would come to be called, more portentously, the United States National Road.

The first contract for building the Cumberland Road was with a man named Henry McKinley, and it was written in January 1812 by the indefatigable Swiss polymath who for thirteen years was Jefferson's (and then James Madison's) treasury secretary, Albert Gallatin. The detail of the thing amazes, even at this remove. Just the briefest of sections will illustrate the attention that was being paid by White House officials to a creation that all knew would ultimately be of lasting significance to the nation:

> . . . the trees to be cut down and cleared the whole width of sixty-six feet, according to the fourth section of the act above mentioned; the slumps to be grubbed, and the bed of the road to be leveled thirty feet in width; the hills to be cut down, the earth, rocks, and stones, to be removed, the hollows and valleys, and the abutments of all the bridges and culverts, to be filled, so that the whole of the road on the aforesaid width of thirty feet, to be reduced in such manner, that there shall not in

*any instance be an elevation in said road when
finished, greater than an angle of five degrees
with the horizon, nor greater than the gradation
fixed by the commissioners who laid out the road,
and so that the surface of the said road shall be
exactly adapted to the marks or stakes, made or to
be made, by the person appointed superintendent
for the said road by the President of the United
States. Where the earth is to be raised, the sides
are to slope at an angle not exceeding thirty
degrees.*

The actual building, starting with the grubbing of
the slumps, had begun in May 1811—before the contract was formally signed—at a stone on a marked lot of
land in Cumberland, at the confluence of Will's Creek
and the north branch of the Potomac. From this point,
surveyors measured out the road westward in miles
and perches, the latter being an old road measure-
ment of five and a half yards: sixteen feet six inches.
McKinley's contract price was $21.25 for every con-
structed perch.

It was a monumental task, not least because it was
performed until its very late stages without any kind of
powered mechanical help. The final section of the road
was completed after twenty-eight years of laboring, at

the town of Vandalia, Illinois, having passed through the towns of Brownsville and Washington in Pennsylvania; Wheeling in West Virginia; Zanesville, Columbus, and Springfield in Ohio; and Richmond, Indianapolis, and Terre Haute in Indiana. The project took long enough for four more presidents to be involved in its construction—Martin Van Buren was in office when the last load of gravel was rolled onto the roadway in central Illinois.

The road was not simply the first; it was also the first to be made in a way that was likely to endure. And that is primarily because the War Department, no less, began to get interested in creating good roads along which it might send troops, allowing soldiers to be dispatched at great speed across the country. The generals used their clout in Washington to persuade the road builders to incorporate into the highway's construction new ideas that were coming across the Atlantic, particularly those emanating to widespread professional interest from an aristocratic Scots highway maker named John Loudon McAdam.

The Scotsman was to be seen as a savior. For until now the roadways were, to put it mildly, shoddy, inferior things. To judge from his growing reputation in Britain, McAdam could do a great deal better. Orders went out that henceforth the laying of the kind of inferior roadways that had been built east of the Ohio River would no longer be tolerated. If any degree of federal funding was to continue, then the newly proved techniques that were now spreading their fame from across the Atlantic would be employed in the United States. All of a sudden, John McAdam was to became an American highway hero.

Road building was for this remarkable Scotsman an adolescent hobby that turned into an obsession. As a youngster, he was made a trustee of a southern Scottish turnpike, and he became fascinated by the way the road had been built and maintained. He insisted on making improvements, as he saw them, to the running of the pike—so many that in 1816 he published a booklet broadcasting his ideas. It ran to nine editions over the next decade, and the ideas he promulgated became so biblically accepted that his name soon entered the language. Roads made of *macadam* were likely to survive.

What John McAdam had done was to throw out the existing rule book, which traditionally held that a road ought to be made of big slabs of rock, the bigger and tougher the better. His new creed, instead, had all to do with compaction. He demonstrated that to give a road the chance of a long life, its topmost two inches should always be made of small stones, small enough to fit into a man's mouth. They should always be significantly smaller than the four-inch-wide wheels of the wagons that used the road.

Building then became simplicity itself. The hired laborers could easily measure their stones: they could sit in shifts beside the roadway, piles of rock dumped by them, and with small hammers they could chip away and shape these larger stones into more-or-less spherical objects that weighed less than about six ounces. They would then pop the finished stones into their mouths to make sure they were the proper size, and finally toss all the suitable candidates into a basket to be carried away and laid on the surface.

Beneath this vital top layer would then be slightly larger stones, three inches or so, but not the gigantic boulders that had been used in earlier road-building efforts. Moreover, nothing—no binder, no cement, no gravel, no dirt—should ever be inserted into the fabric of a macadamized road, decreed the Scotsman. The simple weight of passing vehicles would squeeze the stones together, crushing them into one another's angles, solidifying the roadbed as a result.

In addition—a peculiarly American addition, it has to be said—there was a financial incentive for the users to keep their road surface intact. The tolls were cleverly calibrated to favor lower-impact users. If you had a flock of sheep, deemed very light-footed, then they could pass through Pennsylvania between Brownsville and Wheeling for six cents for every twenty animals; for cattle, which had sharper hooves and so did more damage, one paid twice as much; a cart pulled by four horses would be charged eighteen cents, and so on.

The only disadvantage was dust, particularly when it came to the passage of wheeled vehicles. The lowered pressure beneath a carriage body as it sped over the road* drew up great choking clouds of the stuff, a problem that was not to be solved for the better part of a century. In the 1920s, a Welshman named Edgar Hooley decided to spray tar on John Macadam's crushed-rock surface—creating tarmacadam, or tarmac, in America called blacktop. He was of course too late for both Jefferson and Martin Van Buren: the National Road and many of its immediate successors had to make do with simple original unadorned macadam, dusty but highly durable.

Before the six hundred miles of the Cumberland Road had been completed, Congress had already approved a means of funding an extension, across the Mississippi River and as far along the course of the Missouri River

---

* This lowering of the pressure of high-speed air is an example of the Bernoulli effect, a principle of physics that became central to the much later design of the airfoil and the invention of heavier-than-air flying machines.

as Jefferson City. The route it eventually followed then stuttered across the country during the latter half of the nineteenth century, according first to local needs, then to national planning. And even though during the summer the roadway became a thousand-mile fog of nearly impenetrable dust, the National Road was a great nineteenth-century success. As the WPA Guide for Illinois had it, the highway fast turned into:

> . . . the most traveled thoroughfare in the nation . . . [and] bore a never-ending load of traffic. Great freight wagons lumbered over its length, crammed to overflowing with manufactured goods for the frontier, and returned laden with raw materials for the Eastern seaboard. Travelers of every description ate, drank, sang, and cursed in its roadside taverns and stage houses. Andrew Jackson, William Henry Harrison, James Polk, Henry Clay and John Marshall rubbed elbows and exchanged a passing word with teamsters, actors, settlers and soldiers of fortune. The east-west mail was carried on the Cumberland Road; in 1837 it took about 94 hours to travel from Washington to St. Louis.

It would all soon get a great deal faster.

Rain, Steam, and Speed

It is an unassailable physical fact that the invisible and very hot gas known since the eleventh century as steam occupies 1,600 times the volume of the water from which it is made. The difference in volume is key. Light a fire under a pot of water and bring it to a boil, and the immense amount of steam that the laws of physics compel it to become can be made to perform work—a very great deal of work.

Once inventors had determined various ways of usefully harnessing this volume change, a slew of steam-driven devices were created. Among them were engines with levers and rods and cranks and pistons, pushed and pulled and turned and compressed, that could perform a million miraculous tasks. When applied to the business of roadways and waterways in particular, steam power instantly made obsolete the slow labors of traditional animal-hauled vehicles and boats. Steam engines, uncomplaining, untiring, and unyielding, could henceforward be made to perform all of the labor. All that would be needed from now on was fuel to feed the engines' fires.

In America industrial steam had many champions. One of the earliest and best-known was the very Irishman who in 1789 had produced the country's first highway map, Christopher Colles. Fifteen years before, shortly after arriving in America from Limerick, Colles had built a steam pump for a Philadelphia distillery, and then a much larger clanking monstrosity of an engine for the New York City water pumping system. Neither was a success; the former performed only intermittently, and work on the latter had to be abandoned because of the War of Independence. And though neither machine was designed for transportation, the irony is inescapable: a man whose renown was to be very much defined by his mapping of American roads was also toying with the machine that would soon be, with the invention of the railway engine, the early roadways' temporary undoing.

Colles's first engines were in no way portable. But such contraptions were indeed coming and had been for a while. Ten years earlier, James Watt had invented the condensing steam engine. In doing so, he had opened the way for all manner of more complicated and flexible ways of designing wood- and coal-fired boilers to produce steam to push pistons and turn cranks and drive rotary engines of one kind or another and thus make things move along a road, a waterway, or a specially designed track—a railroad.

Waterways were the first to benefit from the newly discovered physics of steam. John Fitch, a Connecticut button maker, watch repairer, and silversmith, took an early interest in the waterborne side of things. He proposed in 1785 that "there might be a force governed by steam," promptly built a paddleboat with a James Watt–type engine inside, and chugged up and down the Delaware River in 1787, little more than a decade after George Washington had rowed across it on the way to war. But he could never interest investors; he faded into obscurity, a figure of pity and ridicule. In 1798 he killed himself by taking an overdose of opium pills, believing that his pioneering would one day be recognized.

It never has been, cruelly. Instead the honors go today to a much better connected, more amiably disposed, and more ambitious Pennsylvanian, Robert Fulton, who purchased an engine from Watt's company and installed it so as to rotate a pair of fifteen-foot paddles in a steamboat he had built for him on Staten Island. In 1807 he floated it out into the Hudson. He called it the *North River Steamboat of Clermont*, later just the *Clermont*, invited paying passengers aboard, and ran the craft on a regular schedule between New York City and Albany, a journey that took just thirty-two hours, regardless of wind or weather.

*The success of Robert Fulton's 1807 paddle-wheeler
the Clermont led to a shipbuilding boom, with vessels becoming adored icons of the age. Here is another
Fulton creation, the Chancellor Livingston, plowing
through a poetically rendered Hudson riverscape, on
dinnerware. Mr. Fulton himself had to make do with
appearing on postage stamps.*

His creation became an overnight wild success and
was the herald of an entirely new form of waterborne
travel. Ten years later, a steamboat pushed its way successfully upstream along the Mississippi, then turned
right and proceeded up-current along the Ohio and in
due course reached the city of Cincinnati. Just as on
the Hudson, steam-powered travel along the country's
greatest river caught the attention of all: within two

years there was a fleet of no fewer than sixty stern-wheeled steamboats plying a hectic trade between Saint Louis and the sea.

Steam technology on America's waterways then advanced at breakneck speed: in 1817 it took twenty-five days to travel up from the sea to Louisville, Kentucky; ten years later, it took little more than a week. The boats became sleeker and more suited to the waters, many with a draft so shallow that it was joked that they could float on a heavy dew. Of all this the Connecticut craftsman John Fitch—despite the cruel injustice of history—deserves to be remembered as the true pioneer.

But now, following Fulton's success, there was no stopping steam. The changes it wrought came at warp speed. Within fifteen years, these two entirely new forms of powered transportation, the steamboat first and the steam train next, came to dominate the business of carrying cargoes of all kind, especially raw materials, and then in due course, once they became less timid and less in dread of speed, people. Steam under high pressure is dangerous, of course. People were killed (140 in a gigantic boiler explosion in Charleston in 1838, for instance). Cassandras denounced what they saw as a troubling new American tendency to recklessness and blind ambition: "Go Ahead!" seemed to one diarist of the time to be the watchword of the age, "regardless of the consequences and indifferent to the value of human life."

The development of steam spelled the end, at least for a while, of the idea of a national highway system. It ensured instead the beginning of a series of battles among a new and very different set of rivals in the railroad business, battles that would culminate in the midcentury linking of the two ocean coasts of America, not by macadamized roadway but by steel, along which new and mighty creatures powered by heat and flame would travel.

The Annihilation of the In-Between

The first working freight train shuddered out of its station in northeast England in late September 1825. In the same year, an elderly and very wealthy New Yorker named John Stevens made and ran America's first steam locomotive. However, because it was never given a name,* never carried cargo, certainly never carried

---

*   Unlike the British engine, which was named *Locomotion No. 1* and for its historic first journey hauled seven hundred people in thirty-three coal wagons, taking two hours to cover the twelve miles between the mining town of Stockton and the port city of Darlington.

people, and never went anywhere except around and around a small circular railway track in Hoboken, New Jersey, it has somewhat faded into history.

But all technologies must these days have a parent, and the fact that this doughty little engine performed its task before anyone else in America had accomplished anything similar, and in only the fifth decade of the nation's existence, has caused John Stevens to be reckoned the father of the American railway.

Stevens was seized with the idea after hearing the melancholy story of John Fitch and his 1787 experiment with his sad little steamboat chugging around on the Delaware. According to the *American National Biography* entry for Stevens, "From that moment until his death he devoted himself and his fortune to the advancement of steam-propelled transportation both on water and on land. Immersing himself in the science of steam, he designed boilers and engines on paper. His influence with Congress helped pass the first U.S. patent law in 1790, and he received one of the first patents. He improved the design of the vertical steam boiler and invented an enhanced version of Thomas Savery's early steam engine, both conceived with steamboats in mind. He also applied steam to the working of bellows." Steam, in other words, became John Stevens's religion, in which he had unwavering faith.

Initially he was not much more of a success than Fitch. He had a special dedication to the Hudson River, as it passed beside Manhattan. He experimented with the idea of running a ferry across it, between New Jersey and New York City, and when that plan went nowhere, he drew up schemes for building both a bridge and a tunnel to cross the Hudson. All exist today: Stevens was a century and a half ahead of his time.

But as he aged, he grew more vexed and bored with boats, steam or otherwise. Innumerable disputes among his various rivals frankly wearied him. Instead, in his dotage, he turned his attention to the less contentious field—since they had not been fully invented—of railroads. He imagined—rightly, as it turned out—that, rather than excavate the Erie Canal, it made more economic sense to build a railway between New York and Lake Erie along the Hudson-Mohawk Gap. And around 1815 he dreamed up plans for a line made of timber rails bearing carriages and wagons having flanged wheels, "the moving power to be a steam engine." He had sufficient faith in his idea to obtain from the state a license to run a railway. A skeptical legislature, however, declined to advance any money.

Unsurprisingly the rejection frustrated Stevens— who by now was seventy-six years old, a man as crotchety as he was impassioned. That is why, in a last-ditch effort to convince the lawmakers of the wisdom of building steam-powered railroads, he created a working model of a proposed engine and had it run around a circular track that he had his men build on Castle Point, his great Hoboken estate overlooking the fast-amassing buildings of lower Manhattan. But the legislators to whom he showed his invention, whom he saw as conservative and blinkered, never bit.

However, his demonstration and his idea did win admiring comments, and in short order they also sparked imitators by the score. John Stevens may have failed to impress, but his idea most certainly did not fail at all. Within just two years of his showing his toy train, a number of proper full-size railways, all of them powered by steam, were starting to spring up throughout the American East. A mania, one that would develop on a far greater scale than that which accompanied the building of canals, was soon to grip the nation. The annihilation of distance, the erasure of the in-between, became a new calling among the country's premier engineers and the more visionary tycoons.

The old man himself lived on for a further thirteen years, dying in 1838 at a time when all of his railway dreams were fast starting to be realized. On July 4, 1828, Charles Carroll, the last surviving signer of the Declaration of Independence, was invited by the directors of the newly formed Baltimore and Ohio Railroad to turn the first sod for a steam railway. He said, "I consider this among the most important acts of my life, second only to my signing the Declaration of Independence, if even it be second to that."

John Stevens's two sons* eventually went on to become directors of a railroad that won the monopoly for the phenomenally busy route between New York and Philadelphia, amply swelling the family's riches. These days their great house in Hoboken memorializes the patriarch in a manner of which he would have approved: he, after all, had once famously declared that "good morals and good government in a republic are only attainable and maintainable by knowledge and information pervading the whole mass of society." Thanks to provisions in the will of one of his sons, the estate was bequeathed to found and develop the Stevens Institute of Technology, a place where young men and women learning the niceties of new kinds of machine making now stroll on campus lawns, the very

---

* One of whom invented the cowcatcher, a uniquely American railroad symbol.

lawns where in 1825 the first American steam train performed its practice runs and changed the face of the nation.

The speed of that change was dizzying. It began in 1828 and roared through the country for the next forty years, until briefly stopped by the Civil War and then again by the Long Depression of 1873–79, which was itself triggered in part by the frantic post–Civil War boom in railroading. Afterward, the rail-borne changes resumed in a fury. One telling illustration symbolizes the early days. When Tennessee's Andrew Jackson was elected president in 1828, he traveled to the White House in a horse-drawn carriage; when he stepped down from his second term, just eight years later, he left for home aboard a steam train. Less than twenty miles of track had existed in the country when he took office; there were nearly three thousand when he left.

There was some throat clearing. When the Baltimore and Ohio Railroad formally opened in late 1828, the locomotives had not been built, and the wagons and carriages had to be pulled along by horses. Most early engines were imported from Britain, which had already embarked on its own railway-building mania. Within a decade, though, American factories had gotten the idea, and by 1839 they had turned out more than three hundred locomotives of varying designs and sizes. There were six different gauges; the current classic gauge of four feet, eight and a half inches was not fully and formally adopted until 1886.

This popularity of railroads prompted a swift and sudden upheaval in the economy. By 1870, the railroad industry had become the country's second-biggest employer, after agriculture. Soon the dominant railroad companies became the country's biggest corporations, some of them almost the size of governments. Most of them were started with enormous government loans or vast government land grants.

Almost half of the money behind the early American railroads came from the public purse—an exceptionally wise investment in a myriad of unanticipated ways. American railroads helped move metal ores and wagonloads of grain around the nation, benefiting the miners and growers. Once the trains began burning coal rather than wood, new mines opened and flourished from Maryland to Tennessee. With the construction of new factories supplying everything from rail lines to signals to carriages and wagons and engines, it can fairly be said that the railway spawned and underpinned the industrial revolution in the American East and started the country on its way toward global economic supremacy.

Railroads brought about lasting social effects, as well. The companies' ruthless attention to keeping time impelled passengers to carry pocket watches,* and led to the eventual establishment of time zones. A rail traveler initially had to contend with whatever time was chosen by the railway. The Penn Central kept to Altoona time; five other railroad times were shown on the many-faced, many-handed clocks at Pittsburgh station. The American East had fifty different railroad times posted on its various lines. The railroads also prompted the learning of entirely new skills—those of the mechanics, boilermakers, foundry workers, and civil engineers who built the bridges and cut the canyons across and through which the lines ran. Whole new business management ideas and models were tried, tested, and inaugurated. The creation of railways also led to the idea of vacations, for it had all of a sudden become easy and cheap to get away, then perhaps to build a holiday home. Resorts started to be built, along with spas and hidden-away country lodges with stations close by.

On the grander scale the placement and prosperity of whole cities—Chicago not the least—were hugely

---

* The wristwatch was born on the World War I battlefields from the need for officers and senior NCOs to synchronize their "over-the-top" orders in the trenches.

affected by the building of railways and all the junctions, switching centers, and classification yards, where wagons were sorted and sent on to their final destinations. Besides all of these constructions, people themselves had to live and be serviced and fed, housed and watered.

On the other hand, those regions where the railways did not go inevitably suffered. This suffering was felt most painfully in the early days in the American South.

Amid the whirlwind of construction in the North and East came a swift realization: it was much more attractive and economically prudent for a manufacturer in the Midwest to send his goods to New York or Boston or Baltimore on nonstop freight trains rather than to continue to use the locks and slow waters of the southbound rivers. The South's early railways were small, local affairs, largely unconnected to the great new lines of the East. This helped exacerbate the feelings of separation and otherness that led to the Confederacy and the Civil War. The existence of an extensive and technically sophisticated Northern railroad network also became a factor in the outcome of that war, giving the Union generals a major advantage for rapidly moving their troops. So while it is indisputable that railroads played a key role in the uniting of America generally, they also played a not insignificant part in the profound spasm of disunity of the 1860s, when America was riven apart, brother from brother, in the shameful horrors of the War between the States.

In the decades thereafter, railways were frantically built from Maine to Charleston, from Chicago to Cincinnati, from New Jersey to Nebraska. Within only thirty years, the American East and Midwest had been knit together by twenty-seven thousand miles of iron rails (later made of steel, seventeen times more durable). On these, an enthusiastic poet, R. C. Waterston, told his audience in Boston, "gliding cars, like shooting meteors run, / The mighty shuttle binding States in one."

But what was bound here into one was really just the East and the northern Midwest—the America that lay between the Atlantic Ocean and the Missouri River. To forge a real link and bring the entire continental nation together needed a concerted act of will, a practicable route, a formidable design, a great deal of money, and more hard work than the nation had ever witnessed. The transcontinental railroad, which would allow and encourage travel from ocean to ocean—with a promised journey time of days, not months—was the prize above all others.

The first plans for a cross-country railway were offered in 1838, a mere decade after the first trains began running on the B&O. The first serious surveys were commissioned by Congress in 1853, though politics prevented the selection of any one of the five projected routes. Much of the political argument revolved around the issue of slavery, a scourge not then visited upon the states of the "New West," and because most Southern senators demanded that half of all new states allow slavery and wished to use the railroad to export the practice, matters were confounded and complicated beyond endurance. The final accomplishment and the start of actual construction of a three-thousand-mile ribbon of steel from coast to coast is largely to the credit of another remarkable figure: Ted Judah, a Connecticut visionary who was first seen, as many visionaries are, as deranged and even unutterably mad.

The Immortal Legacy of Crazy Judah

Theodore Dehone Judah, a preacher's son from Connecticut, had surveyed the terrible majesty of the Sierra Nevada in 1860, and after much time and thought and whole seasons of climbing and mapmaking, he decided that the Donner Pass, the notorious site where almost half of a party of westbound pioneers had died in 1846, trapped by an early-season snowstorm, was the ideal route to take a transcontinental railroad over the most difficult mountain barrier of them all.

I first crossed the Donner Pass late in the 1980s. It was early spring, and I was in a hurry. I had been in Montana and was driving fast to San Francisco to catch a plane back to Hong Kong, where I then lived. I was on Interstate 80 westbound.

After coming down south through the mountains of Idaho on Route 93, I had joined the freeway near Wells and then sped across the desert as fast as was prudent. I got to Winnemucca at dusk, Reno at ten. Though it was cold and starting to rain lightly, slicking the roadway, problems seemed unlikely: the Nevada Highway Patrol's radio station was reporting that the Donner Pass was dry and clear.

Except that flashing lights beside the road soon said quite otherwise. "Chains Required, Donner Pass" they said, with increasing frequency and urgency. As I climbed slowly up the darkening range the rain turned to sleet, then to driving snow. Then there was a barricade of flashing blue lights ahead: a Highway Patrol roadblock, and a sign: ONLY CARS WITH CHAINS PERMITTED BEYOND. A small group of men waited at the roadside: for $75 they'd sell and attach the necessary chains, cash only. Bills were handed over; a police officer shone a flashlight at my wheels, squinted through the blowing snow, waved me through. "Better you than me!" he shouted into the gathering gale.

As I climbed up through the thousands of feet—Donner Pass is a mile and a half up—the snow fell more and more heavily, blowing directly at me in a frenzy of white needles. The road was now covered with thick powder, and it became ever more difficult to steer. Huge cliffs of snow, cut sharp and sheer by a plow, rose beside me, giving me the feeling I was trapped in darkness inside some infernal white-walled canyon. There was almost no traffic—except that once a plow went past in a blaze of whirling lights, and racing in its wake was a huge truck, flying past at eighty, careless of the weather, New Riders of the Purple Sage on the radio, no doubt, setting me rocking dangerously in his slipstream.

Ten miles on and a mile up, the weather worsened, the wind now howling like a hurricane and setting boils of spindrift dancing on the highway ahead, blinding me. I would grip the wheel and pray that the road didn't twist until I emerged from the maelstrom. But then I was on a straight section. I could see the sign showing I was at seven thousand feet. Then there was a huge roar to my right, and through the blizzard I could see, comfortingly, a train—a pair of yellow-and-orange locomotives hurtling alongside me, heading west, imperturbably shoveling the snow over their shoulders and plunging into the darkness, sleek and unstoppable and magnificent.

That moment has lingered in the mind for years. For of course I survived the journey. I breasted the summit and headed down onto the Pacific side, and in time the snow thinned and turned to splashes of sleet in the windshield, and then it became rain, and a thin watery dawn came up, and there were the lights of Sacramento and, in time, the bridges and the skyscrapers and the sea. And somewhere that great train I had seen was also threading its way over switches and beneath gantries of signal lights and would soon slow and squeal to an easy stop in some mighty freight yard beside the container terminal. And the engineers would climb down from the roasting diesels, and someone would ask them how the night had been, and they would say it had been routine, a nasty storm on the mountains, the Donner Pass quite chilly and with a lot of snow, but nothing that the railroad couldn't handle. It was a reminder that the railroad across the Donner Pass is an essential part of what connects Americans, always open, always there.

Ted Judah's life had been entirely wrapped up in railways. He studied engineering in Troy, New York, at a school next door to the Troy and Schenectady Railroad yards. One of his first jobs was to build a railroad bridge in Vermont. He then performed surveys for the Hartford and New Haven Railroad, joined the Niagara Falls & Lake Ontario Railroad, surveyed and built the line down the Niagara River gorge,* was appointed chief engineer of the Buffalo and New York City Railroad, and then, crucially, was invited out west. His task there was to lay out a route and then oversee the building of the line for a new railway up to the high goldfields of California, along the valley of the Sacramento River. The railway—the first to be built west of the Missouri—was for the mundane business of supplying the gold miners with picks and shovels and pit props, all of which would be brought by paddle steamer from San Francisco to the head of navigation.

It was the first time that this callow, twenty-four-year-old New Englander had seen the might of the Sierra Nevada frontier. It took him only a short while to become convinced that this formidable geological barrier to trade, travel, and the railway, behind

---

* The climactic closing scene of Michael Powell and Emeric Pressburger's classic movie *49th Parallel*, released in the United States as *The Invaders*, was shot at Niagara and illustrates Theodore Judah's permanent way handiwork to advantage.

which the magic and wealth of California were shielded from the rest of America, could and should be breached, and that he was indisputably the man to accomplish it. He became seized with the conviction, to a degree some saw as close to lunacy, that a transcontinental line was essential for securing the future greatness of America.

To Judah, all was obvious. San Francisco and Sacramento were the key cities in the West. The Platte River valley through Nebraska provided the obvious route for the rails to follow. South Pass in Wyoming, where the wagon trains had rattled their way west, was clearly where the Rockies could be most easily breached and the divide crossed. Salt Lake City was where food, fuel, and locomotive crews might be found and cached. Then it remained only to find a route across the Sierra and so to get from the deserts of Nevada down onto the coastal plains of California. He planned the route carefully, considered it as possible as it was desirable, and advanced his thesis near-endlessly in long screeds that were published time and again by a tolerant editor-friend at his local newspaper, the *Sacramento Union*.

Crazy Judah, as he became known, fast developed into a familiar figure on California's fledgling society scene. Crazy, maybe, but seldom thought of as a wastrel or dilettante, of which San Francisco had many. He was taken seriously enough to be elected in 1859 as a delegate to the Pacific Railroad Convention held in a city that, though less than twenty years old, was flush with the wealth of the goldfields and longed to be connected to New York by a route more convenient than by way of a ship to Panama, a mosquito-infested train across the isthmus, and then another long sea journey up to the East River landings of lower Manhattan.

The convention was a success. There was near-universal acceptance that San Francisco was indeed the natural terminal city for a transcontinental line, and Judah—so eloquent! so impassioned! so knowing! so delightfully mad!—was made official agent for the cause. Dispatched back east, he pressed it endlessly and with enthusiastic conviction to the lawmakers on Capitol Hill. He visited the aging and sickly President Buchanan at the White House, and to all he transmitted the same message, the same demand, but with one condition that would keep the idea in limbo for the next several years—that California and the states to be crossed by the new railroad line would not permit the wickedness of slavery. Those in Congress who desired otherwise—and with the Civil War looming there were many, most but not all of whom were

Southerners—argued their case bitterly, and for years that delayed matters.

Lobbying and canvassing and lecturing were all well and good. But going out into the field with a team of burros, a chain, and a theodolite and surveying just where the line might go—that's what Theodore Judah did to clinch the deal. The vital moment came one summer day in 1860 when, at his office in Sacramento, Judah received a letter from a man named Daniel Strong, who ran a drugstore in a mountain settlement sixty miles away to the northeast, the so-called Athens of the Foothills, the goldmining town of Dutch Flat. Strong had something for him to see.

Dutch Flat was a wild mining town, just nine years old. It had a post office; the usual range of bars, brothels, and convenience stores; a doctor to attend to the miners' frequent injuries and ailments, as well as Strong's Druggists next door; and a daily stage line down the front range to Sacramento.*

---

* Germans—"Dutchmen" in the local parlance—had settled Dutch Flat in their thousands after a discovery there of placer gold. It was one of the many finds in the Sierra Nevada foothills that followed the famous sighting of nuggets at Sutter's Mill in 1848, the event that can fairly be said to be responsible for the sudden growth of San Francisco and the utter economic transformation of California.

All the town's miners were involved in the brutal practice of hydraulic mining—the sluicing of flakes of the precious metal from the diggings by the use of fantastically high-powered water jets, which tore the cliffs apart and ran the particle-rich water down through mesh gold sluices. Millions of dollars in gold flowed across the assayers' counters here, until the mines played out. During its heyday, this technique was a quick and dirty approach to gold mining, now generally outlawed, which left an indelible environmental stain on the California countryside, from which the region has still not wholly recovered a century and a half later.

But Doc Strong had no wish to show any of this to his visitor. What he had in mind was quite simply to show Theodore Judah what he considered *a stunningly beautiful and very important view.* Strong knew the young engineer was positively aching to build a railway across to Chicago and New York; how could he not know, with articles about Judah in the local paper almost every other day? He lived in the mountains and knew full well the problems of finding a way through them. Maybe he had at last found the answer, and he would take Judah up into the hills and have him gaze at a vista he had seen himself some days before, then let him draw what Doc Strong suspected were the inevitable conclusions.

The two men went to the local livery stable and secured a pair of strong horses. They then started up the trail—unmarked, except for some relic scars of the Donner party and their rescuers of fourteen years before. It took some hours as they hacked their way through the sweet-smelling pines and over small creeks, gaining altitude steadily, the air cooling, the breeze rising. Finally they reached a summit, at eight thousand feet or so, and Doc Strong gestured to Judah to look before him, eastward, and to prepare to be amazed.

As indeed he was. There was no mistaking what he was seeing. The tall serrated ridges of the High Sierra marched to the north and the south of him. But directly in front was a low pass, with steep cliffs running down to the American River on his right side and to the Yuba and Bear Rivers on his left. Between the cliffs was a low, flat plain, a wide notch in the mountains that sloped down as far as the eye could see, terminated not by another crest or ridge but by the sparkling blue waters of a lake. That lake, said Strong, emptied by way of a small creek into the Truckee River, which led down the far side of the mountains, down onto the Humboldt Plains and into Nevada Territory. Surely, said Strong, that could be the way that the railroad could go.

Theodore Judah was first shown the Donner Pass in 1860, fourteen years after the tragic fate of a party of would-be pioneers who were trapped by its ferocious winter weather. Judah saw the pass as providing the obvious route for the transcontinental railroad, nonetheless; it remains today the key Union Pacific main line between Nevada and California.

Theodore Judah saw it in a split second. The pair duly camped overnight in a shepherd's abandoned house, and Judah woke early to make sure his eyes had not deceived him. It was raining, but as he stood on the precipitous ridgeline and gazed eastward into what, symbolically, was a new American morning, the visibility was good enough for him to be certain. He felt sure that this could, should, and indeed would be the route of the railway to the East. By completing this, he would be welding the missing link into a skein of steel that would now bind the nation firmly and permanently together.

Once back in the state capital, he did the necessary calculations. For the benefit of the congressmen to whom he would have to give evidence, he drew a colossal map of the critical segment. It was a chart ninety feet long, showing every proposed cutting, embankment, grade, and tunnel. It was 102 miles from Sacramento to Donner Summit, a total of 140 miles to the flatlands of Nevada. The route he designed would require that locomotives climb the hills to Donner Summit, 6,960 feet up, smoothly and serenely, and in doing so never facing a gradient of more than 100 feet for every mile. The Baldwin locomotives being manufactured at the company's huge factory in Philadelphia could manage such a climb with ease; some of the passes across the Alleghenies were quite as bad. And the thirteen-mile run down the Truckee Valley on the far side would be no more severe. Building the line might be difficult but was far from impossible. It would be simplicity itself for a good-size train to cross the passes once the line had been built.

The saga of the subsequent construction of America's first transcontinental railroad is now painted in the nation's most hallowed self-portraits. In essence, it involves two giant railway companies—the privately financed Central Pacific, based in San Francisco; and the publicly chartered Union Pacific, based in the newly settled town of Council Bluffs, on the east bank of the Missouri River in Iowa. It also involved an act of Congress, signed into law by President Abraham Lincoln on July 1, 1862, at a time when Confederate troops were massing on the very outskirts of Washington, DC, and the president was preoccupied with defending the capital and writing a draft of the Emancipation Proclamation, which he would discuss in a Cabinet meeting just three weeks later.

The Pacific Railroad Act was as long and complex as might be expected, larded with detailed plans for the financing of the railway and peppered with timetables and provisions for rewards and penalties for those who would be involved "in the Construction of a Railroad and Telegraph Line from the Missouri River to the Pacific Ocean." Three crucial paragraphs, sections one, nine and ten, remain in the memory, and even in précis remain sonorous for all time:

Sec. 1. The Union Pacific Railroad Company . . .
is hereby authorized and empowered to lay out,
locate, construct, furnish, maintain and enjoy a

*continuous railroad and telegraph . . . from a point
on the one hundredth meridian of longitude west
from Greenwich, between the south margin of the
valley of the Republican River and the north
margin of the valley of the Platte River, to the
western boundary of Nevada Territory. . . .*

*Sec. 9. The Central Pacific Railroad Company of
California . . . are hereby authorized to construct a
railroad and telegraph line from the Pacific coast
. . . to the eastern boundaries of California. . . .*

*Sec. 10. And the Central Pacific Railroad
Company of California after completing its road
across said State, is authorized to continue the
construction of said railroad and telegraph through
the Territories of the United States to the Missouri
River . . . until said roads shall meet and connect.*

Work on the Central Pacific line was started first
because, though shorter, Judah's route was obviously
going to be a construction nightmare. Optimism ruled
when ceremonial sods were cut in Sacramento on a miserable day in January 1863, and the first few miles east
of town were easily made along the flat alluvial plains
by the river. But as soon as the loggers and excavators

and rock splitters and bridge builders and plate layers arrived at the sierra foothills, matters started to become phenomenally difficult.

The challenges were myriad and mixed. For a start, all the equipment had to come by sea, around Cape Horn. The five thousand tons of iron rail, the eight locomotives, and the eight passenger carriages, four baggage cars, and sixty freight wagons that Judah ordered all had to be loaded onto ships in New York and brought through perilous seas thousands of miles to San Francisco, then driven up-country to the construction sites, where the real problems began.

The terrain, the geology, and the weather presented an unholy trinity of nonstop nightmare. The cliffs were well-nigh insurmountable, the tunnels had to go through volcanic rocks of drill-blunting hardness—it took eighteen months for two thousand men to finish the half-mile Summit Tunnel—and the snows fell fifteen feet and more with drifts that could bury campsites ten times over. Special snowplows had to be built for the trains as the line advanced eastward day by slow and painful day. In places where even digging out the track by hand proved impossible, the builders made huge iron sheds on top of the line and hoped the all-too-frequent avalanches would sweep over the roofs.

The available workforce proved to be severely limited. Despite the good pay on offer, even the toughest Irishmen and Slavs from the San Francisco docks were reluctant to work so dangerously hard for so long in such appalling conditions. It was much easier to forage for gold or build houses for the newcomers to the city than to go up into the mountains and endure privation and danger for months on end.

Eventually someone reminded the foremen that Chinese workers, despite being supposedly limited in stature and musculature, had once managed to build the Great Wall. Soon twenty thousand immigrants from the fishing villages of southern China were pouring off the boats and into the Bay Area, tempted by the promise of pay in gold coins and stoically undaunted by the rigors of the work demanded. Few will forget their triumphs: the ferociously difficult Summit Tunnel, where they first employed nitroglycerine explosives when the drills proved useless; and perhaps most notably the nearby Cape Horn Cliffs,* where volunteers from among them agreed to be suspended by ropes in small baskets over vertiginous drops, to chisel a passageway

---

* Not to be confused with the cliffs on the small Chilean island, named for the Dutch city of Hoorn, near the southern tip of the Americas—but in their own way, quite as dangerous.

into the thousand-foot sheer face and then clamber into the passage and enlarge it until it was wide and deep enough to carry a train. They were eventually helped by ample supplies of black powder and nitroglycerine, but their unfamiliarity with these high explosives caused casualties by the score, and today there is still a scattering of sad little grave sites marking where they fell. The Chinese were inestimably brave to work as they did: it beggars belief that for so many years they were then excluded from immigrating to America, largely because their working practices were believed to depress the wages of less energetic white workers.*

Matters on the Central Pacific eased dramatically once the crews had pushed their way to the Truckee River and then down onto the hot and waterless plains of Nevada. Before long, the workers were laying track at phenomenal rates, many miles each day. The effort required to do so strains credulity today. Teams would place scores of rough-hewn and well-dried cottonwood ties onto the flattened gravel roadbed. Men who would later be called gandy dancers would next swing and heave down fifteen-foot, five-hundred-pound lengths of rail, two at a time, into position

---

* The Exclusion Act, passed into law in 1882 at the height of a wave of anti-Chinese sentiment, stayed on the statute books until 1943.

lengthwise across the ties. Swarms of workers behind would hammer in the spikes and screw in the bolts and secure the fishplates. The target was often as many as ten miles in a single day, and to achieve this, men had to lay and secure two new lengths of rail, separated by the standard gauge of four feet, eight and a half inches,* *every twelve seconds.*

Back east, the Union Pacific's work had started nearly a year later, on December 2. The groundbreaking ceremony in Omaha—the city across the river in Nebraska Territory, reached by ferry from Council Bluffs and the railway's theoretical zero-mile marker— was a somewhat more elaborate affair than the one back in Sacramento: soldiers fired fusillades; brass bands sounded fanfares; a thousand of the Omaha elite were on hand for a mighty banquet.

But the practicalities were just as irksome as those the Central Pacific faced. The construction team was beginning a railway from Omaha that was designed to reach the West Coast, yet at the time there was no physical connection between Omaha and the East Coast, let alone the West: there was no bridge across

---

* Arguments over the gauge took a year to settle: Lincoln signed another Railroad Act in 1863, setting the gauge as the standard used at the time in Britain on the Liverpool and Manchester line, and then everywhere else in the United Kingdom.

the Missouri River. All of the building supplies— hundreds of tons of rail, ties, spikes, tools, shovels, wagons, and locomotives—had to be brought, along with many of the workers, across the great stream by boat. And worse, no rail link to the east side of the river existed either—and none would exist until the Chicago and North Western Transportation Company extended its line down from Chicago to Council Bluffs.

Once work started, progress was reasonably swift, sweeping prairies being much easier to cross than jagged mountain ranges. But there were many places where federal troops had to protect the railroad crews from Indians, who were not unreasonably angered by the white man's continual treaty breaking, violence, and duplicity.

Many Irish were employed on the Union Pacific rails—immigrant ships brought thousands across the Atlantic from Cork to New York, just as such ships also brought Chinese "Celestials" (from a traditional name for China, the Celestial Empire) across the Pacific from Hong Kong to San Francisco. There was much whiskey, too, and once the Civil War was over and the Irish, who had fought for both sides, were demobilized, they came west to the Union Pacific bosses for work, and usually found it—together with the low-level villainy and alcohol-fueled fighting that seemed to go with it.

But there was organization to the process, too. Special construction trains were made, with accommodations and commissary cars and with hunters like Buffalo Bill supplying fresh meat to the cooks. And then, as if this wasn't enough, portable cities of sorts followed the tail end of the last construction trains—each a tented, movable Hell on Wheels, as all were called, with casinos and brothels and dance halls and anything else the workers needed when they came back exhausted each night from the front of the line.

Thus did the workers of the Union Pacific cross the prairies and pass through the tallgrass counties and the High Plains; they managed to hoist their rails at as comfortable a gradient as possible through the famous South Pass of migrant days, then over the Divide in Wyoming. They headed steadily west as the Chinese workers on the Central Pacific were heading east— until that momentous day in the early summer of 1869, when the two oncoming teams glimpsed each other in the distance. They stopped, waved, shouted, sent scouting parties and greeting parties, prepared their celebration parties, and then continued to build their respective lines steadily toward each other.

Finally came the day—weather delayed; it was initially scheduled to be two days earlier—that will always live prominently in American history: May 10, 1869, the day the two lines converged, intersected, encountered one another, met, were connected. From that moment on, an unbroken line of metal has connected the Atlantic to the Pacific, by way of continental America.

It was a Monday. The two railheads had been halted within feet of each other. The world appeared flat and dry here, nearly five thousand feet up in the basin-and-range province between the Rockies and the Sierra Nevada, at a place that has since been called, for the range of mountains nearby, Promontory. This was Utah Territory, run largely by Mormons from Salt Lake City. Suspicious of the encroachments of the outside world, they paid little attention to the ceremonies that day. Brigham Young had been invited to the celebration but declined.

And what ceremonies then unfolded! The Easterners had worked through the night to make a small terminal station where the event would be staged; the Californians had slept through it all. Various trains arrived, carrying onlookers, dignitaries, workers, celebrators. The Union Pacific's gleaming black-and-brass train *No. 119* drew up to the most westerly end of the eastern line; the Central Pacific's great workhorse, the *Jupiter*, drew up, decorated with flags and bunting, to the easterly end of its line from the Pacific. Thousands waited in the sunshine, millions more around the country, promised the news by telegraph the instant that it happened.

The construction boss from California then ordered a gang of Chinese workers to get down into the short patch of dirt between the two sets of rails to scribe and level it for the placing of the ties. Someone noted a difference in the ties' quality: those used by the Central Pacific had been neatly sawed, chamfered, and finished; those made by the Irish lumbermen employed by the Union Pacific were more primitively cut, still much resembling the branches of the trees from which they had been hewn. The tie that would be used as the last actually came from the Californian stock: cut from a laurel tree, it was eight feet long and eight inches by eight square in section—the number eight is the most valued in Chinese numerology—and it had been polished to a high gleam by a San Francisco billiard-table maker.

A number of ceremonial spikes had been prepared to secure the rails to the final tie. One, presented by a wealthy San Francisco contractor, was made with eighteen ounces of pure gold and was inscribed "May God continue the Unity of our Country as this Railroad unites the two great Oceans of the World."

This, another made of pure silver, and a third made of an alloy of precious metals were to be dropped into predrilled holes and then tapped down into place. But only an iron spike would stand up to actual hammering; spikes made of soft alloy, silver, and most certainly of gold would all crumple under the impact of even the featherlight touch promised by the man who would perform the ceremony, the former California governor and senator and founder of the university that today still bears his name, Leland Stanford. So the last spike was iron—even though the event was to be called the Ceremony of the Golden Spike.

This iron spike had a small copper plate attached to its top, with two thin wires connected to the telegraph lines that stretched back to the East and across to the West to San Francisco. Stanford was to use a silver-plated maul to tap the spike gently into place; wrapped around its handle were two thin wires, also connected to the telegraph line. When the maul hit the plate, the connection would be complete. All telegraph offices around the nation were on alert for the moment, though the local duty telegrapher warned his superiors far away that once he saw that the blow had come, he would tap three dots on his own line to confirm what the hammer-and-spike connection itself, its signal so weak, might not.

The time was coming fast. The *Jupiter* and the *No. 119* were eased gently toward each other over these final rails. Photographs were taken, hands were shaken. The gold and silver and alloy spikes were dropped into their respective holes and left alone. Workmen started the two iron spikes into the wood—including the one with the copper plate on top, with its wire connected to New York—and left them standing upright, ready for the final blows. A minister from Pittsfield, Massachusetts, offered a valedictory prayer, no more or less platitudinous but mercifully briefer than the speeches that had been offered in the hours since the two locomotives had come into mutual view.

The chief of Union Pacific, Thomas Durant, then knelt, poised to hit the unconnected iron spike; Leland Stanford, the thin wire trailing from within his right hand, likewise fell into a state of genuflection, raised his maul, and glanced back at the telegrapher. In unison both men flexed arms, then dropped their fists and lightly tapped their iron spikes into the wood, while simultaneously the operator tapped his three dots and theatrically whispered to the two suited and kneeling men, "OK." Around the nation, the word DONE was flashed to a thousand telegraph offices, and a brief bacchanal erupted, inaugurated by the connection, the signal, the event, which in a memorable instant united the rail lines and made a six-month overland journey from New York to California achievable, affordable, and doable in less than a week.

The cannon fire, the bands, the steam whistles, the train horns, the rockets and the mortars, the marching bands, the sudden paradings and the singings, spontaneous or not, of "The Star-Spangled Banner," the church bells and fire alarms and gongs and every train whistle on the entire Union Pacific system— America erupted in such a sound as could have been heard from space. Durant and Stanford shook hands and proclaimed yet more platitudes; the trains edged forward and touched cowcatchers; champagne bottles were opened and foamed stupidly; train engineers reached dangerously across hot metal plates to embrace one another; Chinese suddenly became Irish; and so many people stole the ties that replaced the hurriedly squirreled-away laurel tie (destined, like the precious spikes, for displays, parades, and final glass encasement in museums) that it was feared the line could never be united, because every tie was the last tie, and each was carved up and torn away until guards had to be set beside the line.

There is a particular poignancy to the story, though. For even though he arranged so much of the detail that led to the road's creation—legislative (for there had to be a Pacific Railroad Bill passed through a divided and fractious Congress), financial (massive sums had to be raised from skeptical and impatient bankers; this short section in Utah alone cost $12 million), and technical (making precise calculations for the ascent of the Donner Pass, working out the curvature of the tunnels and the necessary gradients beside the Truckee River)— Theodore Dehone Judah himself did not live to see and enjoy the realization of his dream. He was bitten by a mosquito in 1863 on one of his transits through Panama and died that autumn. The ceremony had to go ahead without him—and given the profiteering nature of most of the participants who gathered that day, it did so with no more than a perfunctory mention, the briefest nods of acknowledgment. Stanford, Durant, Lincoln— these are the remembered names. Though he has a mountain at the summit of Donner Pass named for him, and a plaque in a park in Sacramento, Theodore Dehone Judah has nowadays all but vanished. Amid all the sound and fury at Promontory Summit in May 1869, he was barely noticed or acknowledged.

And then the crowds melted away, and the summit became quiet and dusty and deserted once again—until the first trains started chuffing by, en route between, if not New York and San Francisco, then at least Omaha and Sacramento. No one suggested that the achievement of that May afternoon had really been for so under-whelming a connection: potential was what the celebra-tion was all about. But the truth was that it would be a while yet before the full realization of Judah's dream: a clear run from New York to San Francisco. It would be four more years before a bridge was thrown across the Missouri, six more months before a railway line was established between San Francisco and Sacramento. People wishing to go from coast to coast might now do it, for sure; but until March 1873 they had to get down from their train in Council Bluffs and take the ferry across to the terminal in Omaha before resuming what would then be a quite uninterrupted ride to the coast. The bridge was crucial, symbolically. Its third succes-sor still stands; nowadays it groans with slow-growling trains hauling wagons of coal and wheat and corn across the Big Muddy. Amtrak passenger trains crawl across it, too, invariably running late.

There would soon be many other lines built across the country and scores of others constructed to run north and, to a lesser extent, south. Fifty years after the ceremony in Utah, there were 250,000 miles of track that seemingly connected every town and hamlet with almost every other. The smallest of places invariably had a train. There is a famous depot in *Winesburg, Ohio*. In the classic Western comedy *Ruggles of Red*

Gap, Charles Laughton, playing Ruggles the butler, leaves his bags with a kindly stationmaster in Red Gap, Oregon. And when I wandered the country to all the towns called Paradise, my old Official Guide showed me, deep in the Union Pacific timetables, train No. 545, which would bring farmers' wives who had been shopping in Salina back home to Paradise, Kansas, each evening at six o'clock, in time to make their husbands' dinner. Ninety-six percent of Americans, when they went anywhere in the late nineteenth century, went by train.

Freight is the business of what remains of America's railroads today; and not surprisingly it is at the nation's center, the omphalos that is Omaha and Council Bluffs and the flat countryside nearby, where freight has its most vivid expression.

It is a railroad phenomenon unrivaled for size anywhere on the planet—the Bailey Yard, an eight-mile-long, three-thousand-acre conglomeration of hundreds of miles of rails and spurs and sidings, where freight cars from all over are sorted, detached from this train and moved to that train, run down long shallow man-made hills onto faraway lengths of line where new trains are being assembled, with coal trains from Wyoming being undone and some of the wagons sent to power stations in Georgia and others to smelters in Alabama and yet others to waiting coal ships headed for Korea, with wagons groaning under stacks of shipping containers from Chicago squealing past to be put on trains bound for a vessel already waiting to unload in Seattle, with a shipment of two hundred wind turbine blades from China brought by way of San Francisco and now to be placed on a mountaintop in Virginia, with lumbering wagons carrying thousands of tons of corn from Iowa headed to Galveston, or iron pellets from Montana heading off to a new smelter in Nevada—and all the cars carrying these and a thousand more mosaic morsels of a vast economy being shifted and rolled and detached and moved and relocked together in a complex computer-controlled dance that the software makers have destined to be quickly accomplished and efficiently handled in as timely and cost-effective and profitable a manner as possible on the various just-in-time schedules that the owners and customers demand.

This is what goes on in Bailey Yard, and Edd Bailey, the onetime blacksmith who worked in a wagon repair shop in Cheyenne and went on to be Union Pacific president, after whom this giant facility was named, could fairly be said to have been familiar with every task accomplished by the thousands who work there, and could most likely have done all of them himself.

The three-thousand-acre site where this all happens is known in railwayspeak as a classification yard, and it is said to be the largest in the world (as Omaha's Harriman Street control bunker is similarly said to be the largest railroad operations center in the world; everything about Union Pacific is massive, on a global scale). It is also said to offer a daily litmus test of the health of America's economy. The more business Bailey Yard is doing, the healthier the nation's balance sheet.

It entirely envelops the town in which it is sited, North Platte, which in the 1930s was a settlement "with no traffic lights; people and vehicles bustle about in unrestrained, comfortable, small-town fashion." Almost every establishment then sported a portrait of Buffalo Bill Cody, who was essentially North Platte's patron saint. Nowadays almost every establishment sports the red-and-white-striped shield that is the Union Pacific logo. If ever there existed a railway company town, this is it, and as the railroad's fortune goes, so goes that of North Platte, Nebraska.

Trains take passengers—and freight, for that matter—only so far: they travel from station to station, not from house to house. And when Henry Ford created a machine, his Model T, a flivver, that for a few hundred dollars and some stoicism on the driver's part would indeed allow a rider to drive himself and his passengers to and from his very home, that changed everything, once again. So far as human cargo was concerned, the brief supremacy of the train was brought suddenly low by the motor car. Trains required that you travel as the railroad demanded, according to its schedule, along its routes. The automobile, on the other hand, returned to the traveler his freedom to go where he wished, as he wished, and when he wished—and at a velocity never before imagined or known.

Not that early travel by automobile was exactly a picnic. Not long after their invention, cars became fashionable playthings for the adventurous, but the condition of the roads over which they were obliged to travel often rendered the adventure more like a serious expedition, not to be undertaken lightly.

By the time America's roadways had extended their tendrils clear across the nation, it became fashionable among the adventurous, with their early motors cars, to attempt to ride along them—not so much as pioneers or settlers, nor for the lure of religion or gold, but for the sheer pleasure of travel. For such explorers, there was much advice on offer:

*To begin with, limit your personal outfit to a
minimum, allowing only a suitcase to each person,
and ship your trunk. Use khaki or old loose
clothing. Some wraps and a tarpaulin to protect
you against cool nights and provide cover in the
case of being compelled to sleep outdoors are
essential. Amber glasses, not too dark, will protect
your eyes against the glare of the desert. You will,
of course, want a camera, but remember that the
high lights of the far west will require a smaller
shutter opening and shorter exposure than the
eastern atmosphere.*

*Carry sixty feet of ⅝-inch Manila rope, a pointed
spade, small axe with the blade protected by a
leather sheet, a camp lantern, a collapsible canvas
bucket with spout and a duffle bag for the extra
clothing and wraps. Start out with new tires all
around, of the same size if possible, and two extra
tires also, with four extra inner tubes. Select a tire
with tough fabric; this is economical and will save
annoyance. Use only the best grade of lubricating oil
and carry a couple of one-gallon cans on running-
board as extra supply, because you may not always
be able to get the good oil you ought to use.*

*And, mark this well, carry two three-gallon
canvas desert water bags, then see that they are*

filled each morning. Give your car a careful inspection each day for loose bolts or nuts and watch grease cups and oil cups. Carry two sets of chains and two jacks, and add to your usual tool equipment a coil of soft iron wire, a spool of copper wire and some extra spark plugs.

West of the Missouri carry a small commissary of provisions, consisting of canned meat, sardines, crackers, fresh fruit or canned pineapples and some milk chocolate for lunches. The lack of humidity in the desert sections, combined with the prevalence of hard water west of the Missouri River is liable to cause the hair to become dry and to cause chaps and blisters on the face and hands as well as cause the fingernails to become brittle and easily broken. To prevent this, carry a jar of cream and a good hair cleanser. Use them every night.

The want of pomade and cold cream notwithstanding, what the coming millions of American car owners really needed were proper roads—and being Americans, the best roads in the world. In 1919, Dwight Eisenhower journeyed along the stuttering web of highways that existed between Washington and San Francisco, and during his presidency forty years later, he delivered to American drivers the road system they wanted. In doing so, he would helped bring to an end the golden years of the railway. But at the same time, he succeeded in binding the nation even more closely together, this time by car.

Major Eisenhower's Epiphanic Expedition

That dark evening in Gettysburg it was raining heavily, and the pages of the diary were getting wet. I had managed to keep my Exxon road map dry, however, and in the dim light, I could just about see on it the way to go. From the hotel on the town's central square, I was to turn right and head out on Chambersburg Street, drive for a mile to the junction with West Street, and there jog up a little to the north, then straighten out and head back out west again. In three thousand two hundred and some miles, a few weeks of none-too-hard driving, I should reach the shores of the Pacific Ocean.

This was possibly what the young Eisenhower said to himself here on a scorching hot Tuesday afternoon in the summer of 1919. It was also what I was betting on, on that dank autumn evening ninety-two years later, as I turned the Land Rover onto the roadway and started to follow Ike's exact directions, just as they had been typewritten onto the now dampening pages of my copy of his ragged old diary.

He had written his journal—a slender volume, not thirty pages in all—very much in the clipped military vernacular that was expected of him. It was terse, matter-of-fact, amply salted with abbreviations, acronyms, and paragraphs of technical jargon, and with not an ounce of romance—despite the adventure's being, and for many still remaining, a thing of almost unbearably romantic association. He was crossing his country, coast to coast, and all of it by road. Most Americans then saw this as the stuff of barely imaginable dreams.

In 1919 the now fast-mechanizing America was becoming a country whose people seemed preternaturally inclined to travel. A social change was in process, with plenty of available money and new transportation technologies fueling it eagerly. Gertrude Stein wrote of America of the time as a space, and a space of time, that was always and forever *filled with movement*.

There was ample reason. World War I was over in Europe, and demobilized soldiers were back from the front, flush with hoarded cash. A Model T Ford or a Chevrolet 490 cost less than $400, little more than three months' pay. Four million cars were already on the roads, Ford was selling six hundred thousand of its machines a year, one American in eight already owned a car, and the number would be up to one in six by the 1920s. Farmers were buying small trucks—a quarter of a million were in use in 1916—to haul their produce to market and to collect their fertilizer from the train depot. The stagecoach had all but vanished—although there were still twenty million horses used for conveying people and goods of all kinds—and bus services were beginning to transport the less wealthy from city to city. The taximeter had been invented, and motor taxicabs were available in most cities. The concept of the joyride was quite new and being enthusiastically tested. Industries were springing up close to sources of iron, coal, or water; workers were needed from all over to man them, and migrations were encouraged. The roar of the Roaring Twenties was as much the thunder of internal combustion engines running at full tilt as ever it was the screech of the jazz band on the dance-hall floor.

America's roads were at the time a national disgrace, but there were plenty of them, with nearly three million miles in use. Only a tenth of these, some 369,000 miles in 1919, were paved with any kind of lasting surface. The rest were made of dirt and in a generally appalling condition—with miles of chassis-deep mud the consistency of horse glue, with hundreds of broken bridges, with break-back mountain passes of solid rock, with faint trails that merely sifted their way through blowing desert sands and then quietly vanished, leaving the traveler utterly lost.

Despite the very obviously growing demands being made on them, the country's roads were simply not keeping up. Their dire state presented a perpetual hindrance to trade, an abiding nuisance to agriculture, and a profound inconvenience to the traveling public. It cost an American farmer almost three times as much to haul a ton of produce as it cost his *fermier* colleague in France. A congressional report of the time noted drily that to move a peach twenty miles from a Georgia orchard to Atlanta by road cost every bit as much as it did to move one three thousand miles by rail from California to New York.

Lobbying groups of enthusiastic car drivers and automobile makers sprang up in Washington to complain and to press the government to intervene. They wanted a properly funded federal roads plan, which would prevent this authority from being left entirely to the states, with their lack of oversight and the malign local influence of cronyism and corruption. The League of American Wheelmen was among the first to complain. The title of its regular publication said it all: *Good Roads* was what it was called and what it demanded.*

But these complaints had little to do with the dispatch of the expedition that summer of 1919. It came about because high officials in the War Department were troubled that they might not be able to use the country's roadways to move soldiers about in the event of an international conflict.

After the successful conclusion of World War I, American generals were starting to draw up plans to ensure that any forthcoming military confrontation would be handled, logistically speaking, with similar dispatch. One of the many scenarios dreamed up in the war college involved an imagined attack on the Pacific coast by an Asian enemy—unspecified but presumed with some foresight to be the Japanese, as the treaty that ended the first great war of the century, between Russia and a victorious Japan, had been brokered by America and signed in New Hampshire.

The war game advanced by the theorists had this putative enemy attack or invade the coast of California,

---

* With articles such as "Convicts on New York Roads: 250 Prisoners in 9 Camps Building Highways" and a less-than-prescient "Roadside Gasoline Stations a Menace to Traffic, Says NJ Highway Chief," together with ads for the Solvay Company's National Road Binder and Barrett & Company's Patented Paving Pitch, it was a rollicking read for many.

Oregon, or Washington.* The war gamers needed to know just how quickly fully equipped soldiers could travel from the big army bases on the East Coast to the presumed battlefield in the West.

Railroads were quickly ruled out because they could not carry the amount and type of equipment and matériel required, including the newfangled battle tanks, of which the army was much enamored and which were too heavy for both the trains and the tracks. It was unlikely that sufficient trains could be assembled quickly enough to carry the necessary divisions out to the West, and besides, if the army already had plenty of wheeled and tracked vehicles, why not have its soldiers drive them across the country themselves? There were still a fair number of skeptics in Washington's military establishment—most of them die-hard old cavalrymen—who clung to their lifelong faith in the horse and the mule. But by dispatching an expedition by road, having men of the motor transport corps drive a convoy of the kind of military equipment and personnel that might be needed to repulse a West Coast invasion, the department could prove once and for all

---

* The only successful modern invasion of the land territory of the United States came in 1941, when Japanese forces took a thousand miles of the Aleutian Island chain in Alaska. The contiguous United States has been attacked—on September 11, 2001—but never invaded.

the superior worth of wheel over hoof, and at the same time determine if the roads were better than railways. Indeed, such an expedition would be able to ascertain what America's roads were truly like.

When Major Eisenhower heard mess-table talk of the trip, he was frankly intrigued. He had long been fascinated by all things mechanical and had trained young men in tank warfare at a camp in Pennsylvania. The experience had shown him a good deal about the design and maintenance of these powerful new machines. Moreover, he also had an abiding love affair with the automobile. He had been born in Texas and grown up in Kansas but knew little of the republic west of the Great Plains. So he volunteered to accompany the expedition as an observer, just for "a lark," as he put it in a letter; he promptly began the diary that I took along with me when, almost a century later, I took off time to follow in his wheel tracks.

The journey left an indelible impression on the young man. It would also in time leave a truly indelible impression on the country that he traversed, for what the young man learned and remembered from the agonizingly slow progress of the column across America that year led inexorably to the creation, on his watch as the country's thirty-fourth president, of America's Interstate Highway System—the greatest engineering project in world history, a vast network of high-speed roads built with the sole purpose of uniting the corners, edges, and center of this vast nation by road.

The convoy gathered at a monument—then temporary but soon made permanent, of granite—by the South Lawn of the White House. The column was three miles long. There were seventy-nine vehicles: thirty-four heavy trucks, oil and water pumpers, a mobile blacksmith shop, a Caterpillar tractor, staff observation cars, searchlight carriers, kitchen trailers, a mobile hospital, and other wheeled necessities. Nine of them were wrecked en route, and 21 of the men were lost as well, leaving 237 soldiers, 24 officers, and 15 observers—Dwight Eisenhower among them—who pulled into Lincoln Park in San Francisco sixty-two days later. They had behaved en route as if the Asian enemy was dogging their every footstep. The condition of the roads—the essential nonexistence of roads west of the Missouri until they crossed into California— meant that, had they actually encountered an enemy along the way, the Americans would most likely have lost any battles that might have been fought.

The expedition was enlightening, a series of teachable moments. It was also, in the simple practical terms of getting a large contingent of soldiers from one coast of the country to another, a complete failure, one that

Eisenhower made a part of his life's work to ensure would not happen again.

As I pulled away from Gettysburg on that wet September evening, I was bound west along a small and uncrowded two-lane road. For most of the coming days, I would stay on small roads much like this, passing steadily through farmland and villages or stopping briefly at traffic lights in the center of scores of the small overlooked cities that the roads connected. The Interstate Highway System might well have been born as a consequence of this old expedition, but the mighty expressways weren't anywhere to be seen nearby. The entire route of the 1919 adventure seems today to have been preserved in the amber of its own history.

After leaving Washington in 1919, Major Eisenhower took twenty-two days to reach Omaha; after I set off to follow on the same small roads his expedition had taken, in the late summer of 2011, it took me ten days, none of them plagued by any particular difficulty. I had none of the mechanical problems suffered by the military convoy. A Mack truck ran into a ditch in Chambersburg; a Dodge blew out a cylinder head in East Palestine; dirty gasoline in Bucyrus forced the drivers to stop to adjust their carburetors; a Cadillac truck got a puncture outside Fort Wayne; a motorcycle crashed while going over the Mississippi. Setbacks litter the pages of Ike's diary, but all are mentioned without drama, in the manner of a soldier.

By the time the expedition reached California, the future president had recorded 230 accidents and emergencies of one kind and another. According to the summary report, "forty-two hours were spent in the most heroic efforts in rescuing the entire convoy from impending disaster." Back in 1919, the roads ran out at Omaha, where the myriad troubles and trials of the expedition properly began.

The young Dwight Eisenhower traveled as an observer on the U.S. Army Transcontinental Convoy of 1919, sent to find out how quickly soldiers could get across the country by road. West of Omaha the troops met one mishap after another on nearly nonexistent highways: it took more than two months to reach San Francisco. This, in Ike's mind, was when the interstate system was born.

For me, it was all a little different. The approaches to Omaha, for instance, had a certain sweet nostalgia to them.

Some forty years before, I had spent a couple of summers on a farm on the plains of central Iowa, near the town of Ames. (Ike's diary records that when the convoy passed Ames, "Garford's connecting rod cap bolts sheared off," but things got better when a local church "served ice cream.") The Judges, the farm family, had generously housed and fed and watered me; Tom Judge taught me how to run a combine harvester to help him bring in the 1975 corn crop. I assumed that they had retired, moved, or worse. But why not look for them? It was just a short detour from the Eisenhower route, and I was in no particular hurry. So I left the country road and spent the next hour driving dustily up and down miles of section roads, along wide avenues lying between rows of genetically modified corn eight feet tall, to see if I might recognize the old property. I never did; the farms all looked the same, all built with severe Midwestern regularity on the square, with corn hiding everything.

But then I stopped at a post office and asked. Sure, said some old-timer sitting in a shaft of sunlight by the window. Pat and Tom Judge? They're in the same place they've always been. George Washington

Carver Street they call it now. No more than a mile away.

George Washington Carver's is a story to melt men's hearts. He was born into slavery in Missouri circa 1864, was given his owner's name, was kidnapped, was turned away from school because of his race, then homesteaded and plowed untold acreage himself and steadily became America's greatest expert on alternative crops—peanuts, soybeans, sweet potatoes—that might replace the destructive cotton monoculture of the Deep South. He taught agriculture for forty years at the Tuskegee Institute in Alabama, is said by many to have invented peanut butter, and has been honored as prolifically since his death in 1943 as he was during his long and decent life.

It was entirely proper that the Judges lived on a road that memorialized so great a man, for they were die-hard Democrats, something that never came up when I stayed but was on very evident display when I knocked at their door all these years later. They looked just as one might expect retired farmers to look: nut-brown, strong, slim, fit. Their fields—a half section, 320 acres—were now rented to a neighbor; they doubted that it would be practicable or profitable to farm so small a lot these days, with many of the neighbors having 5,000 acres or more and living their lives more as industrialists than as dedicated farmers, steering their combines with GPS and listening to Rush Limbaugh as they harvested.

The Judges were glad they had left farming when they did. They worried about their country and thought that maybe they had lived during its best times. They loathed the industrialization of farming yet thought organic farming financially unsustainable, and they recognized the cruel paradox. They refused to use Roundup pesticides. They admired frugality; when Tom Judge's mother and father died recently and he auctioned his parents' possessions, as farming families do in these parts, they realized less than a thousand dollars. "And when we sold up when we retired," said Tom Judge with a sardonic grin, "what all our stuff made wouldn't buy us a new tractor."

Next on the Eisenhower route was Council Bluffs, then the Missouri crossing, and Omaha the following day, but the old road had taken the 1919 expedition to a small western Iowa town named Denison ("Sunday rest period. Baseball game in afternoon: *Denison 15, Convoy 1.* Band concert in courthouse in the evening, also movies at the Opera House for the Convoy personnel. Dirt roads. Heavy dust.") I wanted to go to Denison—not because of the opera house or the sound of remembered music in the courtroom but because a girl named Donna Mullenger had been born there and I had been in love with her for all my adult years.

She had been born in Denison in 1921 and had first wanted to become a teacher but had no funds to pay for school. She took off instead for Los Angeles—the familiar dream—and once there performed onstage, had a screen test, and was signed by Metro-Goldwyn-Mayer in 1941. She was given the stage name Donna Adams, but after the moguls found the name tricky to say, she was renamed Donna Reed. The rest is all trajectory.

In 1946 Miss Reed starred as Mary Hatch, along with James Stewart as George Bailey, in Frank Capra's *It's a Wonderful Life*. As part of the story, she married, and as Mary Bailey she became the most beloved American film star of the day—and to my fevered mind, the most beloved of all time. And her adorable nature had been crafted and nurtured in Denison, Iowa, where Dwight Eisenhower and his brother officers had happily listened to a Sunday band concert in the summer of 1919.

Would that Denison had been as favored as its most famous child. It is a pretty enough place, with a scattering of well-crafted old wooden houses—and on a mantel in one of them stands Donna Reed's Oscar (which she won not for *It's a Wonderful Life* but for *From Here to Eternity*). The old German opera house has been turned into a Donna Reed Performing Arts Center. A soda fountain first installed in 1907 inside Fat Moe's Deli in Chicago was trucked down to Denison and reinstalled, a perfect replica of the one in Capra's fictional Bedford Falls.

But that aside, Denison stinks. Or at least it did on the day I visited. Everywhere I went there was a malodorous sickness in the air. In places it was visible: a low yellow mist hung over the Union Pacific tracks, drifting in from a vast cattle feedlot nearby. It was a miasma of chemical sprays, cattle waste, powdered feedstuff, and atomized drugs. It suffused the town, lying over it in a fetid slump, like an old dog blanket. Feedlots are just one aspect of the industrialization of American agriculture, so derided by Tom Judge back on George Washington Carver Street in Ames. They ruin and despoil the countryside, and where they merge with towns, they despoil and pollute them, too.

There were floods up and down the Missouri; most of the bridges were closed. The floods were caused by the melting of enormous snowpacks up in the Rockies, as well as by heavy rains in the northern plains. The Army Corps of Engineers, which superintends the dams, had opened many of the gates to release the impounded waters in the great man-made lakes of the

Dakotas, and the consequent rush of floodwaters had caused immense damage and inconvenience, in places forcing people to drive a hundred miles to cross from one side of the river to the other.

I had planned to cross a sleek new concrete span with an ancient triple-arched iron railway bridge beside it, both leading to a small Nebraska town named Blair. The bridge was closed, however, and for the first time since leaving Washington, I had to drive on Interstate highways—I-29 southbound to Council Bluffs and then I-80 high across the river into Omaha. The detour turned out to be happily counterintuitive, for although both Omaha and Council Bluffs are cities with histories firmly wedded to the railway—and, as will be seen, to the new digital future—they can fairly be said to have played an essential role also in the birth of the new generation of roads that connect America so comprehensively today.

Back in 1919 this was the place where Dwight Eisenhower's transcontinental nightmare began, and with it the epiphany that forty years later would change the face of America. His diary records the moment. It was early August. The party had crossed the hundredth meridian just outside the small town of Cozad, and were now in one of the driest parts of America, beyond the grassland of the Great Plains, in a place of sand, bones, and tumbleweed. On August 3, they left Gothenburg, a town founded by a Swede who had worked for the Union Pacific and for some reason, despite growing up in a mountainous country close to the sea, had opted to settle in a prairie as distant from the ocean as possible. The weather was cloudy and cool; the roads were sandy with, ominously, "some quicksand." The convoy arrived in North Platte on August 4, the men so tired and the vehicles so overworked that all agreed to spend a day resting— "forced to suspend movement," the future president wrote. They left town the next day. There is much technical talk in the day's diary entry, but the gist is clear:

> *Nine miles west class B Machine Shop #414319*
> *sank in sandy road and was pulled out by 3 class*
> *Bs. One mile beyond Class B #48043 sank in soft*
> *sand, both right wheels and differential being*
> *buried. After five unsuccessful attempts this truck*
> *was finally rescued by the combined efforts of the*
> *Militor and the Tractor, the cargo having first*
> *been removed. All trucks, except the FWDs and*
> *some of the Class A trucks had to be pulled thru*
> *this 200 yd stretch of quicksand. The Militor*

*handled 16 trucks, the Tractor 10, and in 8
instances the combined efforts of both the Militor
and Tractor were required. Delay 7 hours 20
minutes. Five small bridges were damaged during
the day. Between Paxton and Roscoe 2 smaller
sand holes were encountered, one on an up-grade,
only the FWDs and Militor going through
unassisted. Civilian automobile ran into Mack
#51482 east of Ogallala. Fair and warm. Roads
soft, sandy gumbo. Made 53 miles in 16½ hours.
Arrived Ogallala, Nebraska, 11 pm.*

The army and Dwight Eisenhower had discovered
what back in the capital they had long suspected, that
the American West had essentially no roads worthy of
the name.

And yet the road they were on had in theory a long
history and, as roads went, was supposed to be a good
one—it had once been the Oregon Trail. For little
more than a year, it had also been the route of the Pony
Express. But now it was a mere byway for men and
horses, with the trains on their permanent way just a
few yards off, the locomotives snorting contemptuously
past, the engineers sounding their whistles at the soldiers in Ike's party in friendly derision. This was not
a track ever designed for motor vehicles. It was not

suitable anywhere in Nebraska or Wyoming, and it was worse than useless in Utah. Most of Nevada was a near-trackless waste where the expedition got itself hopelessly lost as some of those waiting at the journey's far end briefly suspected dire happenings.

Only when the party reached California did matters improve, and beyond the state capital of Sacramento, the roads swiftly became so good—with a macadamized surface, proper drainage, traffic police, rules of the road, gas stations aplenty, tire-repair depots—that the morale of men who had been lauded as "the flower of the motor transport corps" began to soar.

They were feted and feasted after they ground noisily into Sacramento. They were treated to a triumphal banquet with the governor and a small armada of dignitaries. The menu featured olives and almonds, a chowder of razor clams, Sacramento River salmon, country-fried chicken, corn and sweet potatoes, Turkish melon, Overland ice cream, coffee, California nuts, fruits, and raisins, and cigars. There was entertainment. ("Uncertainty of time of your arrival made arrangement of this programme a bit difficult. There will be these, and more!" said a note with the menu.) The San Francisco Jazz Trio played; a soprano trilled happily; the soldiers were said to have enjoyed a ukulele player; there was a performance by the Whistling

Doughboy; the Allabads did a turn called "Just a Touch of the East"; and the men were offered the services onstage of one Violet MacMartin, who was billed with a mixture of triumph and uncertainty as an Entertainer De Luxe.

But all this cut no ice with Eisenhower. This great convoy, called into action to deal with a supposed threat to the country's West Coast, had crossed 3,251 miles of the country at an average speed of only 5.6 miles per hour. The vehicles were in fine shape, Eisenhower concluded. The men were brave, intelligent, and snappy. But the roads were execrable. If nothing else, he wrote, the experience of the expedition should spur the building, as a national effort, of a system of fast, safe, and properly designed transcontinental highways.

By the time the soldiers got back to their bases and Ike returned to his desk at Camp Meade, Maryland, the seed of the idea of the Interstate Highway System had been planted. Today it is essentially complete, formally named the Dwight D. Eisenhower National System of Interstate and Defense Highways. Because exasperation is often the trigger of a solution, it can fairly be said that it was most probably in or near the town of North Platte, Nebraska, that the seed was first sown.

The Colossus of Roads

It may have been Ike's idea, but it was one Thomas MacDonald who, a quarter of a century later, fashioned the billions of tons of concrete and steel in such a manner as to make the notion whole.

Thomas Harris MacDonald, who is a half-forgotten man today but whose unforgettable legacy was to give Americans of all stations the ability to cross their country by road at speed and with ease, was by all accounts not an easy man. He was diminutive, cocksure, thickset, proud, austere, always scrupulous and conservative in dress—he wore a tweed coat and patternless tie even while fishing or riding horses—and he demanded absolute respect from his social inferiors, from his colleagues at work, and from all within his household. He had his wife call him Mr. MacDonald and his brothers and sisters address him as Sir. The only hint of levity he permitted was allowing his staff to call him Chief. Woe betide anyone who during his entire adult life—he was born in 1881 and died in 1957—addressed him by his Christian name.

Tom (we can risk his ire today, perhaps) was born in Leadville, Colorado, the rough-and-tumble frontier town where Oscar Wilde once gave a lecture in a bar and famously saw a sign on the piano pleading with customers PLEASE DO NOT SHOOT THE PIANIST. HE IS DOING HIS BEST. It was the only rough-and-tumble time of MacDonald's eventual seventy-six years. He was still a child when his father moved the family to the flatlands of Iowa and set up a modest business selling corn and cords of cut wood. They lived in a place called Montezuma, a tiny town so full of Protestant Irishmen immigrants that each July 12 they celebrated the 1690 victory over the Catholics at the Boyne, but otherwise so empty that the roads were uncared for, thick with mud from a soil of an almost velvet blackness, albeit rich and nutritious. Flat, frugal, fastidious Iowa shaped Thomas MacDonald. And it was after he had seen so many times his poor father's difficulty driving his lumber truck or his corn cart through the winter quagmires to the nearest station of the Minneapolis & St. Louis Railroad that he vowed to learn the craft of building highways and eventually laid the groundwork for the Interstate Highway System, which he would not live to see.

His youthful interest coincided with the beginning of a keen public awareness of the state of America's roads. The League of American Wheelmen was initially a cyclists' lobbying organization. By the 1880s, the dangerous-looking velocipede had become hugely popular, soon replaced by the safety bicycle: $18 could buy you the freedom to tour the country—and if you were a woman, to do so in skirts and bloomers, a sport so liberating that Susan B. Anthony approvingly backed it. But the roads were so execrable that a hundred thousand bicyclists went to Washington to complain. There was little point in owning a set of wheels, they cried out in unison, if wheels could not be used.

Near the century's end, pressure from the bicyclists was augmented by protest from a tiny but swiftly growing number of automobile makers, two of whom had been bicycle mechanics themselves, who set about their mission with the eagerness of poachers-turned-gamekeepers. They were brothers, Charles and Frank Duryea, and in 1893 they had slung a one-cylinder gasoline engine under the frame of an old horse wagon and driven it along the back streets of Springfield, Massachusetts—in the dark, so that if the machine broke down, no one would see. But it didn't break (it didn't brake, either, for the first model was not so equipped), and the Duryea Motor Wagon Company was swiftly founded, made fifteen cars, raced one of them creditably against a German Benz in a contest held on Thanksgiving Day in 1895, and helped spawn a new American business: making and selling automobiles. Ransom Olds and his Oldsmobile came next, then Henry Ford, the Model T, and mass production, and soon the sound of lobbying by the two-wheelers faded to a dull roar while that from the four-wheelers of the new automobile industry rose in a crescendo and became deafening.

In 1894, the American government seemed to be listening. Officials were sensitive enough to make the connection between roads and the needs of men like old Mr. MacDonald, who couldn't get his goods from his farm in Montezuma to the railroad station. Buried deep within that year's budget for the Department of Agriculture was an allotment of $10,000 to establish the modestly named Office of Road Inquiry. Though the federal government had been involved in road building since Jefferson's time, this was the first proper road agency set up with an office and a staff. It had a simple mandate: "Get the farmer out of the mud."

The officials did this in a most imaginative way. They assembled a number of trains; filled each with engineers and road-building equipment; sent them out to small, isolated towns throughout the country; and built small half-mile-long sample roads, for demonstration. They were known as object-lesson roads or seed roads, examples of what could and should be done. In the words of a senator who watched one of these Good Roads Trains pull into Lynchburg, Virginia, "titanic machines" spilled from the railroad cars, were cranked into life, and immediately began building a hundred feet or so of "good road over bad road."

The results were predictable. Scores of local people would bring their bicycles or cars or farm trucks and would bump along through the ruts and mud pools until they reached the edge of the new-built seed road—and then sailed serenely across something that suddenly felt as smooth as velvet, as easy to ride along as if it had been made of glass.

To advertise the benefits of well-built and -maintained highways, the government dispatched trains, loaded with mechanized road-building equipment, to show distant communities how to construct them. The Illinois Central sent a train on a journey between Chicago and New Orleans, stopping every few miles to build half-mile sections of roads, to advertise how it could be done.

Meanwhile, over in Iowa, Thomas MacDonald had confirmed his interest, had refined his skills, and had acquired a job. Having once remarked that road building was "a calling of the greatest public responsibility" that mightily increased "the possibilities of enjoyment and happiness of life [more] than any other public undertaking," he had gone to farm school in Ames to study engineering and had advertised his eagerness with sufficient energy to be handpicked on graduation to run a commission to study that state's roads. He took to the job like a fury, riding for weeks at a time on horseback to the farthest corners of the state, advising on building techniques, sacking countless corrupt local officials, and ordering the paving of five hundred of the state's six thousand miles of highways. In doing so, he rapidly came to the notice of men of influence in Washington.

By this time the Office of Road Inquiry had expanded and become the Bureau of Public Roads, still within the Department of Agriculture but now much larger and with a bigger budget. Its first two chiefs were lackluster timeservers, and the bureau had had little impact. It hardly managed to spend its $75 million budget, and during the time that MacDonald had been studying in Ames and then laboring to better the roadway system out in his tiny home territory, it had built less than twenty miles of new road in the entire United States. It clearly needed some ginger.

The bureau got it in 1919, when Thomas Harris MacDonald was plucked out of Iowa, and made its chief, a post he would hold on to like a limpet for the next thirty-four years.

His appointment came at a propitious moment in the nation's history. The Great War was over and the troops were home. A period of prosperity had settled on the country; cars were being bought, and Henry Ford's Model T began to be available. Dwight Eisenhower's cross-country expedition had been concluded, its reports were out, and suggestions for road improvement were on every official table. Men had been crossing the country by car since a Vermont doctor improbably named Horatio Nelson Jackson had done so on a bet in 1903. It took him sixty-eight days; it had taken a Colorado coal-mining cyclist five months, not much longer. Others followed their routes, and by 1919 it could almost be said that the unsigned, unmarked, unpaved, and unattended roads across America were beginning to be populated by a people suddenly aware of the possibilities.

An iron-willed, curmudgeonly martinet, Thomas "The Chief" MacDonald ran the U.S. Bureau of Public Roads from 1919 until his retirement in 1953, and wielded unprecedented influence with the seven presidents he served, in supervising the creation of the nation's present highway system. As well as the 3.5 million miles of road built under his authority, he also oversaw the building of the Alaska Highway.

Now, with MacDonald at the helm, change started to settle on the highway system, change so profound that almost all of the details of America's present highway system owe their beginnings to decisions made by this humorless martinet during his legendary and seemingly endless Washington career.

His initial achievements were prodigious. He persuaded the Treasury and Congress to release hundreds of millions of dollars and began a program of building roads financed directly by the government. They were all fine roads, though not overbuilt. He did not favor the idea of having them all surfaced with ten inches of concrete but instead believed they should meet the demands that traffic made upon them. He wanted them to be smooth, well graded, well drained. "My aim is this," he said. "We will be able to drive out of any county seat in the United States at thirty-five miles per hour and drive into any other county seat—and never crack a spring."

In 1925 he arranged for the country's major roads to be given federally designated numbers that would extend beyond individual states, reducing the chaos of the existing randomly named, randomly signed roads and giving long-distance drivers—who that year had access to twenty million cars—the opportunity to stick with one road with its own number for a trip of many hundreds or even some thousands of miles.

The system he proposed was elegantly simple. North-south federal roads would have odd numbers; east-west, even. All signs on these roads would be the same: a plain white shield edged with black and painted with a black route number. The lowest odd number, Route 1, would be on the East Coast; the highest would be Route 101, now the maddeningly congested commuter corridor in California. So far as the ocean-to-ocean routes were concerned, Route 2 was at the north and Route 90 down near the Mexican border. Roads ending in the number 5 were major routes, as were those ending in zero, routes 30 and 40 becoming of legendary importance.

The most famous road of them all, originally known as the Great Diagonal Way, running between Chicago and Los Angeles, was given its eventual celebrated number as a result of a bitter argument involving a state—Kentucky—that was nowhere near it.

It is a droll tale. When MacDonald's numbering system was announced, Kentuckians were mortified to see that they were going to get only the unmemorable highway number 62. Instead they wanted the big, grown-up number 60, which had first been allotted to the Chicago–Los Angeles highway. They felt that Kentucky should have 60, because 60 would bring business through its bluegrass counties, which had been bypassed by all other numerically notable roads. Its governor petitioned the stern Mr. MacDonald, saying the state wanted nothing of his nastily forgettable number 62.

Since the entire numbering arrangement required the unanimous concurrence of all forty-eight states, Kentucky's petulance had to be addressed. The first plan was to give the diagonal highway the number 62 and give Kentucky 60, but the states through which the Chicago–Los Angeles road passed wanted nothing to do with the much-loathed 62 either. Stalemate.

Then MacDonald's chief engineer in Oklahoma remembered that number 66 had not been claimed. If the Great Diagonal Way could be designated Route 66, then Kentucky could certainly have its way with US Highway 60. MacDonald agreed, Kentucky said yes, and the best-known highway in the Western world, 2,297 miles of star-dusted asphalt, was formally born. Route 66 remains today, but only just. John Steinbeck's "mother road" is now a relic, ill repaired, half broken, and much diverted, scarcely traveled except by the curious, but still a beloved monument to that moment when the steely reserve of Thomas Harris MacDonald briefly let go, and he grinned.

Matters became much more serious a decade later, toward the end of the Depression, when MacDonald was summoned to a meeting at the White House. He had been called in to see President Roosevelt, the fifth president under whom he had served, the man whose New Deal had—along with the European war—started to bring the long economic slump to its merciful end. Roosevelt had long been a highways man,* a keen supporter of all the various good-roads movements that had flourished in the country, and he had long admired MacDonald's tenacity, foresight, and drive. But now, in 1937, it was fast becoming clear that even the existing new and numbered roads could not cope with the exponentially increasing number of vehicles. The Bureau of Public Roads simply had to do something extra.

President Roosevelt took MacDonald over to an Oval Office table where there was a large map of the United States. The president, a man given to big thoughts and sudden decisions, picked up a crayon and drew six broad lines on the map. Three of them crossed the country coast to coast; three more ran up and down

---

* Roosevelt had, for instance, overseen the planning of the Taconic State Parkway, the beautiful, winding 125-mile road that connects New York City to the farms of the Hudson Valley—which happens to be my preferred road to take to and from Manhattan. The design of the picnic tables, my family always remarks, was FDR's.

The utopian idea of a paved paradise: this was what the National Highways Association believed would guarantee a prosperous future for America, and the organization lobbied intensely to win public support.

Maps like this, produced in 1925, presented a vision for an interconnected and thoroughly asphalted America that was to be broadly realized within the following decade.

the country, border to border. A while later, he drew two more north-south lines on the map, making a total of eight. To MacDonald he then essentially said that these eight roads, wide divided highways with limited access and without any intersections or stoplights, which would enable vehicles to travel for hundreds of miles at the highest of speeds, should be constructed to link America together properly, once and for all. How much will it cost? Where exactly should the roads go? And what will be their effect?

It would be twenty more years—World War II and unending domestic battles over budgets would intervene—before the first sod was turned. But this hastily drawn crayon-on-a-map plan of 1937—lost to history, sad to say—was the exact moment of origin for the Interstate Highway System, and Dwight Eisenhower, though renowned by association, had nothing to do with it. He had been the first to recognize the need, back in 1919. And in 1945, he famously saw in the autobahns of defeated Germany the kind of road that should be built. But he did not originate the roads that now bear his name; Roosevelt did. And once given that Oval Office directive, MacDonald and the staff at the Bureau of Public Roads began the planning immediately.

MacDonald did not agree with Roosevelt's plan entirely. FDR wanted tolls; MacDonald did not (an argument he won, the roads eventually being financed by public money, much of it raised by a new tax on gasoline). FDR wanted the roads to help the future economy; MacDonald believed that roads should be built only where they were currently needed. The long winding strips of concrete that eventually drifted with apparent pointlessness across the unpopulated plains of North Dakota and the deserts of Nevada were proof that Roosevelt was right: such roads both provided work for local men and in time helped the local economy grow in other ways.

MacDonald and Roosevelt also disagreed about the projected size of the highways: MacDonald's numbered highways had all been two-lane. FDR wanted his to be four-lane, divided, and in places even bigger than that. The president got his way.

The final triumphal results of MacDonald's work, which would be completed by the hands of many others long after he had retired, came twenty years later, in June 1956, with the passage of the Federal-Aid Highway Act, the landmark legislation that finally approved the building of this vast new highway system. Dwight Eisenhower signed it with much enthusiasm and showmanship.

In September 1956, there then appeared another MacDonald legacy—the plan. It was a hundred-page book of maps, with only three sentences of text, that

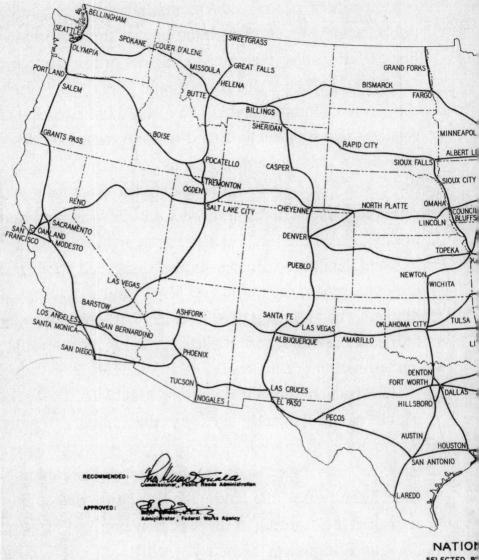

RECOMMENDED:
Commissioner, Public Roads Administration

APPROVED:
Administrator, Federal Works Agency

NATION
SELECTED B

C ROADS ADMINISTRATION
EDERAL WORKS AGENCY

TEM OF INTERSTATE HIGHWAYS
ON OF THE SEVERAL STATE HIGHWAY DEPARTMENTS
S MODIFIED AND APPROVED
NISTRATOR, FEDERAL WORKS AGENCY
AUGUST 2, 1947

Within a year after the famous Yellow Book of 1955, with its map of the intended routes of the main Interstate Highway System, Congress had passed the legislation that set in train the greatest public works program in world history.

was sent to every member of Congress, titled *General Location of National System of Interstate Highways Including All Additional Routes at Urban Areas.* The book's cover was the color of butter; that is why this document, which laid out the exact directives for the building of the Interstates, immediately became known as the Yellow Book. It was the road-planners' bible; it told everyone what to do and where to do it in the biggest public works program in the history of the world.

There would be forty-one thousand miles of highway—with the longest, I-90, running without a single traffic light the 3,020 miles between Seattle and Boston. Roads would climb as high as two miles above sea level in Colorado and fifty feet below by the Mexican border in Southern California. There would be fifty-five thousand bridges and tunnels, sixteen thousand exit and entrance ramps, and thousands of cloverleaf crossings—junctions where in time new kinds of hotels* and venues for refreshment and amusement would grow up, changing in no small way the face of the nation and the habits of all.

---

* Holiday Inn was one of many. Kemmons Wilson opened his first motel in Memphis in 1952. The standard appearance of the Interstate highways reflected his view that all roadside motels should be similar also, and he began planting them at busy intersections as fast as the roads were built. He had fifty Holiday Inns in 1958; ten years later, a thousand.

Most of the roads were brand-new, requiring the acquisition of enormous acreages of land. The construction would in places be brutally difficult. Bitter winters, deep swamps, high mountain ranges, vast deserts—all would be swept across by ribbons of an immense building experiment, the builders learning mile by mile, making it up as they went.

Once the cash spigots were opened, work began in many widely separated places, and in a matter of only weeks the new highways began to appear, growing outward toward one another like *Penicillium* mycelia in a petri dish. The race was to the swift, and though there are at least three separate claimants to being the first finished (two of them in Missouri), the generally accepted fleetest of all were the team at a ribbon-cutting in Topeka, Kansas, on November 14, not five months since Ike signed the bill. The men had widened, straightened, and paved, according to Washington's Yellow Book specification, eight miles of the old US Route 40 and had transmuted it into the beginnings of Interstate 70, which runs now between Baltimore and Utah.

The project was supposed to take thirteen years to complete, but instead it took thirty-five. The first phase was finished in 1992, with its total cost estimated at $430 billion in modern money. Curiously, the completion was achieved with a section of the very same I-70 that had been the first road started, which passes through a particularly rugged valley of the Colorado River known as Glenwood Canyon. This valley long provided the route for the Union Pacific Railroad, and later it was touted as the most scenic portion of the *California Zephyr* route. A memorial to the dome cars that made the *Zephyr* so beloved had to be moved to make way for the highway.

Month by month, new sections of high-speed, limited-access national highways opened. Three states lay claim to be first, among them Kansas in 1956, with the I-70, with four fully paved miles outside Topeka. This ceremony near Waukesha, Wisconsin, held two years later on the I-94, was typical of the times, but such occasions swiftly became too routine to merit celebration.

Thomas MacDonald had meanwhile left government and had gone off to Texas to teach engineering. He never lost touch with "the greatest public responsibility" of building highways. He was well aware of the moment, seventy-six years after his birth, when the ribbon-cutting ceremony was held for the beginnings of I-70 in Topeka. But he would know precious little more of the story. The man who essentially started it all, midwife to the greatest highway system in the world, died in a Texas restaurant in April 1957, as his forty-thousand-mile memorial was just beginning to coil its way around the nation.

Montezuma, Iowa, the town where he had grown up, once had a railway station; it was his father's difficulty in getting his goods to this tiny depot over the mud and ruts of the early-twentieth-century roadways that led Thomas MacDonald to his obsessive interest in building good roads. There is now an Interstate highway, I-80, twenty miles north of Montezuma. There is no longer a railroad connection of any kind: the tracks have long been torn up, and the depot is no more.

One final reminder of the many legacies of Thomas MacDonald came to me recently when I was driving through the Yukon, in far northwestern Canada, and was stopped for speeding. I was heading north on the Alaska Highway—a 1,700-mile international, mostly paved, but still extreme road between Dawson's Creek, British Columbia; and Delta Junction, Alaska. I was tooling along, minding my own business. I had Lake Teslin to my left, spruce forests to my right, and what seemed an empty hundred lonely miles of gravel road ahead of me—when suddenly, out of nowhere, the scene was filled with more flashing blue lights and sirens than the New Jersey Turnpike. A police cruiser was pulling me over. I was *busted*. In the *Yukon*.

It turned out to be a patrolman from the Royal Canadian Mounted Police, a Mountie. He strode over, adjusted his big Mountie hat, and struck an imposing attitude. He was all smiles. He glanced at the backseat, checked my New York license plate. "Don't see many Land Rovers in these parts," he observed by way of introduction, and then: "Welcome to Yukon Road Safety Week! You were doing one hundred twenty-five in a ninety zone. You were speeding. Radar said so."

He was talking in the metric system, of course. I had been doing eighty miles per hour in a fifty-five zone. "But even so," I said, "there's nothing to hit." No people, no cars, no towns, no houses, nothing.

"A moose," returned the Mountie. "If you hit a moose at a hundred twenty-five, you'd be in pretty poor shape." He then explained that a Yukon Road Safety Week ticket—the only one he had given so far—was for much less than the price of a modest lunch; it wouldn't go on my record back home, the insurance company would never know, and here was an 800 number to call once I got to Whitehorse, a day's more driving ahead. So far as tickets go, it could scarcely have been a more pleasant experience.

And when I called the number, the lady who answered sounded as if she was quite well on in years and was doing her knitting. "Have you been a naughty boy?" she asked, and when I allowed that I sort of had, she asked me for the ticket number, the credit card number, and then announced that I was free to go. Except, she asked—who had given me the ticket? I squinted at the handwriting. Sergeant R. Smith, I read out to her. "Oh how wonderful!" she exclaimed. "Rolly Smith himself? Knew him when he was a little boy. Isn't he an *absolute sweetheart?*"

The Alaska Highway was built in 1942. It took two teams of US Army soldiers just six months, one team working up from mile zero in the Canadian south, the other moving down from across the US border in Alaska in the north. They met at a place now called

Contact Creek, six hundred miles from anywhere. They began work in May, the road was open in October, the public got to use it in 1948, and it has never closed since. An American road passing through a foreign country—much like the Manchurian Railway, built by the Japanese but passing through China.

And the father of the Alaska Highway, the man who back in the 1920s dreamed of one day creating something tangible that would link Alaska to the forty-eight states, was Thomas Harris MacDonald. So far as the corps d'elite of men who united the states is concerned, MacDonald can fairly be said to be one of the very few who lured number forty-nine into a physical connection with the rest of the Union. That makes it even more remarkable that he has been so widely forgotten, overlooked, and dismissed in just about all the places he managed to bring together.

And Then We Looked Up

It was the second Tuesday in September 2001, and I was climbing up the Sierra Nevada, heading for New York, breasting the Donner Pass. It was the most beautiful of days—at more than a mile above sea level, the few deciduous trees that stood beside the Interstate were already starting to display the purples, golds, and reds of their autumn exultation. The weather was perfect, cool and crisp, the air like champagne. The sun was low behind me, and the sky ahead was the palest of blues and perfectly, perfectly clear.

Clear and quite empty. Some birds were wheeling and soaring above the trees, but apart from them, nothing. I think I would have noticed, even if I hadn't known the reason. Normally there were at least a few contrails lacing the sky—a delicate tracery of die-straight white lines, usually in pairs, showing where a high plane had gone past, most likely preparing for its landing in San Francisco, a couple of hundred miles behind me.

The smaller, more local aircraft, those running between places like Reno and Sacramento, say, would not be flying high enough to leave a trace, though the planes themselves would be seen glinting in the sun. But the big transcontinental jets would still be climbing, heading east as I was, or else going west, their captains spooling back their engines, starting their deliberate loss of altitude, heading for the comforting arms of the glide paths into Oakland or San Francisco. They were the ones that would usually leave behind these strange cloud signatures, either because of the water in their engine exhausts or because their wings would alter the physics of the thin air six miles high and turn its suspended water vapor into ice crystals.

But there were no glints in the sky this Tuesday afternoon, no contrails visible whatsoever. There had not been any seen over California or Nevada or Utah, or indeed over anywhere in the entire United States or in the skies over Canada (in Mexico matters remained unaltered) for at least the previous ten hours. It was a noticeable lack, a visible emptiness, a strange hollowness in a usually differentiated sky. And it had happened because a man named Ben Sliney, an experienced air traffic controller who by chance happened to be on his first day in the chair of national operation manager in the country's air traffic control center in the Virginia suburbs, had issued a nearly unprecedented order.

The banks of monitors before him had told an awful story. The nation was under attack and in chaos. The president was on the run. The vice president was hidden underground. The Cabinet was in emergency session.

Such of the world as was awake was stalled in horror; in much of the world that was asleep, millions were being shaken into horrified consciousness. Because of what had already occurred in New York and Washington and Pennsylvania, every one of the aircraft crossing America's skies at that moment was to be regarded as a potential flying bomb, a weapon of vast power that could be unleashed at any of a score of targets. The Golden Gate Bridge. Fort Knox. An atomic power station—or all sixty-five of them. A crowded high school. Harvard. A biological weapons lab. The United States Capitol. The White House.

At 9:45 a.m. East Coast time, Mr. Sliney had issued, in direct response to the gathering threat, an order of formidable implications. He had already forbidden any aircraft to take off from any airport anywhere under his vast national jurisdiction. He had already closed the Atlantic and Pacific approaches to the United States, and planes were already turning around high over the oceans, returning to where they started, or diverting to some safe haven nearby.

But now at 9:45, Ben Sliney instructed that word of a strange and seldom heard procedure— SCATANA—be broadcast to every one of almost five thousand commercial aircraft that were in the air. *"This is not a drill!"* was then broadcast immediately afterward, many times, for emphasis. The announcement was a formal, legally and militarily enforceable order for which all aircraft officers were in theory well prepared. It was an order made under the terms of the regulation for which it was the acronym—the plan for the Security Control of Air Traffic and Air Navigation Aids—and it required all aircraft then in flight to land immediately at the airport closest to where they happened to be.

It was obeyed, masterfully. Every pilot appeared to cooperate; none of significance appeared to balk. And as one could then see from the monitor screens showing the aircraft images bleeding away, second by second, the icons of their presence seeming to evaporate like spray on a hot plate, it was effective, in double-quick time. By 11:20, an hour and thirty-five minutes later, every plane was safely down, somewhere, on North American ground. Airport aprons were crowded. Passengers bound for Oregon fetched up in Oklahoma; others, wanting to reach New York, found themselves in the cornfields of Nebraska. Most famously of all, thousands were marooned in Gander in Newfoundland and were made more welcome and at their heart's ease than they could have ever imagined. Millions were inconvenienced.

But the intent had been achieved. The American skies were empty. A few fighter jets patrolled. A small number of prisoners and deportees were flown in planes that had been given brief waivers. Some organs due for transplant were allowed to be flown, as well as vitally needed medicines. But otherwise America's skies were quiet. And my portion of the deep blue sky, the wheeling and soaring birds aside, was as empty as anywhere between the sand dunes of California and the cliffs of Maine.

The tragedy of that September day briefly united the hearts and minds and souls of most Americans, and yet it did so at a moment when, oddly, the threads of cross-country jet routes that had for so long bound the nation together physically had been so abruptly sundered. It remains a supreme moment of double irony that the machine that had done so much to bond the nation into one had been employed by an enemy to try to do the opposite.

The jets started flying again three days later, as the country did its best to resume its life, now so utterly changed. The jets that are taken so much for granted these days—often cursed, indeed, so routinely unpleasant has the business of long-haul flying become—are the legatees of one lonely journey that was taken in the winter of 1911. It was a journey made by air from New York to California just eight short years after the aircraft had been invented on an American beach and ninety years before this magical invention was first used as a weapon on American soil.

An American had tried to cross his country by air, for a $50,000 wager. He lost the wager but succeeded in

crossing the country; then, shortly afterward, he managed to get himself killed. He made history, and like so
many of those who helped to knit the nation together,
he has almost wholly passed from the scene.

His name was Calbraith Perry Rodgers. Although
throughout his brief life he was a flamboyant, cigar-
chomping showman and in no sense the sort of fellow
who would ever wish to become an American civil servant, in one respect he was very much like the sober-
sided traffic controller in Virginia, Ben Sliney: both
were men entirely new to their tasks.

The day that Ben Sliney made his decision to close
American airspace in September 2001 was his first day
on the job. When Cal Rodgers took off on the adventure
that essentially first opened up that same American
airspace in September 1911, he had only just learned
to fly. He had taken his first and only lesson in June,
a ninety-minute session given to him by no less than
Orville Wright himself. He had been given a flying
license, the forty-ninth ever issued; his brother flew for
the US Navy and had an earlier one. Cal had taken part
in an aviation meet in Chicago and had won third prize
for staying aloft for the longest time. But despite that
small victory, he had flown for a total of less than sixty
hours. He was a beginner. Both men, in more ways
than one, were pioneers.

The Twelve-Week Crossing

Philip Danforth Armour, the nineteenth-century butcher-baron who famously remarked that he did not love money so much as the acquisition of it, spent a lifetime trying to prize everything it was possible to sell out of every kind of customarily edible animal in America. He was a meat packer, founder of Armour & Company, which for a while was the nation's biggest food and food products maker. He was as rich as Croesus, thanks in part to his early hunch that Chicago (because of its growing nexus of railway lines) would in time beat out Cincinnati as the world center for butchering, preparing, and preserving meat and rendering fat.

He loved to peddle meat, but he was also a great early believer in diversification. Armour & Company did not simply pack beef, pork, mutton, lamb, and veal and send it onward in Armour-invented refrigerated railcars to the grocery stores of the nation. The firm also traded in glue, fertilizer, margarine, lard, soap, gelatin, isinglass, buttons of horn, hairbrushes of bone and hog bristle, and various chemical by-products that went into the making of pharmaceuticals. Philip Armour is said to have liked poking around in the sewers close to his South Side packing plants, eager to see if there were any animal parts being washed away that might have a potential commercial value.

Philip Armour died in 1901, two of his sons continuing to run the business. In their eagerness as ingenues, they may have gone too far. In 1910 the company, diversifying rather further than normal from hoof and horn, began selling a soft drink. It was a lurid pink syrup, supposedly made out of grape juice, carbonated water, and a slew of secret ingredients. Because it was a sort of aerated nonalcoholic wine, it was called Vin Fiz. It was best drunk cold or, in the opinion of many, not at all. It was wildly unpopular. It was said by some consumers to have a decidedly laxative effect, and others, presumably competitors, said it tasted "like a combination of river sludge and horse slop." But J. Ogden Armour, the older son and company president, was no slouch and was eager to sell whatever his firm made as widely as possible, so in the best American tradition, he looked for inventive ways to market this otherwise unmarketable product. It was then that he heard of Cal Rodgers and his plan to try to fly across the entire United States, and the company promptly hatched a plan to bring the name Vin Fiz to the tongues of the nation.

The Armour people decided that Vin Fiz would sponsor Rodgers, in every imaginable way. They would name his plane. They would festoon it with advertising signs. The plane would carry an oversize bottle of the stuff between the wheels. They would christen the craft with a bottle that would be broken (carefully, by a pretty girl) over the fuselage. They would provide Rodgers with a railway train, also festooned with advertising slogans, to trundle beneath his flight path carrying every spare part he might ever need. The firm would also pay him. A rate was set: $5 for every cross-country mile he managed to fly east of the Mississippi, where Vin Fiz sales were more brisk than elsewhere, and four dollars a mile over the unpopulated west, where the drink had yet to catch on (and, in fact, never would).

William Randolph Hearst, in the name of his morning daily newspaper the *New York American*, had offered a prize* of $50,000 to the first flier to

---

* Aviation prizes did much to stimulate early flying. Lord Rothermere at the *Daily Mail* got fliers to cross the Channel and then the Atlantic; the owners of New York's tony Brevoort Hotel and of the Bulova Watch Company both gave the prizes that were won by Charles Lindbergh when he reached Paris single-handed; and Tokyo's newspaper *Asahi Shimbun* offered $50,000 for the first nonstop crossing of the Pacific. American fliers won the contest, but were paid only half because they were not Japanese.

cross the United States, between New York and Pasadena, going either way, within thirty days. Many accepted the challenge; three men eventually set out. The first was a race-car driver, heading west; the second a jockey, heading east from the California coast. Neither of them made it; their engines were too wanting in power, their aircraft were too fragile, their navigation was too poor, their energy too easily sapped.

And then came Calbraith Rodgers in the *Vin Fiz*, the third and only other birdman, as fliers were often called back then, to enter the contest. His plan was to leave Sheepshead Bay in Brooklyn, near Coney Island, at dawn on September 17, 1911, and head west for some three thousand unnavigated miles. Because Hearst's offer was due to expire exactly one year after he first made it, on October 31, Rodgers had just forty-three days to complete the entire trip. If he managed to reach Pasadena by Halloween, he would not only pocket a much needed $50,000 but would also earn an extra $20,000 from the Armour Company.

In the end, he failed. He made it to Pasadena nineteen days too late to claim the money. He pressed on, however, and dipped his wheels in the water at Long Beach, thereby becoming the first person ever to fly from one coast of America to the other, taking a total of twelve weeks to do so. But contest failure or not, his first-ever journey turned out to be so rich with colorful misfortune and gay amusement, and he himself so jolly and good-tempered through it all, that he accreted a folk-hero reputation that grew steadily as he went. By the time he was halfway across the country, he was known to everyone—as was Vin Fiz, just as the Armour brothers had wanted. At the end of 1911, Cal Rodgers—tall, half deaf, impeccably dressed, ebullient, broadly smiling, and interminably brave—had become the best-known, most celebrated man in America.

His journey began with what would be an all too characteristic mishap. His team of mechanics had mistakenly assembled his Wright Flyer not at the Sheepshead Bay racetrack, where a makeshift runway had been put together, but in a field a mile away owned by a local farmer. They had gotten themselves lost, and the farmer kicked them out his meadow, afraid for his cows. "Never mind," said Rodgers, as he lit yet another cigar and went back to bed while they moved his biplane to the right place.

Eventually, the contraption, assembled under the Wrights' supervision in their workshops in Dayton, Ohio, and sent to New York by rail, in boxes—took off twelve hours late. A pretty and rather buxom young woman from Tennessee had poured a bottle of Vin Fiz over one of the lower wings—or planes, as they were then still called—having been told quite sternly by Rodgers not to try to break it, because it would most likely break the plane instead. Someone had lashed a small pouch of mail beneath the aircraft—with 263 letters, including notes from the mayor of New York to his counterpart in Los Angeles. Ten gallons of gasoline were poured into the tank.

And then Cal Rodgers, his fifth cigar of the day clenched between his teeth, his brown tweed suit clashing somewhat with the painted bunches of pink grapes on the rudder and flying surfaces of his tiny craft, climbed aboard. He told the waiting crowd to step back, shorted the magneto, pulled the choke cable. The propeller began to turn, then roared into invisible life; black smoke poured from the exhaust; members of the team pushed the flimsy-looking machine onto the racetrack and pointed it into the wind; and with a roar, the release of a brake, and the pull of another accelerator lever, the craft bucked down the track, tilted crazily, bounced up into the sky, and rose, smaller and smaller, silhouetted against the late-afternoon sun. Within minutes the speck in the sky was gone from view.

William Randolph Hearst started his stopwatch and clutched his wallet nervously.

The daily log tells its own story. Rodgers had a good first day, heading north across Brooklyn to the East River, swinging left over Manhattan and the Hudson to Jersey City, then veering northward to Paterson and Suffern before making for the wooded western New York State hills around Middletown, and settling in there for the night. The following morning, he crashed on takeoff and totally wrecked his plane.

For most, that would have been the end. But not for Cal Rodgers and certainly not for the makers of Vin Fiz. In their three-car support train, they had duplicates of every piece from which the Wright Flyer had been made. They had a completely assembled second plane. They had a car, a sporty and reliable Palmer-Singer, which would take mechanics to the crash site if it was far from the railway line. And at Middletown, New York, they did just that: by lunchtime they had the aircraft repaired, as good as new, and Rodgers took off once more, bruised but in no other way compromised and not the slightest bit reluctant to continue.

A staccato tale of disaster unfolded. Crashed on landing, Hornell. Wrecked on takeoff, Cattaraugus.

Calbraith Perry Rodgers, a flamboyant pioneer of early aviation, made the first transcontinental airplane flight between New York and Pasadena in the fall of 1911. It took seven weeks, interrupted by dozens of mishaps and crashes. His trip was sponsored by the Chicago-based makers of Vin Fiz, a generally undrinkable pink-colored soda, which was taken off the market soon after the flight.

Broken skid. Cracked skid. Ruin! said the newspapers. Catastrophe! Disaster!

But he kept on going. He was across the Mississippi by October 19. It had already taken him thirty-two days, so Hearst's money was safe. But there was no hesitation: the flight should go on, just because the nation *needed* to be crossed, it needed now to be united by air. People were now coming out each evening in their thousands, summoned by the sight of the black speck in the eastern sky and the distant thrumming roar of an aircraft engine. These were sights and sounds utterly unfamiliar to most, in a country that had never seen anything smudging its skies other than birds. Here a man and a machine would appear in the emptiness, suddenly and without warning, the watchers below told that he was on his way from New York to California and the Pacific Ocean—incredulity abounded, and the need to see and feel and touch and understand him and his likely achievement swept up thousands—and before long, millions—as he passed roaring by.

In truth, though, there was more stuttering than roaring. The catalog of mishaps grew. Broken this, cracked that, wrecked on takeoff, wrecked on landing, lost, delayed. He cracked a cylinder in Muskogee. He was given a baby jackrabbit as a good luck token in San

Antonio, but two days later, he crashed close to today's art-world gathering place of Marfa and contemptuously tossed the animal away, deciding it was more a curse than a charm. The entire engine then blew up in his face in the Texas cow town of Waco, and despite being rebuilt, it exploded again in the hot and lonely California town of Imperial Junction. A magazine writer named French Strother caught this sensational moment:

He was flying west from Arizona, intending to go on to Banning, Cal. He had flown over Imperial Junction, in the solitary waste of the Colorado desert, and was speeding above the Salton Sea at an elevation of 4,000 feet, when suddenly, without a moment's warning, the No. 1 cylinder of his motor blew out, completely wrecking his engine and filling his right arm with flying splinters of steel. An instant's hesitation would have meant sudden death; a false move with his injured arm, which controlled the warping lever, would have tilted him down sideways and sent him hurtling down 4,000 feet to destruction. The aeroplane made two lunges downward before Rodgers could control it; and then he began a long, easy, graceful spiral glide, descending in loop after loop of

diminishing radius, six miles in all, judging his
distance so nicely that he landed only a short space
from the station at Imperial Junction. I saw the
remains of this engine in Pasadena, and a man
could literally put his head into the hole that had
been blown out of it.

By now he was only a week away from finishing. His journey thus far had described a weird, more or less diagonal route across the nation, northeast to southwest, and weird because there were so many places—in eastern Indiana, Chicago, around Phoenix—where he went around in great circles. In northern Texas he went around in no fewer than three circles, lost and bewildered, so that his eventual track, followed as best they could by his intrepid supporters in their three-car railway train, looked like an intricate filigree of needlepoint.

But then came the warm afternoon of Sunday, November 5, in Pasadena. Thousands were waiting in Tournament Park, where the New Year's Day Festival of Roses is held. A band played jaunty music; a polo match was staged to keep everyone amused. Astronomers at the Mount Wilson Observatory trained their enormous telescopes over the shimmering desert landscape to the east. He was coming in from the town of Banning, though word spread that he had put down in Pomona to check the oil. A great tension gripped the gathering, everyone aware of the historic portent.

Suddenly there was a flash of white light from a heliograph on the top of the mountain. The band started on a quickstep. The thousands rose to their feet. And then a small boy cried out, "There he comes!" and the tiny confection of string and sealing wax, baling wire and cloth-covered aerofoils appeared through the mountain haze. Rodgers tilted danger-ously at one point, the crowd silent with horror, but then he righted himself and his image grew larger and larger. The sound of the engine changed from a hum to a growl to a roar—and then he *refused to land.* He performed ten minutes' worth of stunts for the crowds, wowing everyone with rolls, dives, spi-rals, and passes so low he set the female bystanders clutching their hats. And then at last he turned, cir-cled the field twice, came in low, and bounced down on the strip. A tire burst. The skid cracked once again. The engine finally shuddered and died, steam gushing from another hole.

But suddenly these problems did not matter one whit. He had done it. He had crossed the country and was now officially done with the contest. He gave the pouch of precious mail, the first *airmail*, to an official of the post office. He held up the bottle of Vin Fiz, neither knowing nor caring that the product was doomed and would be sold in stores for only a little while longer. He gave a brief interview to the Associated Press, and when the reporter asked him if he minded not having won the Hearst money, he retorted, "Never mind about the money. It don't amount to much that way—but I did it, didn't I?"

He was then swept from the field to the Maryland Hotel, where he signed the register: "Cal. P. Rodgers, New York to Pasadena, by air." America had been crossed by flying machine, by a device powered by fire, a device potentially so much faster than all others. It had been done once. It could now be done again and again and again. Millions would follow in the aerial footsteps of this majestic adventurer, the country now bonded more closely together by yet another indissoluble band of human achievement.

The next few weeks were all footnote. He took his plane to Long Beach, fourteen flying minutes away, and taxied it into the sea, wetting its wheels in the surf so that he could say he had traveled from one ocean to the other. He then returned home to New York, by train—taking six days, whereas his flying had taken eighty-four.

And then the story ends, on April 3, 1912, when he came back to California, to Long Beach. The train was still there, the backup aircraft still inside. So he took it up, alone, for one final hurrah, with the famous bottle of Vin Fiz still strapped beneath him as a talisman. Knowing all too well that people were watching from the pier, he performed some antics, well aware of the fleeting nature of his celebrity, which had peaked so spectacularly just before Christmas but which all could tell was ebbing away, fast and irresistibly. Maybe he hoped that he might perform one aerial feat so spectacular that he would be remembered always—but whatever his motivation, it remained unaccomplished, and fatally so. For some unknown reason—a bird strike, suggest many—the little plane nose-dived into the surf, the engine sheared from its bolts and slammed into Calbraith Rodgers from behind, breaking his neck and cutting his throat at the same time. Sunbathers rushed down from the beach into three feet of warm Pacific water. The aircraft was crumpled and broken like a great ungainly monster, and Rodgers, killed instantly, was found bobbing in the waves, under a mass of broken spars and canvas. He was thirty-three years and three months old, and he changed the face and nature of aviation forever.

One small harbinger of a further revolution to come lies buried in the accounts of that first flight. It came while the enormous crowd was waiting on that November Sunday afternoon in Pasadena, expecting Cal Rodgers and his tiny craft to appear in the sky across the wall of the San Gabriel Mountains that separate Los Angeles from the deserts beyond. There was a dusting of early snow on the summit of Mount Wilson. The Carnegie Institution astronomers, whose sixty-inch telescope had just been installed in the observatory, were in for another cold night of stargazing.

But on this afternoon, November 5, the men had another task. They had the wherewithal to spot the flier approaching them from the east, and they could signal the news to the crowds waiting to the west. But how to tell them what to do, what to look for, how to give the news to the mass of excited Californians?

It was done by telephone. The manager of the local company had arranged for a special instrument to be installed in the park, and he established a connection from it to the astronomers six thousand feet up on the mountaintop. He asked these watchers of the skies to tell him when they first glimpsed the tiny plane, and for safety's sake to use their mirrors to flash a heliograph reflection of the setting sun down to the crowds below.

And this is what they did. They sent their coded sun message as planned. But more important for history, the moment that they glimpsed the incoming machine, they called, they *telephoned*, to let the Pasadena crowds know what they had seen.

The call was by no means the first made in America nor the most important. It would be four more years before the first transcontinental telephone call was made, between New York and San Francisco. But this one call, made between a San Gabriel mountaintop and a Pasadena park, hinted to the waiting public at much that was to come. For while one day it might eventually take a person or a piece of cargo just a few hours to travel along the route that it had taken Cal Rodgers twelve weeks to complete, it was now clear that the existence of the telephone meant that other communications, weightless but no less vital than persons or their belongings, might travel along that same route, too, not in days or hours or even minutes—but in an instant, at the speed of light.

That single telephone call suggested the birth of a new idea: that news, gossip, casual conversation, information, intelligence, knowledge, and perhaps even that which passeth all understanding could cross from one place to another, no matter how far apart, not just swiftly, by the power of fire, but instantly, along a wire.

In years to come, such information would manage to pass with equal velocity in quite a formless, invisible, and almost unimaginable manner, through the very air itself. But in the first instance, as here in Pasadena, it would be carried along wires that were forged and spun and drawn from copper or any properly conductive metal.

Some random touch—a hand's imprudent slip—
The Terminals—a flash—a sound like "Zip!"
A smell of burning fills the startled Air—
The Electrician is no longer there!

>—HILAIRE BELLOC, "THE BENEFITS OF THE
>ELECTRIC LIGHT," NEWDIGATE POEM, 1893

Electricity is not in any sense a necessity, and under no conditions is it universally used by the people of a community. It is but a luxury enjoyed by a small proportion of the members of any municipality, and yet if the [generating] plant be owned and operated by the city, the burden of such ownership and operation must be borne by all the people through taxation.

>—TAX LAWYER HENRY ANDERSON, QUOTED IN THE
>*Richmond Times-Dispatch*, October 24, 1905

. . . it shall be said of Jacob and of Israel, What hath God wrought!

>—NUMBERS 23:23. THE LAST FOUR WORDS
>OF THE VERSE WERE EMPLOYED BY SAMUEL
>F. B. MORSE AS THE FIRST TO BE SENT BY A
>PUBLIC ELECTRIC TELEGRAPH, MAY 24, 1844

The Song of the Talking Wire

>—HENRY FARNY, TITLE OF 1904 OIL PAINTING
>DEPICTING AN INDIAN AT FORT YATES, DAKOTA
>TERRITORY, PRESSING HIS EAR TO A TELEGRAPH POLE
>IN THE HOPE OF CATCHING A PASSING CONVERSATION

PART V

When the American Story Was Told Through Metal

1835–Tomorrow

To Go, but Not to Move

The story goes that in 1861 the chief of the Shoshone Indians, whom his white visitors had come to know as Sho-kup, declared his belief that the electric telegraph, whose metal wires were beginning to sprout across his people's lands, was some kind of mysterious and ill-intentioned animal. The chief especially could not understand how this animal was to be fed. The white men told him it ate lightning, but Sho-kup remained suspicious.

The men told the chief he could use this instrument to communicate with his pair of wives (at the time of the reported conversation, one was said to be under the weather) no matter how many miles separated them. But Sho-kup was not as impressed by this as they hoped. He said he would prefer to talk to each of them in the customary way, face-to-face. Nevertheless, he was grateful to them for suggesting the idea, and now he felt a little calmer and somewhat reassured—indeed, so calmed that he now probably could accept the telegraph's inexplicable magic. He decided he could best deal with its mysteriousness by naming it, by giving it a sesquipedelian Shoshone word of his own making. He thought for a while, consulting those of his fellow chiefs who had some linguistic skills, before declaring the suitable word to be *we-ente-mo-ke-te-bope*. An interpreter quietly told the listening white men that this elaborate confection of sounds signified the phrase "wire rope express."

A name was nice, but it was still not exactly what the white men wanted. And so they waited, for they had come on an urgent mission. They were from the Overland Telegraph Company, and long-distance telegraphy was their business. They were trying for the first time to string a line right across America. In places this was not an easy task, the simple engineering difficulties of the project aside. In a place like this remote corner of the Rocky Mountains, for instance, there was still much nervousness among the local native people about the new inventions and what they were bringing with them. The men of the Overland Company needed the local chief's permission to arrange the lines across his people's mountains. They also needed to be assured that their line—here suspected of being an animal, but feared for many other reasons in many other places—would not be attacked, torn down, destroyed.

After some further thought, Sho-kup told the visitors that he by no means liked their wire rope express, nor did he trust its intent, and given his continuing belief that it was some kind of beast, he feared its appetite. But in an act of great magnanimity, he finally agreed to treat it kindly and with respect. Moreover, he promised he would send out an order to his men never to do it an injury. If the visitors truly wanted to build their poles across his land and hang their wire rope express from one pole to another, then that was acceptable, more or less.

After saying their farewells to the old man and his council of chiefs, the white men set off in their small convoy of wagons. They left behind a detachment who then dug deep circular holes to hold the thirty-foot-high poles of stripped and treated pine, with crossbars from which the wire rope could be suspended. Once they had the poles firmly set in place, the detachment galloped out of the rugged lands of the Shoshone, and headed westward toward California, all of them in a confident mood.

Neither Sho-kup and his people, nor indeed the men who put up the poles and stretched the wire between them, had any true idea what would eventually take place. The wire up above them looked so modest and innocent and incapable. It was just a wire. It didn't move, except when the wind made it vibrate and send out a plangent whistling, somewhat akin to the grass on the plains when the breezes ruffle it like the sea, and it bends and waves and seems to sigh.

The all-too-rapid changes afflicting some quarters of American society brought about by the invention of the telegraph were famously caught by the Cincinnati artist Henry Farny in his 1904 painting Song of the Talking Wire, in which a puzzled Plains Indian tries to listen to the conversation supposedly passing overhead.

There was something odd about the metal wire. It had to do with movement. Things that shifted their ways along roads and canals and railways and even through the air above—eagles, for instance—could be seen to proceed from place to place. Such normal, traditional progress was an entirely visible thing, easy to understand. The whole arrangement—whether from a team of horses being lashed along, gouts of steam pouring from a locomotive funnel, the flap of a bird's wing, or in later years the whirl of an airplane propeller—gave the clear impression of the frantic expenditure of some kinetic energy, which roared and thundered and zoomed and collided and perhaps had accidents but still shipped people and goods from place to place on time, or not, as the case might be.

But the wire was something quite different. Small wonder it was so feared, so suspected, so much the target of attack. For it was always quite silent, yet it was said by those who knew that *things were moving along it*—and yet the peculiar thing was that if anything, either cramped and crammed inside its narrow interior or crawling precariously along its insignificant outer surfaces, was moving, it never showed.

To a mid-nineteenth-century person—whether an Indian or a Scottish or German farmer newly arrived on the prairies or a slave survivor of the Middle Passage—the business of the telegraph wire was so unfathomable as to verge on the impossible and the magical.

The telegraph was carrying *information*—and it was doing so in a manner that required no agency of man to transport it. The taking, the sending, the shifting, the transporting—all of this could be done instantly, while the carrier seemed to remain perfectly still. The invention marked a climactic moment in the history of the United States. The implications were enormous, and they were recognized immediately. The moment instant communication was within the grasp of all—banker, baker, merchant, soldier, doctor, farmer, and yes, even a hesitant Indian chief—America was bonded and annealed into an almost unbreakable and indivisible one.

If the ultimate effects of the telegraph were to be of great benefit, its proximate effects were less so—at least, for one particular and legendary corps of ambitious and courageous men. The Central Overland California and Pikes Peak Company—the Pony Express—had begun its heroic runs between Missouri and Sacramento in April 1860, but within months the message-bearing horsemen galloping frantically between the transfer stations could see the poles rising along the roadway beside them, could see the insulators being set in place, the copper wires being strung across them.

The end for the brave little horsemen was as swift and savage as it was inevitable. By October 1861, the cross-country connection of cables had been completed.

Two days after the opening of its circuitry, with code being successfully received in California just seconds after it was tapped out in New York, the final orders went out to the Pony Express riders, too. No further men would be sent out from the stables, and the way stations built every ten miles along the two-thousand-mile route would be torn down or put up for sale. The staff who manned them would be sent home. The riders, Buffalo Bill and Frank Webner and Pony Bob Haslam and their other colleagues in the saddle, would be thanked, paid off, and their services dispensed with. Their fleet little horses would all be put out to pasture or sold off for glue.

The electron would now take over from the spur, the saddle, and the hammered iron shoe, and matters would never be the same again.

The Man Who Tamed the Lightning

Most information is passed between human beings as *conversation*, the six-century-old word that signifies the interchange of thought in words, which is conventionally conducted face-to-face in what has come to be called real time. Over long distances, conversation was seldom possible, save for a bellow across a chasm, a wave from a point on high, a chain of fires, or a series of signals sent by smoke, reflected sunlight, or rattled sticks. For any exchange beyond single-syllable simplicity and for exchanges that were to be made across truly lengthy distances, conversation in a strict sense was not possible. Hand-carried messages and letters had long been the sole alternative.

Until the invention of the electric telegraph, which at last permitted a long-distance version of conversation. With it the transmission of information—gossip, news, or intelligence—could be accomplished, if not necessarily with perfect ease, then at least rapidly, in the blink of an eye.

This immediacy of communication is what made the electric telegraph and its successor inventions so mystifyingly different from what had gone before. It was a development involving a new magic that until then had never been much discussed: *electricity.*

Long known but long uncomprehended, electricity was recognized first as the strange attractive charge that came from rubbing chunks of natural amber with cloth. It was also realized to be lightning, which, as Benjamin Franklin had almost fatally discovered, could be brought from the sky down to earth along a conductive filament like the wet string of a kite. Electricity was also thought to be in some way related to the invisible force that made magnets attract great masses of iron and lift them unaided in a way no man could do.

Then it was found that electricity could be hoarded and stored. A pair of Italian inventors created a *voltaic pile*, so-called, that produced conductible electricity, while a team of French doctors managed to bottle up already-made electricity and then use it, by *discharging* it, to shock patients into what they imagined might be better health. In time, man would generate this strange force, and once there was enough of it and it was massaged in certain ways to allow it to pass across great distances, it would be persuaded to do innumerable things—produce light and heat, turn wheels, calculate numbers, make sparks, help lift weights, and move objects small distances.

In the earliest days of discovery, one possible use of electricity stood above all others. It could be made to power a subtle kind of device that would move information—a weightless thing, after all, requiring little power to transmit it—along conducting wires. If such a thing worked, it could move this information invisibly and very fast.

Samuel Morse, who died in 1872, probably was the first to come up with the idea of an information-transmitting electric device and, if technically beaten by a nose by some other contender, he was certainly the man who perfected it. But he did so only by chance, and the achievement was not at all what was expected of him. Morse had begun work as a portrait painter, a photographer, and a professor of fine arts, yet despite his own high hopes and best efforts (but fortunately for American communications), he is unremembered in these fields, because he was not especially distinguished in any of them.

When he came up with the idea, he was a sad and disappointed man, and not without reason. His descent into melancholy began in 1825, when he had traveled from his atelier in New York down to Washington, ostensibly to paint two portraits of the Marquis de Lafayette.* While there, he was informed by a horse messenger that his wife, back in New York, was desperately ill. By the time he returned home, he was too late; she had not only quite unexpectedly died during the birth of their fourth child, but had already been buried. It is said, perhaps fancifully, that his belief in

---

* The great French aristocrat who had played such a heroic role in the American War of Independence had returned to America as a distinguished graybeard a half century later to perform what was essentially a yearlong victory lap. He was lauded by all. Having his portrait painted by Morse was seen as something of a distinction, but being given twenty-three thousand acres of public land in what is now Tallahassee was probably more welcome.

the importance of high-speed communication—which might have allowed him to reach her bedside while she was still alive—was born at this moment.

Nor was this to be the only tragedy of his year. Shortly afterward, Morse's adored father died, and then a few months later, his mother. Stunned by all this, he fled to Europe, hoping to forget, to win painting commissions beyond the America that was for him now so freighted with misery. He prayed in particular that his artistic labors in Washington would be recognized and he would be invited to paint a great mural inside the Capitol rotunda. But that wish turned to ashes, too. The works he produced in Paris and on his tours to Italy were widely thought to be indifferent at best, and his hopes of winning commissions came to nothing. He then learned—Pelion heaped upon Ossa—that a rival, an Italian named Constantino Brumidi, had been selected by Congress to paint the ceiling of the dome.

In October an utterly dejected and demoralized Morse, embarrassed at the prospect of being shunned by many of his more successful artist friends, decided to come home, an admitted failure. He journeyed to Le Havre in October 1832 and there boarded the French packet ship *Sully*, bound for New York. It was halfway across the Atlantic aboard her that he experienced the epiphany that would help him change the world.

Voyaging through strange seas has a way of bringing together the unusual and the unanticipated, and this passage was no exception. Twenty-six American passengers were on board, not including half a dozen farmers in steerage. Each night Morse dined with the senior American diplomat in France and with a lawyer from Philadelphia, together with the captain, as well as an otherwise forgotten man named Palmer and a Harvard geologist named Charles Jackson. It was his discussions with Jackson sometime during that stormy three-week passage that let Morse see the potential usefulness of electricity.

Both men were far from ignorant about the topic. Jackson's scientific work in France had brought him into contact with matters electrical; and despite his ambitions in art, Morse had been taught such fundamentals of electricity as were known while he was at Yale thirty years before, and he had been interested enough in the topic to take additional classes at Columbia College in New York in the 1820s. He was also no mean inventor: twenty years before, he had made a new kind of pump and in 1822 had designed a machine to cut marble.

Perhaps it was an innocent discussion of the properties of limestone that first brought Jackson and Morse together over an evening drink. Whatever forged the first connection, it kept them together for the rest of the journey. In the sway of the ship's saloon, the two men are remembered for endlessly debating and discussing electricity's mysteries, suggesting various tasks for which it might be best suited.

Lawyers would in due course bitterly debate which one of them first suggested using electricity as a messenger, but most probably it was Morse, as Jackson's later claims were declared the work of an embittered fantasist. (Jackson would later die in an asylum.) Suffice it to say that one of them remarked, "If the presence of electricity can be made *visible* in any part of the circuit, I see no reason why intelligence may not be transmitted by electricity."

In other words, if an electrical current can be made to spark, to do something that can be seen, and to do it instantly over any distance, and if that spark or a number or pattern of sparks can be made to stand for a letter or a number or a word or a name or a piece of data, then the information that these sparks denoted could likewise be transmitted instantly over any distance.

This was a vatic revelation. Morse, who like his geographer father was an avowed imperialist, believing profoundly that America must create "the largest Empire that ever existed," realized in an electric instant the role his imagined invention could play in such a vision. He jumped from the boat the moment it drew alongside, sped to his widower's apartment in New York, and promptly began to draw up designs. It was far from easy. Over the next four years, primarily in his offices in the newly opened university building on Washington Square, he patiently experimented with a number of wood-and-brass-and-wire-and-mercury devices built to do one basic thing: to use something—in his case a movable notched ruler fitted with metal blanks—to open and close an electrical circuit in patterns and sequences representing words and numbers in a code of Morse's invention.

Information was fed into one end and sent down the wires. The circuit was then opened and closed at the other end, signaling that the information had been received. Every opening and closing of the circuit could be signified with a bell or a light or a click.

The complexity and sophistication of the message was up to the user, and the speed at which it needed to be written was up to the user, too. All he had to do was to write, carve, or insert the metal blanks into the ruler according to the prearranged code. In Morse's primitive devices, the code was usually no more complicated than repetitions of clicks, five of them representing the number 5, and three followed by a space and then four more signifying 34. These numbers were then used to signify letters, or whole words in a numbered dictionary. The code, scribed onto a ruler and fed into the machine, would open and close the electrical circuit in the proper sequence. Far away, at the other end of the wire, the circuit would immediately display this same pattern of clicks, to be read as 5 or 34. Let's say a pattern of *click, click-click, click-click-click* was received; a secondary codebook might translate this 123 as ABC.

From there it was a matter of refining the code and the equipment. The words NEW YORK, say, could be punched in at one end, and out of the far end would come a pattern of dots and dashes that could be recorded on a strip of paper and that, to a person able to read the code, would also spell out the same two words, NEW YORK. And crucially, in theory, no matter the distance between the writer and the reader, no matter if the seven letters and their single space were tapped out in half a second or half an hour—all were received an instant after being sent. It was an invention of great simplicity and elegance. It was to be called *telegraphy*, "distance-writing." It was foolproof, exact, and precise. And it was made by Samuel Finley Breese Morse, a painter no more.

In time and with sedulous attention to detail (which included writing to all who had been aboard the *Sully*, and asking if they would swear an oath to what they remembered of his telegraph-obsessed conversations at sea), Morse eventually won all arguments* as to who was first and who invented the instrument, and on June 20, 1840, he won the American patent to it. The file still exists, one of the most significant documents in world history: US Patent No. 1,647, Improvement in the Mode of Communicating Information by Signals by the Application of Electro-Magnetism.

There were problems, inevitably. The electrical current available was too weak to allow messages to be sent as far as theory suggested. Most electricity at the disposal of experimenters was not much more powerful than the sparks that had famously made frogs' legs twitch in experiments performed half a century earlier. To send messages farther than from one side of a room to another—and Morse's original telegraph worked over only some forty feet before its signal petered out—required bigger and better batteries, coils, magnets, and most crucial, the invention of *repeaters* to amplify the signal and push it down the line with rejuvenated energy.

---

* The case was, however, fought out again famously in the US Supreme Court in 1853, in the now classic and widely cited *O'Reilly v. Morse*. Though in the main Morse was declared to be the inventor of the telegraph, the crucial chapter 8 of his claim, demanding rights to the future of his invention, was rejected by the Court as being "too broad"—a ruling often still employed today in cases concerning software inventions and discoveries.

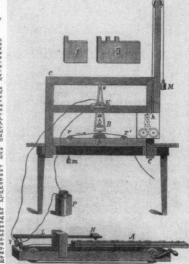

Few patents in world history can be said to have had quite the same impact on society as US Patent No. 1,647, won by Samuel Morse in 1840, signaling what was, essentially, the invention of the electric telegraph. Morse began his career as an artist and turned to electrical experimentation only after being disappointed by being refused a slew of commissions.

Soon other, more skilled inventors came to Morse's aid. Bit by bit, measured foot by transmitted foot, the range of the device increased. The power from one galvanic cup—a primitive battery, zinc and copper plates immersed in a bath of acid—could be easily multiplied; when forty such cups were used, the signal traveled a hundred feet. When copper wires were wound more tightly around the magnet, a message could be moved a thousand feet. (Appropriately, the telegram A PATIENT WAITER IS NO LOSER was used for this test.) By the time a chemistry professor from Oxford came across the Atlantic to see a demonstration, Morse was sending messages half a mile and more, and then, triumphantly, across no less than ten miles. He knew he was onto something big. He started to petition friends in Washington to urge Congress to take an interest.

"We have no doubt," he wrote in 1838, "that we can achieve a similar result at any distance." He felt it vital to the national good that the government be involved in development of the telegraph. "It would seem most natural to connect a telegraphic system with the Post Office Department," he wrote, because he saw the telegraph as another way of transmitting letters.

However, the Democratic Party controlled Congress at the time, and feeling that the telegraph was no matter for government (a reminder of the evolution of party-political sentiment over the years), rejected his overtures outright. Internal improvements to the country were not, the party chiefs intoned, "susceptible to federal aid." It was instead up to the states or to private individuals to throw money into projects like this. The doctrine sharply divided politicians of the time; there was much suspicion about corruption, and no certainty that the majority of taxpayers would win the benefits for which they were being asked to pay. So Congress decided it would offer Morse no money for any public demonstration, nor would it offer any kind of official support for the establishment of a trial telegraph line between two cities.

On the technical side, Morse's scientist allies were almost daily helping him with improvements. One man in particular, Alfred Vail, a priest turned machinist whose family owned a foundry, came up with the crucial invention that solved the problem of the telegraph signal's progressive weakening over distance.

Vail broke the intended circuit into smallish pieces, and at each junction he introduced a device that he called a *relay*. This was another magnetic transmitter that "read" the incoming impulse, no matter how weak; boosted it; and then sent it on its reenergized way down the wire of the next circuit. The magic of the relay was that its work could be performed as many times as relays could be built and installed into the line. The line could be continued endlessly until a chain of circuits circled the entire planet, if that was wished.

In electrical terms, the key was the sensitivity of the relay. Alfred Vail put his wish for it this way: "It matters not how delicate the movement [of the receiver] may be. If I can obtain it *at all*, it is all I want." So he built the device—a tiny magnet that could be triggered by even the tiniest movement at the circuit's receiving end and would then open and close the next circuit. Messages could always be sent as fast as anyone wanted to send them; now, with what was called Vail's "creative engineering achievement of the first order," they could be sent as far as anyone wanted to send them, too.

Morse traveled the world for four years in search of support, all the while continuing to experiment, improve, and hone his invention and to inveigh against its naysayers. Then, suddenly and to universal surprise, the political climate changed, and congressional intransigence melted away like an April frost. Morse was summoned to Washington, where he arranged to string wires between congressional committee rooms to demonstrate his system's abilities. The show seemed to soften the mood of the hitherto ill-tempered politicians sufficiently to prompt Morse to make yet another attempt to wheedle money from them.

This time he succeeded. It was hardly a rout: his plea only squeaked through by 89 votes to 83, but he cheerfully remarked that six votes were as good to him as a thousand. He had been offered a $30,000 subvention earmarked for a trench and a line between Washington and Baltimore, a distance of forty-four miles, to conduct the demonstration of his life.

This was the first electrical engineering project ever undertaken in America, and the technical problems turned out to be legion. The trench was tricky to excavate, especially when winter froze the ground. Not enough wire was available; the lead tubes used to insulate the conductor often broke; not enough batteries could be found. The paperwork required to free the government moneys was wretchedly complicated, and feuds broke out among contractors, would-be inventors, and new claimants to the existing inventions. Finally an exasperated Morse decided that the line need not be underground at all but could be suspended in the air from thirty-foot-high chestnut poles, spaced two hundred feet apart.

This turned out to be relatively easy, and he started to employ such of the line as had already been built to send news of the Whig Party's national convention, which by chance was being staged in Baltimore, back to Washington. He had a colleague at the convention ride out on the train to the closest telegraph pole and report what he had found out to an operator, who would then telegraph it to politicians in Washington. On May 1, he was able to report that the convention had nominated Henry Clay for president. The whole city was suddenly agog with news that was spread about long before the newspapers had it; the atmosphere of febrile excitement that suddenly gripped political Washington generated enormous publicity for the device that transmitted the first official news message ever sent at the speed of light, city to city.

Of that later moment in telegraph history, the famous Friday, May 24, 1844, when the first formal telegraph message was sent, we know rather less than we might wish. Alfred Vail set up his instrument in the offices of a railway station about a mile outside Baltimore. Samuel Morse, well aware of his place in posterity, set his up in no less august a setting than the chamber of the United States Supreme Court, which was then on the ground floor of the north wing of the US Capitol, close to the rotunda whose giant ceiling Morse had once hoped to adorn. But there would be no disappointment now, on this, his second great encounter with the enormous government building.

Morse had taken advice on what the first message should be. Patent Commissioner Henry Leavitt Ellsworth had championed the invention, and Annie Ellsworth, his daughter, had brought Morse the news of the congressional funding vote, so as a gesture of gratitude, he asked her to choose a suitable phrase. We know little as to how many congressmen were present in Washington and how many others were watching and listening in the depot forty miles away in Baltimore.

All we do know is that sometime on that warm afternoon, in the presence of a select group of the nation's elite, Morse sat down before his confection of wires and magnets and cogwheels and mercury vials and levers, and tapped out eighteen letters, the final four words of the twenty-third verse of the twenty-third chapter of the Book of Numbers: WHAT HATH GOD WROUGHT.

The first public message to be sent by telegraph— WHAT HATH GOD WROUGHT, from the Book of Numbers—was transmitted by Morse on May 24, 1844, from what was then the US Supreme Court chamber in the Capitol, to the railroad depot forty miles away in Baltimore.

The words alone formed a simple declarative exclamation, a statement of Samuel Morse's faith, a plangent line of gratitude, and above all, a suitably portentous epigraph for an era of change that now commenced with unimagined speed and unimaginable consequences. It marked the moment when, as Henry Adams would later write in his *Education of Henry Adams*, "the old universe was thrown into the ash-heap and a new one created."

New telegraph lines went up almost overnight. They went up so fast, according to the music-hall wits of the moment, that it seemed new wires were being spliced onto the tops of cornstalks and beanpoles. Lines were sent out strategically, tactically, or randomly, and were planned to run hither and yon, some for highly profitable reasons and others for purposes less commercially wise. Major cities would be connected in a matter of months; distant states within just a few years.

By the mid-1850s, there were dozens of telegraph companies, offering services of wildly variable quality. Many of these companies had been established so quickly and with so little thought of geography that one message often had to pass through several competing companies' hands before it reached its target destination. Between Chicago and New York, for example, a message might pass through the operators of six different companies, with each retransmission occasioning delay and risking inaccuracy. Then a team of investors led by Hiram Sibley, a wool carder, agricultural equipment dealer, banker, and county sheriff based in the wheat-processing center of Rochester, New York, decided to buy up all the competing companies and create a giant near-monopoly. Sibley and his crop-buying business friends needed accurate and up-to-the-moment news on wheat prices, and for that there had to be one telegraph company, and one only.

There might have been missteps. Sibley first backed a rival telegraph system, which was based on keyboards and the Roman alphabet rather than Morse code, and which had been invented by a man with the magnificent name of Royal House. And the giant new company was nearly given the name the New York and Mississippi Valley Printing Telegraph Company. But in the end, Morse's system proved entirely more reliable than the House brand, and the new name chosen for the company was memorable and elegantly spare: Western Union.*

It was under the banner of this Western Union, appropriately, that the push then began to unite the nation telegraphically, coast-to-coast. Sibley pressed

---

* Although Sibley was Western Union's first president, it was a former rival, Ezra Cornell, who first chose the name. Cornell today is remembered less for his involvement in telegraphy than for his establishment of the great mechanical and agricultural university in Ithaca, New York, which bears his name.

Congress as early as 1857 to support building a trans-continental line. Three years later, the politicians passed the Pacific Telegraph Act of 1860, which under-wrote the $40,000 Sibley decided the line would cost.

The act was signed into law on June 16 by President James Buchanan—one of the few notable deeds of a president whose repute remains mired by his having to watch as the slave states of the Deep South seceded and the sides drew apart to begin the first fighting of the Civil War. His decision in signing the telegraph legisla-tion was fully intended to unite the nation, and yet it was a decision taken even as the country was beginning the most savage period of division and disunity in all of its history.

Within weeks Sibley had set up companies to under-take the various sections of the line from Omaha to Sacramento—the same route, more or less, that would soon be taken by the transcontinental railway. By the time construction began in the early summer of 1861, the Civil War was well under way. The attack on Fort Sumter, the generally accepted starting point, had been in April. It thus took more time and trouble than had been expected to acquire supplies of wire and wet-cell batteries and insulators for the fifty-volt line and other essential material—especially supplies for the west-ern half of the line, which still had to be shipped from

Northern ports, around Cape Horn, and up to San Francisco.

Once supplies were at hand, the business of cutting poles and making holes, tipping the one into the other, nailing up the cross ties and setting into them the iron-and-glass insulators that would carry the wire, then lifting the wire from an adjacent pole or splicing a new one onto the free end all turned out to be none too difficult, and progress was swift. The prospect of the linemen having to spend a harsh winter in the Rockies was a powerful disincentive to delay. The line had reached Laramie in August, Salt Lake City in early October, and then with Sho-kup's generous sanction, the tiny town of Fort Carson later in the month. By October 24, 1861, the line was complete.

The first message was sent out from San Francisco by a high panjandrum of one of the telegraph companies. It went to an American president who was by then already reeling under the first months of the Civil War. "I announce to you," went the message to Abraham Lincoln, "that the telegraph to California has this day been completed. May it be a bond of perpetuity between the states of the Atlantic and the Pacific." No mention was made of the North or the South.

The message was a harbinger: in all the future development of the nation, east and west time and again

would trump north and south, leading to marginaliza-
tion of the southeastern corner of the country—a mar-
ginalization that to a degree remains today.

It was all so quick. It had been no more than a decade
and a half after the first signal messages had gone from
near Baltimore to the Capitol building that engineers
had come inquiring at tepee doors in Idaho, explaining
that they were planning the line to California, and could
they please pass by? And then it was just moments later,
in the greater scheme of things, when all America was
connected. Once the final connections had been sol-
dered together and the last of the knurled brass screws
tightened, when switches were thrown and voltmeters
flicked upward and to the right, quivered and stayed
there, then began the roar of a long-withheld national
conversation.

Borrowers could now talk to lenders; buyers could
negotiate with sellers; stock market trends could be sent
to all corners of the nation; commodity prices could be
relayed to potential market makers; credit references
could be checked no matter how far from home the
applicant was. In short, the whole complicated apparatus of a fully functioning capitalist economy, lubricated
with what economists would come to call "perfect
information"—data that were up-to-the-minute, accurate, and offered disinterestedly to all—was now in

place, ready to be pressed into service. America could now be instantly in touch with itself, and from that moment on, the din has never ceased.

But . . . neither of the progenitors of the system remained much interested in their creation in later life, though both benefited mightily, accruing great fortunes. Samuel Morse tried to go back to painting but found his talent had deserted him, and he died, less content than he deserved to be, in 1872. Hiram Sibley lasted sixteen years longer, and after retiring from Western Union, became a successful dealer in seeds, leaving the technological developments of the age firmly behind him and his company's legacy to others.

Western Union once carried more than 90 percent of all telegraphed messages sent in America, but it eventually dwindled in standing as its technology was superseded. Horticulturists, however, belong to a community that today remembers Western Union's leader for an entirely different legacy. A century ago, Hiram Sibley and Company, seed merchants of Rochester, New York, found and then promoted lavishly a banana-shaped summer gourd that is known still to keen gardeners as the Sibley squash. Though the company's last telegram was sent in 2006, the Sibley squash endures and will presumably continue until the end of botanical time.

The Signal Power of Human Speech

In scientific terms, the telegraph truly transformed the age and broke new ground. It created a profoundly new epoch in human history. Yet it was quite lacking in intimacy and privacy. Ordinary people did not warm to the idea of having telegraph instruments in their drawing rooms any more than they had wanted railway stations in their backyards. Most people could not read Morse code any more than they could shunt steam engines.

But then came the telephone, a device of immense social value, an instrument of personal empowerment. For it was private, intimate, and immediate. The telegraph allowed the transmission of information, but the telephone let individuals have their own long-distance conversations without any need for an intermediary— making it an immediate object of desire.

The telephone became something that was to be owned, to be placed proudly as the centerpiece of a sitting room, next to the aspidistra and across from the soda siphon. And that simple fact—that an electrical device was now made to be installed inside a private house—led to the creation of an entirely new skein of networks, for telephony first but also for an entirely new electrically powered world besides.

Alexander Graham Bell was the genius first seized with the idea of the telephone, in 1874. He was at the time employed in Boston as a teacher of the deaf— both his father and later his wife were afflicted—so he knew a good deal about vibrations in the air. One day he had wondered out loud whether, as he later wrote, "it would be possible to transmit sounds of any sort if we could only occasion a variation in the intensity of the current exactly like that occurring in the density of air while a given sound is made." It was a technological epiphany that impressed many, not least the imperious Joseph Henry, one of the country's greatest electrical engineers, then the head of the Smithsonian Institution.

"You have the germ of a great invention," he said to Bell, who replied pathetically that he did not have the skill to do the necessary work. Henry's stark bark of a reply remains the stuff of legend. "Get it!"

And Aleck, as his family knew him, promptly and famously got it. Less two years later, in March 1876, this twenty-nine-year-old Scotland-born Canadian American teacher and inventor was awarded the famous US Patent No. 174,465, "for transmitting vocal or other sounds telegraphically . . . by causing electrical undulations, similar in form to the vibrations of the air accompanying the said vocal or other sounds." Some weeks before the patent was awarded, he had conducted the equally famous experiment that concluded with perhaps the most important fragment of peremptory speech ever known.

Bell had been experimenting for months with different ways of transmitting "the vibrations of the air" by means of the electric telegraph. Much of his work had been done in a damp basement in Salem, a dire place where the witches' trials had been held back in colonial times. But in early 1875 he rented a much more agreeable and airy space on Court Street in Boston, in the shop attic of the man who supplied him with electrical equipment—a man who also customarily supplied his inventor tenants with assistants. He happened to assign to Bell an equally young man named Thomas Watson, who became a faithful companion and assistant for the next several decades and played a significant supporting role in the history of America and the world.

One day in the summer of 1875, the two men happened to be working in adjoining rooms, testing three vibrating reeds they were using for an experiment on sending multiple signals along a single telegraph line. One of Watson's reeds had become too firmly stuck to a magnet, and the young man pulled it clear, with a twang. At the very same instant, Bell, in the next room, heard the same twang through his own receiver—and realized the vibration had induced a tiny electric current that had traveled from one room to the next *by wire* and had made the reeds on Bell's magnets twang at the same frequency as Watson's, at the very same time. He bent his languid frame over the instrument and cupped his ear: there was no doubt about it. Sound was being sent and received where only symbols had gone before.

It was an unanticipated moment of the purest serendipity, and it stopped the two men in their tracks. This, Bell realized, was how to send those "vibrations in the air" down the line—by making a magnet induce a small electric current that would travel along the wire carrying with it the encoded vibrations, as it were. Bell and Watson refined and refined what they were doing, working on ways that the full spectrum of sound frequencies and timbres could be sent, recognized, and received. And then on March 10, 1876, came the peremptory command that remains frozen in history.

Bell had rigged his instrument with a speaking tube, and the circuits between his own room and Watson's were live. The magnets were working, varying the resistance and producing an undulatory current to mimic the sound that was causing the variation; all other components in the chain were working, too. Now Bell needed his assistant to come to his laboratory to check his instruments.

Whether by accident or design, Alexander Graham Bell spoke formally and sharply into his speaking tube: "Mr. Watson—Come here—I want to see you." And Thomas Watson came as summoned, running. He burst through the door. "I could hear you. I heard what you said!" he exclaimed.

He had become the recipient of the first intelligible telephone message ever sent. Bell's order may have lacked the sonorous majesty of the carefully thought-out WHAT HATH GOD WROUGHT of Samuel Morse's first telegraphic message thirty-two years before, but it came to serve as an epigraph of which he was inordinately proud.

Bell had occasion to recall it nearly forty years later, on January 25, 1915. By this time, the telephone was wildly popular, with millions connected around the country. But now and for the first time, a telephone line had been strung clear across the continent, and it was about to be tested. Alexander Graham Bell was in the American Telephone and Telegraph Company headquarters, on the fifteenth floor of the Telephone Building on Dey Street in Manhattan. Thomas Watson was in the Bell Building in San Francisco, almost three thousand miles away.

The test started. Engineers said all was ready. Watson picked up the telephone on his desk. At first, dead silence. Then came a click, a faint electrical buzz, and after a tantalizing pause, out of the earpiece came a voice.

"Hello, Mr. Watson," it said, in the familiar, scrupulously modulated, tones of an educated Scotsman. "Can you hear me?"

"I can hear you perfectly," Watson replied to his former chief.

"Mr. Watson," Bell went on, a smile playing in his voice, "come here, I want you."

Thomas Watson, still on the far side of a partition, but one now three thousand miles wide, was quick on the uptake. Maybe the moment was rehearsed, for there were many cameras and reporters present. He smiled, and spoke into the receiver: "I could—but this time it would take me a week to get to you."

The gist of America's first transcontinental telephone conversation, in other words, was that most impeccable of social lubricants—a joke. It was an exchange that would set the tone for how the most recognizable and universal technological instrument on the planet would most commonly be employed. It would be used for business, of course, and to transmit the perfect information required for efficient capitalism. But most of all, it would transmit the quotidian ebbs and flows of common chatter, which could now be conducted as easily across the country as across the room.

"Making a Neighborhood of a Nation," said Southwestern Bell's advertisements that announced the new service. But the angelic country-spanning operator who is pictured beside the slogan, holding in each hand a telephone, one on the East Coast, the other on the West, has a single word on her headband, suggesting the real hero of the moment: *Science.*

Alexander Graham Bell, scientist, though a kindly and considerate man and a philanthropist of the first water, was also an adroit businessmen, and he and his heirs became extraordinarily rich. When offered a share in Bell's company, the chiefs at Western Union at first scoffed at his invention, dismissing it as no more than a toy. A year later, they realized their mistake and offered $25 million for the patent rights. The offer was too late and, by then, far too small. The telephone was fast becoming a success of epic proportions. Everyone wanted one. By the time Bell died and was laid to rest in Nova Scotia, there were thirteen million in America alone. Today there are more telephones in the United States than there are people. And while it would be idle to suggest that telephony is universal—billions of people, in Asia and Africa most commonly, still have little access to clean water or basic sanitation, let alone to even the plainest of telephone instruments—the statistic is undeniable: there are currently six billion cell phones in the world, a number that is increasing rapidly as long-distance conversation becomes almost a basic human right.

As soon as the telephone had been invented in 1876, offering people the opportunity to have conversations from home with people far away, the instrument became fantastically popular. Small forests of poles soon sprang up in cities, to hold up the wires running between subscribers and exchanges, as here on New York's Broadway.

Small wonder that US Patent No. 174,465 is said to be the most valuable patent ever awarded, though it might not have changed the universe quite as thoroughly as did Morse's Patent 1,647. No one would say of the telephone, as people had said of the telegraph, that it caused the old world to be tossed into the ashcan and a new one to be born. For many years, the telephone's newfangledness was feared by those who came to it late—especially, for some reason, Midwestern farmers, who saw it as an instrument that brought only bad news, usually in the middle of the night.* More commonly, though, the instrument came swiftly to be seen as seen a device with lasting commercial class and mercantile clout. Alexander Graham Bell has long been regarded fondly as an archbishop of the capitalist cathedral, especially by the millions around the country who were prudent or prescient enough to own telephone shares. As the glum Western Union chiefs must have said to themselves a year after they turned down his offer, *Some toy!*

---

* In 1979 N. W. Ayer & Son came up with the slogan *Reach out and touch someone* for AT&T, in an effort to lessen the public's wariness toward the telephone. It became one of the most famous advertising campaigns in Madison Avenue history. Almost overnight the late-night ring of a telephone in a Kansas farmhouse went from a fearful "It must be the hospital" to a welcome "That must be Jill!"

With Power for One and All

It was a crisp blue morning in mid-September, and I was standing on Battle Pass, in southern Wyoming, mugging for the camera with one foot on each side of the Continental Divide. There was a dusting of early frost on the peaks of the Snowy Range, fifty miles to the east, but here, even at 9,900 feet, it was still comfortable, and the air was scented with warming pine needles. For a few moments, it was serenely peaceful, with just the occasional eagle swooping in the thermals. But then from behind a rock, with the tinny sound of a lawn mower engine, there appeared a stripped-down tractorlike vehicle, with two men aboard. "Hunters!" I cursed. *So nice a day! Let's go out and kill something!*

But they turned out to be a perfectly pleasant father and son from Laramie, the lad a poster boy for the joys of hunting and paternal bonding. The pair had skipped their respective work and education for the day and had come up into the high hills for the formal start of the antelope season. When they heard what I was doing—researching the old American Western trails—they both pointed down to the west, where a small gleam of silver shone in the morning sun. "Battle Lake," they said in near-unison. "Where Edison invented the lightbulb."

Well, perhaps. Not quite. There is a plaque on the roadway above the lake, which is on the western side of the Divide, and it suggests rather modestly that Thomas Alva Edison had indeed come to Battle Lake in July 1878 and that being there had something to do with inspiring him to think of using burned bamboo to help make light for the nation's households.

He had traveled to this corner of what was then Wyoming Territory, along with a team of astronomers, to observe an eclipse of the sun while standing on the very same high pass where I had encountered the father-and-son hunters who told me the story. Once the phenomenon was duly observed and done with, he went downhill to fish in the lake, and, so the story goes, he slipped and let one of his bamboo fly rods fall into his campfire. The bamboo was reduced to fronds and filigrees of filaments, which glowed brilliant white with heat. Edison wondered if such a filament could perhaps be made hot artificially, with an electric current running through it, and by becoming hot emit a fine white and useful light. The filament would be fragile, of course; but its lifetime could perhaps be extended and preserved by enclosing it in a vacuum in a specially blown glass bulb.

Thus was born—allegedly, supposedly—the idea of the incandescent lightbulb, in the up-country wilds of Carbon County, Wyoming Territory, in the summer of 1878. But skeptics abound. Most suggest that the nation's inventor-in-chief experimented in his laboratory in Menlo Park with scores of potential illuminating candidates—strands of burned baywood, boxwood, hickory, cedar, flax, and bamboo among them—before finally settling on the carbonized cotton thread from which he made his famous first-ever patented lightbulb in 1879. Bamboo was but one of some six thousand vegetable products that he tried. To find the longest-lasting filament, "I ransacked the world," he said.

But whether or not Wyoming is due any pride-of-place laurels, Edison's incandescent lightbulb spawned a need that neither the telegraph nor the telephone had created: a need for power. The telegraph and the telephone drew their sustenance from battery cells in the exchanges, but the electric light was different. Its sheer popularity presented the world with a brand-new challenge, for there were soon far too many of them to power in any conventional way.

Once working models of lightbulbs were put on the market, they quickly began to replace the oil lanterns and gaslights of Victorian households—but the number of bulbs required to light even a modest house or the smallest town demanded a change in the way power was distributed. It made no economic sense to equip each bulb or each household with a battery pack, nor could any battery array at some central plant possibly illuminate as many bulbs as were now hungering for power.

It was necessary that electricity be made, or *generated.** Once made, it would then be necessary to distribute it along wires to the waiting communities, and from there send it on to businesses and individual houses, lighting the bulbs in the streets and inside the buildings. Thomas Edison worked out just how to do this. He was a man of many achievements, with no fewer than 1,093 patents to his name, from the phonograph to the stock-market ticker to the movie camera to the electric chair, and of course the incandescent lightbulb; but his creation of the new science of electric

---

*   As so often happens, it took some while to find the proper word. The use of the verb *to generate*, in an electrical sense, appears first in 1895, and when the *Times* of London illustrated the first recorded use of *generating station* in 1878, its chosen sentence showed another phrase that was being tested: "The way in which it is proposed to 'lay on' electricity to any given district or town is somewhat similar to the arrangements for gas. There would have to be, in the first instance, one or more electrical generating stations." The secondary phrase survives minimally in British expressions like "Do you have electricity laid on at your place?"

power generation and distribution is the signal unifying achievement that arguably outshines them all.

It all began on Christie Street, in what was once the town of Raritan, New Jersey. (Since the early 1950s, it has been the township of Edison.) The street Edison chose as the first in America—or anywhere in the world, for that matter—to be lit by incandescent electric lamps ran for half a mile outside his laboratories. The lamps were switched on, connected to batteries inside the laboratory, on New Year's Eve, December 31, 1879.

Today most of the houses on Christie Street are just as one might expect of a prosperous bedroom community—neat, mostly ranch style, with clipped lawns, basketball hoops outside the two-car garages, garbage containers ranked outside according to the latest recycling rules. There are a good number of sodium vapor lamps to illuminate the street, all of them now powered not by batteries but by solar panels. Then, at the street's southern end stands something quite phenomenal: an enormous model of a domestic lightbulb, 13 feet high and made of segments of Pyrex, sitting on top of a tower nearly 120 feet tall that is decorated with eight columns of concrete mosaic, with each topped by a floodlight aimed at the great bulb in the sky.

Every evening the light, a gigantic model of Edison's first domestic lightbulb, beams out over the suburb, illuminated from within by a giant incandescent filament (soon to be replaced by light-emitting diodes, which the great man did not invent) and lit from below by the floods. Huge loudspeakers set into the concrete then crackle during the daytime with recordings of Edison's words or else with barely discernible music, mostly sounding like wasps trapped in a milk bottle, spun from his early gramophone recordings. On his birthday each February, the speakers sound with encomiums to the man who, as they say in these parts, "invented today."

The motto of Edison Township is "Let there be light," and not without reason. During the summer of 1879 he saw to it that lamps were erected along the byways of the township's thirty-six acres of Menlo Park, where he had sited his laboratory. They were an exhibition of his abilities and his vision, an exhibition he would employ to persuade those who mattered in New York to allow him to use the city as his first test market for lightbulbs and for the generation and distribution of the electricity to illuminate them.

Not that Manhattan was exactly wanting for electric light. For the previous decade, many of the city's streets, parks, docks, and factories had been lit by thousands of arc lights, devices which poured cascades of brilliant, unforgiving, harsh white light from between a pair of pointed carbon electrodes. They emitted gas that caused headaches, smelled bad, and left a grimy, oily residue on ceilings. The electrodes also burned out rapidly, the drain on batteries was immense, and the quality of light was intolerably extreme. Their makers, knowing they were unpopular, offered the argument that by erecting tall arc-light towers from which illumination could be directed across entire neighborhoods, light would be offered to all equally and democratically, in a very *American* way. But it was an argument that cut little ice with New Yorkers. Although everyone agreed that the security the lights offered to businesses and people late at night promoted the twenty-four-hour economy that still defines Manhattan today, no one liked arc lighting, not one bit.

Edison hoped New Yorkers would turn instead to his smaller, softer, more human-scale incandescent vacuum-tube illuminations—bulbs his company promised would offer "milder" light. He consequently invited all manner of grandees over to Menlo Park to demonstrate what he had in mind.

It was quite a show. On his thirty-six-acre spread, he had laid out whole streets, each lined with wooden poles topped with glass lanterns, inside each of which was an incandescent bulb. Imaginary houses, designed to look like those in lower Manhattan, were also staked out, and they were lighted, too, and this whole unreal New York City was connected to an array of batteries with feeder cables (which took the power to the streets), mains wires (which took it into the houses), and service wires (which went to the individual house lamps). Edison's basic plan for electric distribution—generator, feeder, mains, service—remains today the standard model for all distribution everywhere.

Once he threw the switches, his display burst into a frenzy of glitter. It immediately and mightily impressed the city's Blue Book visitors. The Vanderbilts, prominent from their railroading fortune, were the first to back Edison's efforts. Then a portfolio of barons—J. P. Morgan, Baron Rothschild, and Henry Villard among them—followed suit, commingling between them sufficient funds to allow Edison to forge ahead.

Early in 1881, Edison gave a formidable dinner party, catered by Delmonico's—with Sarah Bernhardt performing—where he showed New York's aldermen and commissioners what he had in mind: a power station in lower Manhattan and conduits with cables running beneath the streets from it to connect buildings to his new electric service. It must have been a wild party, because less than three weeks later, the

Thomas Edison's eagerness to have electricity generated in power stations and distributed by wire to private consumers led to the sudden popularity of outdoor "electric shows," where lights were erected in fields and parks and switched on and off to the amazement of visitors. Soon, Edison promised, power would be so cheap only the rich would burn candles.

city's new mayor, the Irish chemical magnate William Russell Grace, granted permission for the Edison Electric Illuminating Company to build its first power station in a dilapidated industrial building at 257 Pearl Street.

Edison somewhat grumpily paid $155,000 for the property, which he complained was a slum and smelled richly from the fish market nearby on Fulton Street. City bureaucrats initially opposed his plans for digging up the streets, threatening to tax every foot of digging. But he argued fiercely for the system's efficiency and before long began a positive orgy of trenching, his teams working through the hot August nights to place fifteen miles of thick copper mains lines deep in the macadamized roads, insulating them with beeswax and linseed oil and thick asphalt that had been especially ordered from the tar pits in Trinidad.

Six giant steam-powered generators then arrived from Philadelphia, weighing thirty tons each and known as Jumbos. They were coupled to the devices that actually created the electricity, strange-shaped dynamos known as Long-Waisted Mary Anns; following this, giant water tanks and boilers were bolted to the floors, chimneys were raised, and fuel supplies were gathered in. Finally, on the evening of September 4, 1882, the stokers lit the fires, steam began to push pistons, the engines started to turn, the dynamos began to run. The armature of coils turned within its blanket of magnetism, and in accordance with a basic principle of physics, direct current began to flow into the wires, snaked out of the Pearl Street building, poured along the underground copper cables in their conduits, surfaced out of manhole covers and into the mains cables of some forty households, and from there was sent to the server circuits and terminals of the first eight hundred bulbs that had been installed.

In one golden instant, lower Manhattan came ablaze with a soft, uninterrupted light, all coming from vacuum bulbs that neither polluted nor glared nor gave headaches to those who read, dined, walked, or dozed beneath them. The promises of illumination made across the Hudson River in Menlo Park, New Jersey, had now been made good in New York City, and the success of the trial took hold with a speed that astonished even the city itself.

Scores of plutocrats became early adopters; mansions lit up from Central Park to Chambers Street. The Stock Exchange installed electroliers, as they were known, massive chandeliers with sixty-six bulbs each, to throw light across the trading floor. Industry of all kinds suddenly sprouted in the city, with machines that sewed clothing, spun sugar, carved pianos, and printed books and newspapers, all of them suddenly illuminated in a way that no longer carried, as previously, the terrible risk of fire. Theaters put up advertising marquees studded with bulbs—enough of a blaze for Broadway to be christened the Great White Way. The electric elevator was invented: with heights no longer limited by the number of stairs their occupants could reasonably be expected to climb, buildings began to rise high into the sky.

Electric billboards were hauled up into that same sky—one of them a four-hundred-square-foot monster with fifteen thousand bulbs that flashed on and off under the direction of a man sitting in a hut on a roof nearby. H. J. Heinz put up an electric sign with bulbs all painted green in the shape of a forty-five-foot pickle. Nightclubs flourished, in which pretty girls waved customers to low-lit tables with electric-powered wands and winking lights created seductive effects on the dance floor. A prostitute would advertise her wares, her availability, and her address with a small red electric lamp. The elevated trains now had stations gleaming with incandescence, and the docks and boulevards beside the water also gleamed. People who came to New York loved what they saw: the today that Edison promised electricity would bring was now taking Manhattan by storm and would soon spread like an uncheckable fire from New York City throughout the entire country beyond.

There were small problems. The generators broke down or speeded up for no apparent reason and exploded. J. P. Morgan had his walls and carpets scorched by a sudden outrush of power. One of the Vanderbilts got so cross with short circuits that he went back to gas, although the rest of the family kept the faith in their investment. Once in a while, electricity leaked into the street. A man peddling tinware directed his horse across an intersection near Wall Street, whereupon it seemed to go suddenly mad, its ears stood bolt upright, its tail rose skyward, and it ran crazily off toward the East River; a few moments later another team of animals crossing the same spot all fell to their knees and refused to budge. Workmen, it turned out, had punctured one of the copper feeder lines, and the manhole cover was live with current.

In 1888 a terrific blizzard coated the thousands of miles of overhead wires with ice, bringing down innumerable live wires, which fizzed and snaked and blazed on the icy streets, and had to be captured, cut, and turned off by men with axes, who dealt with them as they might deal with hordes of poisonous snakes.

But there were bigger problems, too, and one bigger than all the rest. It soon became apparent that the current Edison favored for New York—direct current, in which the electrons all move in unison in one direction along a wire—could not be easily transmitted along supply lines that were longer than a mile or two. This meant that more generating stations had to be built, creating noise, pollution, and expense. Yet at the same time that Edison and his immediate rivals and copycat inventors were installing their wires and lights, it was discovered that quite another kind of electric current— alternating current, in which the electrically charged particles move back and forth many times each second and essentially go nowhere at all—was, miraculously it must have seemed, not at all limited in terms of distance.

Thus began the so-called War of the Currents, the decisive outcome of which would eventually lift electricity from purely local use and make it a truly national utility. It would also become, in common with such seemingly unrelated things as knowledge, fresh air, and lighthouses, a public good to which all should have access.

One of the features of Edison's direct current was that it needed to be generated at more or less the same voltage as that required by the devices it was powering.

In the early days, almost all of these devices were electric lightbulbs. And because tests had shown that a carbonized cotton filament in an incandescent lightbulb survives longest and shines brightest when a current of about 100 volts passes through it, the generators at Edison's power stations were designed to generate electricity only at 100 volts.

Two problems resulted. The first, already mentioned, concerns the three-wire copper transmission lines taking this power from the generating station to the lightbulbs. When the wires warmed up as a direct result of carrying the 100-volt transmission, the current's power dribbled slowly away. The longer the lines were, the more power they lost. Lightbulbs that were nicely close to a power station burned with a pleasing bright white light ("soft, mellow and grateful to the eye" wrote the New York Times). Bulbs that were in offices or houses half a mile away, on the other hand, burned with a yellowish, sickly look. A mile farther on and they were dimmed with a kind of invisible soot. More than that and the feeble glow of the filament wavered and faded until it was quite dark. DC power was indeed useless over any useful kind of distance.

The second problem involved direct current's limited ability to carry electricity to devices other than lightbulbs. As electricity became publicly available,* scores of inventors came up with uses for it that went far beyond the bulb, world-changing though it might be to have illumination at the touch of a switch. The electric motor, for example, found a myriad of uses— Edison himself came up with a motorized electric fan, which suddenly made the sultry American summertimes of the 1890s bearable; and before long there were electric clocks, vacuum cleaners, hair dryers, and water pumps. The success of the motor never wavered: today half of all the electricity produced in the world is used by electric motors of one kind or another.

But what is true of motors today was true in the 1880s: different motors required different voltages of electricity to perform the different tasks, and in those early days it proved tiresomely difficult and costly for engineers to transform Edison's direct current into the different voltages that were needed. It was not impossible: a rotating device had to be attached to the

---

* Edison distributed electricity free at first, in part to work out the teething problems of supply, in part to allow people to become familiarized and then hooked. He made some money from lightbulbs, for which he charged the substantial sum of a dollar apiece. He then started charging for electricity itself in 1883: the first recorded bill went to the Ansonia Brass & Copper Company in January in the sum of $50.44.

generator to do this; but like all rotating devices of the day, it was expensive, it wore out, and it needed to be replaced. The whole process was formidably uneconomical.

So arguments started to be put forward, most notably by George Westinghouse and his business colleagues, in favor of the significantly different alternating current, whose voltage could be easily altered up or down—transformed by a *transformer*—to power a tiny light-bulb or a massive heavy-duty electric motor, no matter how far away each might be. It was not long before the AC lobby, in spite of Edison's charismatic, near-heroic standing in the America of the time, steadily began to gain ground. It was late in 1887 that the lines were drawn for a brief and vicious little electrical war, which would decide the future of the electrification of the United States.

Although the name of George Westinghouse is the one most commonly associated with the currents war—Edison versus Westinghouse is how the bout is cast in present-day shorthand—he actually had little or nothing to do with the creation of alternating current. His expertise was in mechanical engineering, and he had made a fortune designing brakes for railway trains and making signaling systems and gas pipelines. He was quick to spot an opportunity, however, and late in 1887 he bought for $60,000 the patents for seven crucially important inventions in the field of AC power generation and distribution, all of which had been granted to the true pioneer in the field, Nikola Tesla, a young Serbian inventor and polymath who had come to New York in 1884 to work for Edison.

The adoration of Nikola Tesla—some call him the father of the electric age—has recently made him almost a cult figure. It is easy to see why. He was tall, handsome, supremely clever, impeccably dressed. He was a fastidious gourmand who liked weighing his food and dining alone. He was shy, celibate, polite, soft-spoken, courteous, and kindly; a melancholic man blessed with a photographic memory; a man affected by phobias of such things as pearl earrings, spherical objects, accidental contact with women's hair, and the imagined dirt on knives and forks. He liked to play billiards, chess, and poker and to feed the city pigeons—for one of which, its feathers pure white, he built a $2,000 splint when nursing it back to health after it broke one of its gray-tipped wings. He was also afflicted with an obsessive-compulsive disorder; he refused to shake hands, he would feel a need to read all the books written by any author he encountered, and he had a peculiar devotion to the number three and to any other number divisible by it.

Though the Serbian-born inventor Nikola Tesla, clever, fragile, and an eccentric showman, was until recently widely overlooked, it is now generally accepted that he made vast contributions to the development of alternating electrical current and the invention of radio, long before Marconi. He has lately won legions of new admirers, mostly young, who see him as a forgotten hero of American science.

Nikola Tesla was, in short, the classic exemplar of the mad scientist, and the fact that he made dangerously interesting inventions by the score resonates still with today's imaginative fans of the far-fetched. He made devices like the Tesla coil, a step-up transformer that could raise current into the tens of thousands of volts and then generate Niagaras of gigantic electrical sparks. Smaller versions of the coil were used as scalp massagers and violet-ray-emitting devices for amusing the clients of beauty salons. He drew up blueprints for a gun that could shoot out thousands of tiny particles of tungsten. He promised the US Army a lethal particle-beam weapon, which was immediately seized upon by headline writers and comic-book editors as a death ray. He persuaded his backers to erect a huge iron pylon at a place called Wardenclyffe on Long Island, from which he promised he could beam waves of electrical power directly through the air, just like radio, and by doing so end the need for transmission lines and in theory connect the whole world electrically.

Tesla is also said by some to have held Thomas Edison in spectacularly low esteem. The supposed antipathy went back to his first job at Menlo Park, which he had won after a brief interview soon after his arrival (with just four cents and a book of poems in his pocket) from the Balkans. He had reportedly told Edison he could remake a DC motor that Edison had designed and make it more efficient; Edison promised him $50,000 if he succeeded. But when he showed the successful result to Edison and asked for his money, the notoriously stingy inventor is said to have remarked with withering superciliousness that Tesla clearly had no understanding of American humor, and offered him instead a $10 pay raise. Tesla resigned immediately and remained at daggers drawn with Edison for the rest of his life, making the war of the currents a highly personal crusade as well as a battle based on science and economics.

The first demonstration of public distribution of Tesla's generated alternating current was made in the hill town of Great Barrington, Massachusetts. A transformer manufacturer named William Stanley, backed by Westinghouse and basing his techniques on Tesla's designs and patents, had established his small factory on the banks of the Housatonic River, which flows through the town on its way to Long Island Sound and the ocean. It was beside the river, in an old rubber mill on a short road named Cottage Street, that Stanley made his experiment.

He had erected a generator, driven by the fast-flowing Housatonic water, that would produce 500 volts of single-phase AC electricity. There was nothing very difficult about this, but Stanley knew that if he simply supplied this voltage to the customers in town, their lights would grow dimmer the more distant they were from his generator. So he did what it was not possible to do with Edison's DC current: he connected a big, heavy, wire-wound Tesla-style transformer into the circuit and stepped the output up to 3,000 volts, a hitherto unheard-of current.

This he then sent out by wire to Main Street, hanging the cable from the giant elm trees (now replaced by flowering cherries) that lined the roadway. He connected this big, high-voltage wire to a series of six much smaller step-down transformers designed by Tesla and brought the voltage down to a manageable 100 volts. He connected these small lines to a chain of domestic lightbulbs. On March 20, 1886, he threw the switch and the lights began to glow.

Every light came on and stayed on. Every house, every business that lined Main Street, from the small banks in the north to the piano factory in the south, suddenly had power. They were recipients of the first distributed AC electricity ever offered to any community in the world. And it all worked flawlessly.

As a result of Stanley's success, Edison was almost ready to acknowledge defeat. He railed against what he wanted the public to see as the dangers of Tesla's AC. The back-and-forth current was peculiarly suited, he suggested ominously, for killing people, and even much larger creatures with what some might think were electrically impermeable skins. To the dismay of many and with lasting damage to his historical standing, Edison in 1903 demonstrated the use of a prototype AC-powered electric chair by fitting copper shoes onto a rogue circus elephant named Topsy and electrocuting her in public on a steel plate at a Coney Island zoo.

Topsy was a disagreeable elephant, no doubt; she had killed three of her keepers, though perhaps understandably in the case of one of them, a man who had tried to persuade her to eat a lit cigarette. Few today believe she deserved electrocution, however, even though some at the time believed that "riding the lightning" was, as zoo visitors were told, quick and humane as these things go.

Topsy is seen in a film, standing on the sheet of metal. She had already been given a quick pre-execution snack of cyanide-laced carrots, just in case. The current is switched on. Her legs immediately start to emit white smoke, as if they were stumps caught in a fast-moving forest fire. After no more than a second or two, she falls over on her side, her trunk flailing, her legs in wild seizure. She is dead within a minute. It is a dreadful few seconds of film—evidence of a cavalier attitude to cruelty, which when added to Edison's reputation for plundering others' ideas and for general irascibility, left an indelible stain on his reputation. That he was the first man to connect America by electric wire tends to get forgotten, despite the best efforts of the museum that stands today in Menlo Park, in the shadow of his monster lightbulb on a stick.

One last contest, an event far less macabre than the public assassination of Topsy, proved to be the hinge of fate that decided the nature of the electrical system America would formally adopt before the transmission lines started uncoiling across the land. Companies offering the two competing systems were invited to bid to light the Columbian Exposition, the yearlong Chicago World's Fair dedicated in 1892 to mark the four-hundredth anniversary of Christopher Columbus's arrival in the New World. It would be the first electrically lit exposition in history, and whoever illuminated Chicago would illuminate the nation.

Money, rather than the relative utility or safety of the two competing systems, turned out to be the deciding factor.

Edison had made an early pilgrimage to Chicago to sell his ideas for DC electrical supply to the exposition's chief planner, Daniel Burnham. He certainly convinced Burnham to install incandescent rather than arc lighting, and up to that point, the meeting was cordial. But once the matter of how best to supply electricity to the lamps came up and the corporations became involved, all pretense of courtesy fled. Edison was no longer fully able to speak for himself in these matters since J. P. Morgan had become financially involved and Edison Electric had become a different and much grander organization, the General Electric Company. Such was its swagger, even in those early days, that it had no problems suggesting to Burnham that his exposition could be properly illuminated by DC-powered lightbulbs for $1.8 million.

The backers of the Chicago show were indignant, suspecting they were being made fools of by tricky New York bankers. "Extortionate!" they choroused. They then welcomed the arrival of George Westinghouse, who asked to be considered and claimed he could install a more reliable and less costly system using AC. The organizers asked for new, sealed bids from the two companies. General Electric revised its opening bid downward to the more sensible figure of $554,000. But when the board members opened the envelope from Westinghouse, they read a quoted figure of $399,000. There was now no further question: the alternating-current system would be used to light the greatest world's fair in American history.

DC would continue to have its uses. Subway systems, New York's most prominently, still employ direct current, and very new technologies these days allow for ultra-high-voltage transmission lines to carry DC across long distances also. However, AC promptly became the dominant electrical system in the United States, as it has been ever since the end of the nineteenth century.

Lighting the Corn, Powering the Prairie

The first transmission line in the country was opened in 1889, giving Portland, Oregon, a constant 4,000 volts of AC from a generating station built beneath the Willamette Falls, fourteen miles away. With 4,000 volts, people found they could do more or less as they wished— they could light streets and houses, run tramways, take X-rays, pump water, cool milk, cut hair, show movies, weld pipes, cut stone. Power companies sprang up like weeds; in 1892, Chicago alone had twenty of them. The output of their generators and the capacities of the wires that carried their electricity kept increasing. By 1907 certain transmission lines carried a hitherto unheard-of 110,000 volts—the first of these lines opening between a hydroelectric generating station under a waterfall in Michigan and the modest city of Grand Rapids.

By the outbreak of World War I, more than fifty major electrical firms were making and distributing power, as were any number of smaller companies and municipally owned utilities. Come the 1920s and the number was more than six thousand, though most were controlled by much larger holding companies, of which there were about a hundred. Electricity, which everyone now wanted, was a means of making an almost limitless amount of money, and if the business was profitable, then big companies wanted as much of it as they could possibly acquire.

That, however, was the problem. Not every aspect of the electric business was profitable. It was a good moneymaker in the cities: it cost next to nothing, relatively speaking, to throw up a shed full of boilers and turbines, string wires out a few miles to customers' houses, and then send them a bill each month for the power they consumed. It was not much more difficult in the suburbs or in the smaller towns beyond: the generators might have to be a bit larger, the transformers a little more powerful; the transmission towers and the lines they supported might have to be measured in tens or even hundreds of miles rather than in yards or city blocks. It was not as profitable to supply such places with power, but it still made money.

However, there was no money at all to be made out in the farm country of the American Midwest. The pitiless arithmetic of capitalism did no favors for those who grew corn or alfalfa or soybeans out on the prairies or whose cattle grazed in the foothills of the Ozarks or up in the high country of Idaho. These people could be left safely disconnected from the electrical world. They could remain, though not of their own choosing, steadfastly outside the American dream. They could be reliant only on the wind and the sun and their own muscle and grit to give them the energy that they and their farmyards needed. The electric power was there, ready and waiting and straining at the leash, to give them relief and hope, but in the 1930s, the chiefs of the utility giants judged it as being too costly to bring to their doorsteps. So their hardscrabble lives were to remain that way for much longer than seemed the right of every other American.

In 1932 Franklin Delano Roosevelt, in running for the presidency, campaigned against what he saw as an inequity, attacking in particular "the Ishmaels and the Insulls" of the electrical industry, "whose hand is against every man's." His particular target was Samuel Insull, a Briton who had worked for Edison and had subsequently created a vast electrical monopoly from the Dakotas to Maine, a highly leveraged, precariously balanced corporate monster of a kind more familiar in modern times. Insull lived lavishly, spent freely, conducted his business recklessly, and was eventually charged with federal mail fraud (but later acquitted) and fled to Europe, dying penniless in a Paris Metro station and being buried ignominiously in Putney. Orson Welles said that his *Citizen Kane*, popularly supposed to have been based on the newspaper baron William Randolph Hearst, was in large part modeled on the equally megalomaniacal Samuel Insull.

Insull's electrical companies had an established policy of not doing business with the faraway farms. Almost none sent their wires out onto the windswept prairies of America's heartland, a fact that sorely tried Roosevelt and led him, once he was securely in the White House, to act swiftly to apply government right, as he saw it, to a monstrous capitalist wrong.

Everyone in the Thirties—and especially, from mid-decade onward, the readers of *Life* magazine, which took a keen interest in America's dispossessed—seemed to know that with the combination of Dust Bowl and Depression, the lot of the people of the prairies was grim indeed. Everyone also knew their lot could be mightily improved by the arrival of electricity. Only one farm in ten had access to mains power in 1934—a statistic that made a mockery of the claims that the telegraph, the telephone, and electricity were unifying the country. Electricity might be doing great things for Manhattan and Malibu, but for tiny settlements in Iowa and Nebraska and Kansas, the lack of electric power meant that America was divided still, into those on the grid with electricity and those who survived beyond it.

For those without, life could be a trial. The government published a catalog of the woes of the powerless, a lengthy book that began

> *Because there was no electricity, a farmer could not use an electric pump. He was forced not only to milk but to water his cows by hand, a chore that, in dry weather, meant hauling up endless buckets from a deep well. Because he could not use an electric auger, he had to feed his livestock by hand, pitch-forking heavy loads of hay up into the loft of his barn and then stomping on it to soften it enough so the cows could eat it.*
>
> *He had to prepare the feed by hand: because he could not use an electric grinder, he would get the corn kernels for his mules and horse by sticking ears of corn—hundreds of ears of corn—one by one into a corn sheller and cranking it for hours. Because he could not use electric motors, he had to unload cotton seed by hand, and then shovel it into*

the barn by hand; to saw wood by hand, by swinging an axe or riding one end of a ripsaw.

Because there was never enough daylight for all the jobs that had to be done, the farmer usually finished after sunset, ending the day as he had begun it, stumbling around the barn milking the cows in the dark, as farmers had done centuries before.

Washing, ironing, cooking, canning, shearing, helping with the plowing and the picking and the sowing, and, every day, carrying the water and wood, and because there was no electricity, having do so everything by hand by the same methods that had been employed by her mother and grandmother and great-great-great-grandmother before her. . . .

Because there was no electricity.

And then came FDR and his memorable promise of July 1932: "I pledge you—I pledge myself—to a New Deal for the American people." And one by one there then came from Washington the ingredients for the great alphabet soup of federal agencies, the mass of government bodies, known still by their initial letters, that were created in a fierce whirlwind of anti-Depression activity mounted by Roosevelt during his first hundred days in office. There were as many as a hundred such bodies, some of them so small and hidden from view as to escape government audit; some massive, with gigantic budgets that were rammed through Congress by presidential fiat (and later found to have been unconstitutional—except that by then they had done their job of helping lift America out of the Depression).

The bigger of the alphabet agencies ranged from the AAA, the Agricultural Adjustment Administration, a vital part of FDR's farm-relief program, down to the WPA, the Works Progress Administration. For eight years, the WPA provided gainful employment for millions of jobless Americans, undertaking all manner of public works. Writers wrote government-supported books, poets performed in government-backed slams, and artists and musicians were commissioned to beautify hitherto unadorned corners of federally administered property.

Buried deep within this thick catalog of big government is the one agency that amply settled the hash of such profiteers as Samuel Insull. Founded in 1935 as the REA, the Rural Electrification Administration, it became a godsend to the families of the prairies. It was the government body that lit up those parts of America that capitalism forgot.

The key player, long forgotten, was an engineer named Morris Llewellyn Cooke, who'd had a bee in his bonnet about electrifying America's farmland since the 1920s.

Just like his new president, he thought it wildly unfair that the very people who made the nation's food were so ill served, without access to the one utility that could make their lives easier and their work more efficient. Not only did the power companies refuse to connect faraway farms; they charged extortionate rates to anyone who dared to live in the countryside where lines did exist. Such imprudent inequity, Cooke declared, would have to stop. Roosevelt himself had long been particularly vexed; when he got his own first electric bill for his small cottage in Warm Springs, Georgia, he was irritated beyond measure to discover that the local power company was charging him four times the rate he paid in New York.

Cooke did his best to alert the president and his Cabinet as to what he had in mind to solve the Insull problem. He prepared a formal proposal, now renowned in bureaucratic history for being boldly bound in black-and-white zebra stripes and illustrated with attractive watercolors of bright red barns, with a note on the cover stating, "This report can be read in 12 minutes."

Cooke's flamboyance did the trick, as did his compelling argument. There were currently six million farms in America, Cooke wrote, and of them only 650,000— one in ten—was connected to electric power from the national grid. However, it should be possible to electrify almost all of these farms quickly and at almost no cost to the American taxpayer, providing that a network of cooperatives were set up, which farmers would pay to join. Each one would buy power from existing makers, and it, with government help, would build the lines to supply its farmer members.

The secretary of agriculture promptly read the document in the allotted twelve minutes and hired Cooke—"the big boy from Philadelphia," as he came to be known—to direct the program, then arranged for Congress to pass the necessary legislation to get the show on the road.

On May 11, 1935, under the authority of Roosevelt's historic Executive Order 7037, the REA was in business. It took offices in a Victorian mansion in central Washington that resembled nothing so much as the Bates Motel in *Psycho*. Six hundred administrators were hired. Orders went out to hire many hundreds of unemployed electricians, as well as thousands of junior technicians in overalls, and to buy scores of thousands of white-pine utility poles, millions of yards of wire, and tens of thousands of porcelain insulators. Agents then fanned out across the country, checkbooks to hand, ready to lend REA moneys to rural co-ops that were willing to build the lines, connect the farms, negotiate to buy the power from the generating companies, and then arrange to distribute it and sell it to the farmers.

The first such plan was put into place in November 1935 with the establishment of the Piqua Municipal Light plant in Miami County, in the flat countryside (and Eighth Congressional District) of western Ohio. The REA loaned this co-op a quarter of a million dollars to start building. Advertisements went up—with images of linemen, the countryside heroes of the New Deal, suspended and safety-belted high up on poles, working to connect the glittering wires. Serried ranks of these poles soon went up, marching away from the Piqua plant in all directions toward the distant Ohio horizons of farmhouses, barns, cattle, and stands of corn, where everything was still worked by calloused hand, where everything was dark, but for a few oil lamps, each time dusk had settled on the plains. "It's Coming!" the advertisements blared. "Electricity for You!"

And farm by farm, quarter section by quarter section, dirt road by dirt road, the co-op won over the customers. They paid $5 apiece. "Sign up and get the REA" was the slogan. The REA salesmen liked to have the farmer's wife around when trying to sell the co-op idea. They were instructed to look meaningfully at her as they explained the advantages, especially if her husband was busily examining the dirt under his nails and protesting that he didn't want to "owe nuthin' to the guvmint." Back in 1935, after all, $5 was no mean sum.

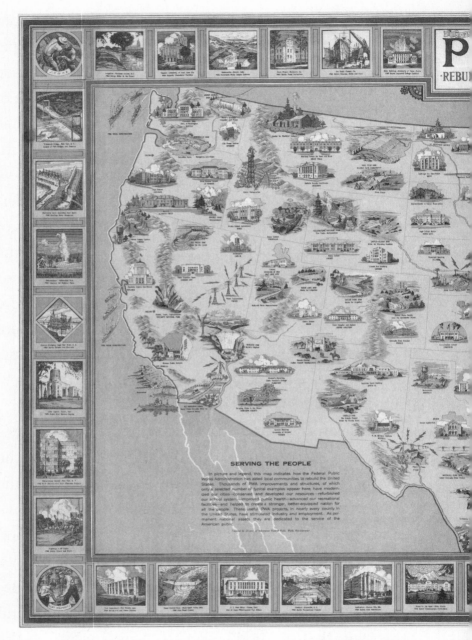

The Public Works Administration, part of FDR's New Deal, promoted its myriad projects with this 1935 map showing the variety of tasks to which it was officially bent.

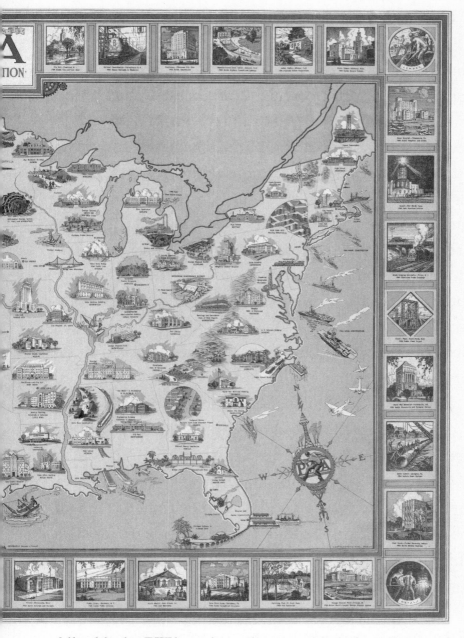

All told, the PWA undertook some thirty-four thousand projects, including dams, bridges, and reservoirs, which still dominate the American landscape today.

Once agreement came—and it usually did—the REA workers moved in, digging holes for tall poles, uncoiling spools of wire, raising it high, then tightening it and linking it together. And then the men of the connection teams came right into the farmhouse, to the bewilderment of all within, and put in the fuse box; climbed up on ladders to fix the ceiling sockets; put in the big rubberized cables for the 60-amp circuit for the kitchen range, then the smaller wires for the 20-amp circuit for the rest of the still-to-be-bought kitchen equipment, and the 15-amp circuits for the lights; then put in the switches, and the sockets on the walls for an occasional table lamp or maybe for that most important piece of equipment that soon would be an even more powerful national unifying force, the *radio*.

Finally, it was all done. The moment usually came at dusk. "I wanted to be at my parents' house when the electricity came," wrote one REA customer. "It was in 1940. We'd all go around flipping the switch, to make sure it hadn't come on yet. When they finally came on the lights just barely glowed. I remember my mother smiling. When they came on full, tears started to run down her cheeks."

The moment is still recalled by old-timers in some of the more distant parts of the country. "The night the lights came on" was as significant a moment as a birthday, often more so. The REA's celebratory volume, published half a century later, is filled with affecting reminiscences, such as this one from Kentucky:

We'd heard the government was going to lend us money to get lights, but we didn't believe it until we saw the men putting up the poles. Every day they came closer, and we realized it was really going to happen. So Dad went ahead, and had the house wired.

It was almost two months later before they finished the job and turned on the power. I'll never forget the day—it was late on a November afternoon, just before dark. All we had was wires hanging down from the ceiling in every room, with bare bulbs on the end. Dad turned on the one in the kitchen first, and he just stood there, holding the pull-chain. He said to me "Carl, come here and hang on to this so I can turn on the light in the sitting room."

I knew he didn't have to do that, and I told him to stop holding it, that it would stay on. He finally let go, and then looked kind of foolish.

It was like that all across the land.

By 1942, half of all American farms were connected to the grid; by 1952, they almost all were, less than a century after Thomas Edison threw that first switch on Christie Street in Raritan, New Jersey, in 1879. And though it still took some time to bring electric power from central generating stations deep into all the American Indian reservations, it can fairly be said now that all who want electric power in America today either have it by right or by right can get it.

It was swiftly accepted that electricity was a public good—a marketable public good, to be sure, a good with a price tag, but something that should be accessible to all as a hallmark of a civilized society. When Thomas Edison threw that first switch, neither he nor anyone else thought electricity would ever enjoy such a status, but now the country was fully connected, united in 110-volt, 60-cycle-per-second concert, city and countryside humming together as one.

The Talk of the Nation

It was a host of Europeans, not Americans, who invented the means of transmitting information invisibly through the air, not with smoke signals or semaphore flags or heliograph flashes but with electromagnetic signals sent wirelessly. Most of the pioneers' names—Maxwell, Hertz, Popov, even the majestically named Italian Temistocle Calzecchi Onesti—are now familiar only to specialists in the field. But Guglielmo Marconi, who cracked the final technical problems and can be called the true inventor of wireless telegraphy, is more properly memorialized. The precipitous cliffs overlooking the entrance to the harbor at Saint John's, Newfoundland, where Marconi first discerned the faint Morse signal for the letter *S* being tapped out by his assistants thousands of miles away across the ocean in Poldhu, Cornwall, have become a pilgrimage destination. To gaze out over the boiling Atlantic waves below and imagine the sound of a radio signal borne across them from the tip of the Old World . . . Signal Hill is a wildly romantic place, in a wildly romantic story.

But what these wireless signals carried were then only clicks, the dots and dashes of Morse code. It was not then possible to employ Marconi's wireless transmitters and receivers to carry the sound of the human voice. Not, that is, until Christmas Eve 1906, when from Brant Rock, Massachusetts, there came over the ether a cacophony of never-before-heard noises, all sounding (over the clutter of atmospheric static) quite unlike Morse code. They announced the arrival of a new star on the radio scene, a Canadian-born former chemistry professor and electrical engineer named

Reginald Aubrey Fessenden, the man who, some claim, made radio come alive. And while it was in Europe that the idea of radio germinated, it was in America that the technology was first employed to transmit the human voice and allow people to begin to talk to one another not just individually, as on Mr. Bell's telephone, but en masse, in their millions.

The basic principle behind the early wireless transmitters was simple enough. It is based on the discovery that each time an electrical spark is generated, it produces an invisible pulse of electromagnetic radiation. The more sparks generated each second, the higher the vibrational frequency of the pulse of transmitted radiation. Tune a piece of receiving apparatus (first made by a Frenchman named Édouard Branly and initially called a coherer) to the same frequency of the pulse that is being generated by the transmitter, and the pulse can be reproduced exactly as it was sent.

Moreover, this reproduction can be achieved—it can be seen on a meter or heard through an electrically driven signal-to-sound *transducer* in a speaker or in headphones—without the use of any wires and across a distance that varies according to the power of the transmission and the frequency of the pulse. A low-powered, high-frequency pulse will go far, and the signal can even be made to bounce off the ionosphere and cross continents. Modifying the signal by briefly opening and closing the circuit with a keyed instrument allows a coded signal to be sent and received by someone far away. This was what Marconi (or Tesla: there remains some controversy) discovered and perfected; the keyed instrument was essentially the same Morse tapper that was used to pass signals over the now old-fashioned telegraph wires. The wireless equipment, now having been tried and tested over the previous decade, was the kind of equipment that was being used by Reginald Aubrey Fessenden when he joined the staff of the United States Weather Bureau in 1900.

The bureau had set up a chain of wireless stations along the American East Coast, and from them it sent weather information to ships at sea. Fessenden's contract called for him to test and improve the system—he had been trained by Edison, so he knew his stuff—and to allow the bureau use of anything he might invent along the way. This he did, in spades. His years working for this American government agency produced a flurry of inventive energy that mightily accelerated the already fast-moving progress of radio technology.

Most notably, he experimented first by sending out a continuous, very-high-frequency radio wave from a rotating transmitter that produced thousands of sparks each second. He then connected an ordinary microphone, just as you might find in the mouthpiece of a telephone, to the apparatus to *modulate* this signal, to increase or decrease what might be called, if it were audible, its volume.

The critical aspect of Fessenden's idea was that the modulation, not the basic signal, contained the information that was to be sent. The high-frequency radio wave simply carried the information that was encoded in this modulation. (Because he was modulating the volume, or *amplitude* of the carrier wave, his system was known as amplitude modulation, or AM radio. The later development of frequency modulation led to the creation of FM radio.) All one did at the receiving end, using one of Branly's coherers, was to strip away or ignore the high-frequency radio signal, the carrier wave, and what remained was the information that was being sent out from the transmitter.

The truly magical aspect of this system was that a modulated signal need not be restricted to the dots and dashes of a code. It could be made to carry anything. It could carry music. It could transmit the sound of the wind or the rain or the crashing of waves—or the human voice. Anything that was sound, and had a measurable frequency, could be turned by a microphone into patterns that could modulate the amplitude of the carrier wave and ride on its back wirelessly out to anywhere.

The first time Reginald Fessenden—a big, bearded, bluff man, quite full of himself and relentlessly self-promoting—is said to have tried doing this was on December 23, 1900. He was at a weather station on Cobb Island, one of the outer banks of the lower Potomac estuary, in Virginia. He started the carrier wave going, attached a carbon microphone, and spoke into it, hoping to attract the attention of a colleague listening in a building about a mile away. "One—two—three—four. Is it snowing where you are, Mr. Thiessen? If it is, would you telegraph back to me?"

And Mr. Thiessen heard this clearly enough to understand it, and he did indeed telegraph back, though history remains silent about the snow. If this account is true, it was the first time that a human voice was ever transmitted wirelessly—by an employee of the United States government in a Weather Service experiment funded by the American taxpayer.

A much more famous experiment took place six years later, by which time Fessenden had left the government's employ (there had been a quarrel over the rights to his ceaseless blizzard of patents). This test was backed not by the taxpayer, but by a pair of wealthy Pittsburgh businessmen, men who were convinced that with his revolutionary technique of modulating a continuous radio wave, Fessenden was going to eventually outdo Marconi in making serious money out of wireless.

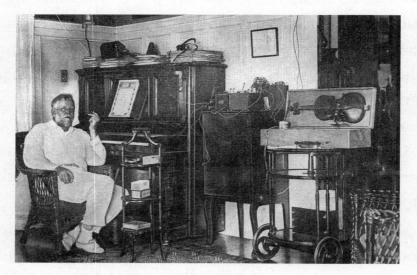

Reginald Fessenden, a Canadian-born engineer with the US Weather Bureau, is generally given credit for making the first broadcast of voice and music over the radio. Until Christmas Eve 1906, ships' radio operators heard only Morse code. Then, from his transmitter on Brant Rock, Massachusetts, Fessenden broadcast music by Handel and read from the Bible—changing the nature of broadcasting forever.

The pair of them financed the setting up of the National Electric Signalling Company, NESCO. They first built a huge antenna for Fessenden on a remote eastern Massachusetts headland known as Brant Rock, a former island now connected to dry land and so named because it lies beneath a flyway for migrating brant geese and because the wearier birds once liked briefly to settle there.

The headland—a place of great historic importance now, though seldom visited—is basically a low, wide mound covered with short springy turf and projecting into the Atlantic surf for some yards. Today it sports a few flyblown shacks; a century ago it was quite isolated from nearby houses and passersby. Its remoteness, as well as its proximity to the clever technical men of Boston, prompted the NESCO engineers to build their very tall experimental radio antenna there. It was slender, alarmingly fragile-looking, a confection of steel mounted on a layer-cake base of cement and insulating ceramics and rising 420 feet straight up. Sixteen guy wires held it steady against the Atlantic gales, and nearby were an engineer's shed and the buildings that housed the transmitting and receiving apparatus.

To make the experiments comprehensive, they also paid to build a second tower on the far side of the Atlantic, at the Scottish village of Machrihanish, in southern Argyll. Early in 1906, technicians reported success in using the Fessenden transmitter to send messages and receive replies, but with Morse code only. Two-way code transmissions had never been achieved before; for that alone, Fessenden would have made history. But he had rather more ambitious plans, for transmitting the kind of voice messages that he knew from his Cobb Island days could feasibly be sent. There were delays; technical problems to be sorted out; and repairs were needed after the Argyll antenna blew down in a gale. Moreover his two investors were getting cold feet and were starting to think seriously about the wisdom of their commitment.

Their skepticism was necessarily short-lived. In mid-December their man announced that he was ready at last to begin a voice transmission. Word went out, by Morse messages sent over the conventional telegraph, to a number of ships then sailing on the Atlantic—most notably to a flotilla of banana boats owned by the United Fruit Company, which had been equipped with NESCO receivers—that they should be ready to listen to the broadcast. Late on Christmas Eve, it was blowing a full gale, the classic dark and stormy night, which only adds to the skeptics' somewhat jaundiced view of the story—when Reginald Fessenden powered up his transmitters, stood before the microphone, and depressed the switch. As he recounted some years later:

> The program on Christmas Eve was as follows: first a short speech by me saying what we were going to do, then some phonograph music. The music on the phonograph being Handel's "Largo." Then came a violin solo by me, being a composition of Gounod called "O, Holy Night," and ending up with the words "Adore and be still" of which I sang one verse, in addition to playing on the violin, though the singing of course was not very good.
>
> Then came the Bible text, "Glory to God in the highest and on earth peace to men of good will," and finally we wound up by wishing them a Merry Christmas. . . .
>
> We got word of reception of the Christmas Eve program as far down as Norfolk, Va., and on the New Year's Eve program we got word from some places down in the West Indies.

The thought of the aria "Ombra mai fu," or "Shade There Never Was," being heard for the first time far out at sea, the idea of so resolutely landlocked a conceit as a Persian king's contemplation of the cool shade under a plane tree being heard by seamen riding out a storm on the Atlantic Ocean . . . Whatever the timescale, whether this was the first or the second time (as some skeptics still claim) that voice broadcast had been demonstrated, whether or not this experiment was in the running at all, the very idea behind the Fessenden story has a magic about it that now positively urges us to believe it.

Yes, the story may well not be entirely true. It is most decidedly odd, for instance, that no ship's log that survives mentions a broadcast that would surely have set most navigation bridges a-hopping with excitement, for not once in the past would anyone aboard have ever heard a sound coming through the wireless receiver, other than that of a Morse tapper. And yet here suddenly there was a human voice and sweet music coming to them out of the thin air! To most it was quite incredible and surely worth mentioning in at least one shipmaster's logbook.

So maybe Fessenden is out by a year or so; surely no matter, really. What *does* matter is that sometime around 1906, inventions made by this great bear of a man made such things possible.

It was an achievement that did little for Reginald Fessenden. His wealthy partners fought with him and sacked him from his own company. He spent fifteen miserable and costly years dragging them through the courts, without satisfactory results. Though eventually he got a cluster of gold medals for his work and was said by many great men and biographers to have been radio's greatest pioneer, he became embittered, turned his back on his invention, wrote an obscure and little-noticed book on antediluvian civilizations, and died in Bermuda in 1932, largely unsung.

His invention, however, went from strength to strength, and at an appropriate lightning speed. Initially the listeners—or perhaps *hearers* would be the better term, for what was received was in the early days the result of random and unplanned transmissions—were rank amateurs, "hams," who saw the airwaves as a magical democratic free-for-all and began exchanging all kinds of information, usually in code, very occasionally by voice, from all kinds of sources to all manner of places, about all kinds of topics, commonplace and arcane.

A small number had the patience, skill, and wherewithal to broadcast voice and music. Charles Herrold, a professor (in the Herrold College of Wireless and Engineering, which he founded in 1909) built a transmitter on the third floor of a bank building in San Jose and began a regular Wednesday-night program of songs and chatter, which he called the *Little*

Hams Program. He had a wind-up Victor phonograph (complete with the little dog Nipper listening to "His Master's Voice"), and he borrowed 78-rpm records from Sherman Clay, the local music store. Positioning himself carefully in front of the microphone, which, because it was wired into the same electrical circuit that caused the transmitting sparks, would get searingly hot, "Doc" Herrold would introduce the music—Sousa marches a specialty—and then read out columns from the local papers. And when he played music, he would tell his audience where he had bought it. The lines of customers outside the Sherman Clay store the following morning testified to yet another new and hitherto unimagined phenomenon: broadcast advertising.

At first maybe fifty local ham operators listened in to his 9:00 p.m. hour-long show. But then in 1913, Herrold invited his young wife, Sybil, to be an announcer. That ploy, together with his adroit plan to stage contests and offer as prizes small galena crystals—which were known to be tunable to specific radio frequencies and so could be used to make the new "crystal set" receivers—led his listenership to mount steadily, until soon several hundred were tuning in. Most of his listeners were on the Peninsula; but on some atmospherically favored listeners as far away as Oregon and New Mexico were able to hear the music, interspersed with

Sybil's sweet made-for-radio voice, swooping and soaring above the crackle of static.

Fan letters started arriving at the bank. Listeners would telephone to ask Sybil for their favorite tunes. Doc Herrold's show, which ran for five years, unregistered, informal, but regular and reliable, was the first properly broadcast radio show in America and possibly the first anywhere.

For a long while, however, code remained king. Young men in attics and garages from Maine to California, with their trinkets of galena and other tunable crystals, built crude radio receiving sets—and the bolder and cleverer ones made transmitters, too—and spent countless nights scanning the ether for streams of code, much as radio telescopes do today, hunting for signs of deep-space life. The tradition of ham radio was born, with scores, then hundreds of people experimenting, a brotherhood of radioheads chattering across the nighttime miles, their encrypted letters and numbers rattling through space by the millions, the invisible pixels of a huge unseen national portrait.

The US Navy, which by now was also enthusiastically using Morse and radio telegraphy to communicate with its warships at sea, frowned on this sudden burst of landlubberly interference. Admirals protested, saying that safety at sea was threatened when urgent maritime traffic was made unintelligible by the constant racket of amateur gossip. Civilian enthusiasts countered the sailors by arguing that the newly discovered airwaves were a realm quite beyond the control or jurisdiction of the state. It was a realm that should be democratized, allowed to be employed at will by the people.

The result of the inevitable standoff was inevitable too—regulation. Urged on by powerful naval lobbying, Congress passed the 1912 Radio Act, which suddenly required anyone operating a transmitter to have a license. And while the government could not refuse anyone who wanted one, it could from now on set time limits on transmissions, and it could order amateurs to cluster in certain corners of the frequency spectrum, while reserving the better ones—frequencies less troubled by whether it was night or day, for instance—for the armed forces.

Then came World War I, and once America had joined battle, the entire world of amateur radio was ordered shut down. Doc Herrold and his small number of followers were told to cease and desist. Attics and garages were raided by police, who seized and destroyed the equipment of anyone whose signals were still bleating out into the night after the order had gone out. For more than a year, the American airwaves fell silent.

Only the chatter of naval vessels and weather stations could be heard by those still listening in, hoping for nighttime music.

But the moment the war was over, an enormous wave of enthusiasm swiftly gathered, and applications for new broadcast licenses poured by the hundreds into the Commerce Department. It was as though the country suddenly wanted to start talking to itself: magazines throughout the land began to hail the arrival of broadcasting—of the human voice, music, sounds of all kinds—as a new means of giving reality to the concept of nationhood. Radio was all of a sudden seen as a possible way to help further unify the country, to ensure that the stories of the Down Easter were made familiar to a listener in Nevada, that the accents and thoughts of a man from Alabama or a woman from Arkansas could be heard and appreciated by a listener with an entirely different manner of speaking up on the high plains of North Dakota or down among the sierras and the arroyos of New Mexico. Radio, one writer said, could give everyone into the fold; it was a democratizing influence that would give those isolated from the mainstream of American culture—"the farmers, the poor, the housebound, the uneducated"—a sense of communion. Radio was suddenly invested with an ideal—as a free, limitless site for a forum of ideas that could knit the nation together spiritually, intellectually, emotionally.

That is not, however, quite what happened.

Making Money from Air

Since the very notion of *broadcasting* has its linguistic roots in agriculture, in seed spreading, it is entirely proper that the oldest radio station in the nation is in Madison, Wisconsin, and that its first aerials were sited in middle of a nearby field of corn. The station, known at its beginnings in 1914 by the call sign 9XM, first transmitted weather forecasts to farmers in Morse code, mercifully switching over to voice broadcasts three years later.

The era of officially sanctioned radio then got itself properly under way in 1920, when George Westinghouse's company, the Radio Corporation of America, launched the station in Pittsburgh known by its all-letter call sign, KDKA. (9XM up in Madison was awarded an all-letter designation too, being reborn as WHA in 1922—coincidentally the same year that the word *broadcasting* entered the English language as a term of art for the new phenomenon.)

And with this launch of a service that then began to spread inexorably to every corner of the country, the guiding philosophy of American broadcasting began to make itself clear. It was championed by such decidedly non-messianic figures as Herbert Hoover, who as secretary of commerce in the late 1920s directed a series of radio conferences that laid down policy for the medium's regulation.

American radio, the conferees agreed, was to be based from its start almost entirely on the perceived need to employ it to make money. Commercial stations, financed by advertising, and so purposely aimed at winning the best audience by providing programming directed at the spending masses, were to be central. Educational radio stations, and indeed any broadcasting that was deemed to be in the dreamily impractical role of promoting the public interest, were consigned initially to unattractive corners of the wireless spectrum, and given little help and generally short shrift by all.

The American idea of broadcasting was born of the country's firm belief—during the prosperous, economically vibrant 1920s, when all of this was unfolding, along with mass production and a seemingly inextinguishable blaze of consumerism—in the purity and sanctity of capitalism and market forces. Britain, which by the 1920s had already been eclipsed by America as the world's leading economy, was in its approach to the medium still clinging to an almost mediaeval aristocratic model—a "mother knows best" approach, which has caused the BBC to be referred to, even today, as Auntie.

America would have none of this. The idea of radio as being some kind of rallying point for the preservation of democracy was, in the view of Hoover and his government, dangerous quasi-Bolshevik piffle. To the radio bosses at NBC, CBS, ABC, and all the other networks that soon sprang up, the American audience was not a manipulable, educable herd of like-minded serfs, but was composed of consumers whose individual motivations were different and based on self-interest. They needed radio for entertainment and amusement and for a means to participate in the nation's commercial and political life.

American commercial radio did not have any high-minded wish to become a way to bring the country together: it was all about making money. Any notion that the medium might offer what had originally seemed most likely—point-to-point communication between people or communities—was swept away by the realization that radio was best in offering point-to-mass communication, one broadcaster talking to millions. For thirty years, beginning in the 1920s, radio networks skippered by men like David Sarnoff, the Russian immigrant who was deeply involved in directing an empire that both manufactured radio sets (RCA) and then broadcast programs to them (NBC), entered a truly golden age.

Sarnoff had already suggested what he wanted: "I have in mind a plan of development," he wrote in a famous memo in 1920, "that would make radio a household utility in the same sense as a piano or phonograph. The idea is to bring music into the home by wireless." To show that he was as good as his word, he helped set up the NBC Orchestra and persuaded Arturo Toscanini to come over from Europe to be its principal conductor. Or maybe people would prefer swing, he then remarked. Whatever they wanted, basically "we're just the delivery boys."

He promised his customers he would deliver sport as well. Early on he staged a boxing match between Jack Dempsey and Georges Carpentier that had three hundred thousand listeners. Mass appeal was generated with extraordinary speed. In the 1930s twenty million people tuned in to listen each Sunday night to Walter Winchell, the opening of his acidic gossip show invariably the same: "Good evening, Mr. and Mrs. America, from border to border and coast to coast, and all the ships at sea. Let's go to press!" In 1938 the twenty-two-year old Orson Welles broadcast his radio version of H. G. Wells's *War of the Worlds*. His adroit dramatization of imagined news bulletins announcing an invasion of Martians in suburban New Jersey startled millions, prompted a formal apology, and jump-started his long career as a director and actor. Fred Allen, Jack Benny, *Amos 'n' Andy*, the *Champion Spark Plug Hour*, and the *Bell Telephone Hour* in the 1930s and '40s were listened to by a country that was in this one sense entirely united—and united by a form of mass entertainment that could transfix millions, all at the same time. By the end of the Second World War, more than 80 percent of the American public was listening in regularly, to shows that had vast transcontinental audiences.

But then some started to whisper that perhaps the pursuit of commerce was not the only reason for this now vastly significant new national medium of radio. By midcentury, as the dollars continued to flow into the stations and into the pockets of their owners, a number of figures made an eloquently impassioned case for radio, not merely as a machine for printing money but as a nation-unifying force for the public good. Among them was a broadcaster from the upper Midwest named William Siemering, who in the late 1960s would become the principal founding architect of what would become National Public Radio, NPR.

The image of a family gathered around the radio set remains an enduring testament to the briefly unifying power of early broadcasting. Before long, the vast range of choices available meant that the family scattered, each to listen in private, to march to a different drum.

The idea of public-service radio—radio for the public good, not financed by advertising—had been around almost from the beginning of broadcasting. Educational radio stations, as they were first called, had sprung up at many of the country's colleges and universities, especially at the so-called land-grant colleges, which had been established, mostly in the late nineteenth century, on land given for educational purposes by the federal government. In Madison WHA, or 9XM, had, indeed, been one of these stations.

After the Second World War, many of these stations banded together to try to form a cross-country radio network. Though initially their efforts were unsuccessful, they did win one victory through a moment of clever cunning.

In the late 1940s the commercial broadcast industry was mainly interested in promoting AM radio and the new medium of television. The steady growth of FM radio did not concern the industry greatly—and this lack of interest allowed the small number of educational stations to band together and successfully petition the government to reserve the left-hand side of the FM radio dial, frequencies 88.1 to 91.9 MHz, permanently for their brand of noncommercial radio alone. The rest of the FM spectrum could be as much of a free-for-all as the AM dial already was; but from 1945 onward any potential listener who wanted educational radio, whether in Seattle or Miami, Bangor or Los Angeles, knew just where on the FM dial to find it. The reservation of twenty channels of guaranteed spectrum space for public radio remains to this day.

The four hundred stations involved in the first cooperative effort then got themselves more or less organized into a lobbying group, arguing for rather more than the reservation of parts of the spectrum. They wanted recognition, and they wanted some kind of government-sponsored financial help. In the mid-1960s, reports and books like *The Hidden Medium* and *The Public Radio Study* began to be circulated among legislators—prompting the country to sit up and take notice of the revolutionary idea of a national network of radio stations that were designed solely for the common good.

The lobbying proved effective: in 1967, the Public Broadcasting Act was signed into law, creating a non-profit, extragovernmental entity, the Corporation for Public Broadcasting, which would raise money for the creation of public TV and radio networks. It was then that William Siemering's thinking started to be taken seriously.

Bill Siemering was from Wisconsin, and he seemed to have been in radio since birth. His family's house was almost literally in the shadow of the historic aerials from which WHA, at the land-grant University of Wisconsin, had for the previous four decades been broadcasting its deliberately uplifting, useful, locally beloved—and commercial-free—radio programs. He dived into radio as soon as he was able, working in the mid-1950s, part-time during term, full-time in the summers, as a board operator, an announcer, a newscaster. He came to believe profoundly in his station's mission.

By the early 1960s, he had left the small university town of Madison and had become a public radio station manager in the much larger and more racially diverse city of Buffalo, New York. While he was there—"the formative years for NPR" as he was to say later—he changed the station's direction, making it quite unashamedly a community resource available to everyone in town, and highly successful thereby. He also listened avidly to the broadcasts coming from across Lake Ontario from the government-supported Canadian Broadcasting Corporation stations in Toronto—since the CBC was known then, as today, for producing a vast slate of highly original programming.

In 1970, and basing his thinking very much on his eight-year experience in the cold of upstate New York, he famously wrote a mission statement that brazenly suggested that what had proved so good for Buffalo and across the border in Ontario would also be good for the nation. "National Public Radio," as the new system was almost casually named by a radio journalist colleague named Al Hulsen,

will serve the individual; it will promote personal growth; it will regard the individual differences

THE MEN WHO UNITED THE STATES · 577

among men with respect and joy rather than with
derision and hate; it will celebrate the human
experience as infinitely various rather than
vacuous and banal; it will encourage a sense of
active constructive participation, rather than
apathetic helplessness.

Congress had already anticipated the network's creation. It was now decreed that federal moneys would be earmarked for the new system, with broadcasts scheduled to commence in 1971. Intelligent radio would be given an officially blessed leg-up. Under the banner of Siemering's mission statement, with a radio newsmagazine program designed by him, NPR was formally incorporated in midwinter 1970 and broadcast the first edition of *All Things Considered* the following May. That first show began with a twenty-minute report on antigovernment protests in Washington over the Vietnam War, played an interview with Allen Ginsberg about the legality of drugs, ran a portrait of a young girl addicted to heroin, and reported on how a barbershop in Iowa was diversifying by offering to shave women's legs. NPR, it was abundantly clear, was going to be radio of a very different stripe.

Few would dispute that in the forty years since, it has gone on to be become a grand success. NPR is now an

established and essential part of the American broadcast continuum, admired by most who hear it, held in
high esteem both within America and beyond. It has
proved to be formidably successful in broadcasting
intelligent, nonpartisan information to an immense and
ever-growing daily national audience. There is what
one might term an NPR culture in the country: decisions are often made, conversations are often begun,
conferences often commenced, with a simple commonly heard phrase: *I heard it on* Morning Edition,
or more simply, *I heard it on NPR.* Despite reaching
fewer than one in ten of the American public, NPR
seems sometimes, in terms of its influence, to be just
about everywhere.

But can NPR be fairly said to have unified the nation?
Did it—indeed, does it today—help connect the people
of America in the way that the invention of the telegraph, the laying of the railroad tracks, or the making
the Interstate Highway System so unequivocally managed to do? Was that ever the intent of its creators? Was
NPR devised to be both a national bulletin board and a
social sounding board—or was it to be something with
rather more strength, an entity of great size and power
that could employ the metaphor of being a network to
help link the nation together by an invisible skein of
radio waves, and thus forge a bond quite as strong and

enduring as any railroad line, telegraph wire, or high-way vanishing over the horizon to the mountains?

To be sure, the country is all too often gathered suddenly together, as it was in grief at the moment of John Kennedy's killing in November 1963, or on learning of the terrible events of September 2001. Similarly, there can be no more unifying exuberance than was displayed with the broadcast news of the 1969 landing on the moon or the bicentennial in 1976. The part that radio played in alerting Americans to these events was key: people heard the news flash, they gathered around the sets, they remained transfixed, and for days after spoke of little else.

But what annealed the society into one was the *events*—not the medium that transmitted the news of them. On ordinary days American public radio some-how retreats, then subsides. It fades into the background to be ever present, always available, like the staff in a country house or the spigot of a hot-water supply, as a supplier of impeccably turned-out news and analysis, of entertainment of a higher order, of music well chosen and worthy. But it performs no other social function than this, on ordinary days. It does not on ordinary days allow the cranberry picker in Maine to feel any kind of national connection with the computer techni-cian in California. The life of an Iowa prairie farmer is scarcely connected by radio to the quotidian routines of the taxi driver in Manhattan. She is quite foreign to him, and he to her, and the presence of the radio set that each has at home or in the car does little to make their relationship otherwise. A scattering of programs—*This American Life, Story Corps*—attempts a kind of connectivity, but NPR's basic institutional structure, with member-station fiefdoms whose managers are necessarily occupied with their own parochial concerns, militates against that.

This is not quite how NPR saw itself back when it was founded. And it should be recalled that the late 1960s were a time of deep divisions in the country—over the Vietnam war, over race, over the power of youth and the voice of authority—and a time when any attempt to forge a sense of unity should have been blessed, and noble. Siemering was eager to give voice to all in those divisive times, "to celebrate differences," and by doing so to help bring the fractured community together. To construct the *unum* from the *pluribus*, by radio.

To help accomplish this he planned to have as much as a third of the material on the big nationally made shows come from the local stations, so that America could genuinely be heard talking to itself. Had his plan worked it might well have helped create some sense of national oneness via the medium, rather than wholly through the big events that the medium occasionally broadcast. But it was not to be. Back in 1971 few of the original hundred-odd member stations had the staff or the time to contribute stories to the network; most of the programs were eventually produced by an ever-growing corps of Washington staff—and in the process, some of the pan-American vision slipped; some of the magic vanished.

Bill Siemering was eventually dislodged from NPR. Internal politics and stated differences about vision did him in. While his fingertips were still grasping the window ledge he took to pleading to stay on at headquarters, if only to cut recording tape. But his colleagues declined the offer.

In the end he left the capital altogether and took a job far away, up in the great northern plains, managing yet another educational station on the border between North Dakota and Minnesota. He spent a happy five years there, with just a three-person staff, returning to radio basics—finding the news, editing it, recording it himself. Necessity and reality tempered his vision, somewhat. From his new windswept aerie he managed to look both inward—producing a series of radio essays on a scattering of small towns in North Dakota, telling Dakotans more about themselves than perhaps they had ever known—and also outward, setting himself the target of contributing at least one piece of High

Plains broadcasting each week down to NPR head-quarters, and so having a prairie story told regularly to all America. In this modest goal he succeeded—helping thereby to employ radio to knit Dakotans together, and to more firmly cement in place a part of the magical mosaic of America. In a modest way, in other words, he fulfilled the task with which public service radio around the world is entrusted, and in which he so believed.

And meanwhile the commercial airwaves of today are loud with the harsher sounds of hectoring and demagoguery. There are national conversations in process, true: the most popular of all, a conservative rabble-rouser broadcasting from Florida, preaches to as many as twenty million listeners a day and earns $50 million a year for doing so. But his is far from being a unifying conversation: it is broadly seen as divisive and unkindly, sharply separating radio audiences into radically opposed camps.

The radio broadcasting that Reginald Fessenden inaugurated from Brant Rock, Massachusetts, in 1906, has hardly lived up to its promise of creating a kind of continental togetherness. For thirty years, it did: between 1920 and 1950 the cosy image of the walnut-veneered radio in the living room, the family gathered about, all listening in Minneapolis to the same program that was going out to listeners in Manhattan and Montezuma,

Iowa, had a solid truth to it. But since, the dream has sputtered, changed, and faded. Television altered much, of course. But in other countries with a strong public broadcasting tradition—Canada, Australia, New Zealand, Britain—the idea of radio as a unifying force still obtains. Public radio in America is undeniably good radio; but aside from those events of great moment that in and of themselves unify the nation, it devotes itself largely to the service of its hundreds of local audiences, each of them typically isolated from the others. And the commercial radio that did once help bind the country now often divides it, and brutally. Reginald Fessenden and his like would probably not wish to be listening.

Television: The Irresistible Force

In the summer of 1963, when I was still a student, I was hitchhiking north of Los Angeles and was given a ride by a helpful middle-aged man who worked as a technician at NBC television studios in Burbank. Like so many of those who gave me rides, this man—his name long forgotten, I'm ashamed to say—was exceptionally well connected. He knew studio chiefs everywhere and happily admitted me to a closed set at the Paramount studios in Hollywood, where people were shooting scenes for a film to be called *Seven Days in May.*

The director, John Frankenheimer, took kindly to me also and let me sit briefly in his director's chair, which was just as I imagined a director's chair to be. He also introduced me to Kirk Douglas, who was exceptionally warm, told me amusing stories about various Britons with whom he had worked, and signed the brim of the straw hat I had bought in Mexico.

All of which was fine and exciting and would become the stuff of a wealth of stories when I got back to Oxford. But the real deal, as they said in those days, was another promise that my technician friend had made to me: that he would get me onto the set in Burbank where NBC technicians were taping episodes of *The Tonight Show* and that if I was exceptionally lucky and well behaved, I might get to meet the now legendary new young star who was just then briefly hosting the show there, Johnny Carson.

All of which duly came to pass. I met, shook hands with, was bought coffee by, and duly fell under the spell of Mr. Carson. I cannot say I gave any thought to the possibility that he and his late-evening show, screened in bedrooms throughout the land, would tuck millions of Americans into bed for the coming thirty years or that he would become one of the most popular mass entertainers the country would ever know.

THE MEN WHO UNITED THE STATES · 585

None of these things would have even interested me at the time. I was more interested in the technical aspects of the production, in watching in detail the workings of a full-blown American television operation. And what remains in my mind was driving each night with my new friend and two big metal drums in the backseat of the car. We would place these drums securely on an overnight plane bound for Idlewild Airport, New York. They held precious, freshly recorded tapes of *The Tonight Show* and would be handed off to the technicians at WNBC's headquarters in Rockefeller Center and played over the air the next day for the stations on the American East Coast.

The three-hour time difference across the continent had much to do with it. When Carson opted to perform his show in California, and went live on the air in Burbank at 11:30 p.m. Pacific Time, it was 2:30 a.m. the next day in New York, a time when even his most ardent fans were likely to be a-slumber. NBC had at the time the technical capacity to hook up stations for coast-to-coast live broadcasts, but if everyone back east was asleep, there was clearly no reason to. What the network did not have, however, was the ability to send the show down the line and have stations record it for broadcast the following day. Hence the visit to the airport; hence the concern that the tapes were treated royally, being worth millions in potential East Coast revenue.

In its youth, television proved as unifying a feature of postwar society as radio had been in the 1930s. Johnny Carson, a late-night television host, can fairly be said to have sent the nation to sleep each night, though not before presiding over a million conceptions that he never witnessed. By the 1990s, cable TV and other later splinterings of the TV networks' command of the airwaves had diminished the role of such figures.

Our nightly airport excursions were reminders that television's true transcontinental network was still not quite finished. Its great potential as a cultural unifier was still a work in progress. Yet perhaps no other device had progressed from invention to near-omnipresent acceptance with such speed.

The first public demonstration took place on Thursday, April 7, 1927, in Manhattan, at the Bell Telephone laboratories on Bethune Street in what is now Tribeca. It featured the amiable moon face of the secretary of commerce, soon to be president, Herbert Hoover. He was speaking from his offices in Washington. Reporters from the *New York Times* watched him, enthralled, though not a little skeptical.

FAR-OFF SPEAKERS SEEN AS WELL AS HEARD HERE IN A TEST OF TELEVISION, read the headline on Friday's page 1, above six decks of subheads: "Like a Photo Come to Life; Hoover's Face Plainly Imaged as He Speaks in Washington; The First Time in History, Pictures Are Flashed by Wire and Radio Synchronizing with Speaker's Voice; Commercial Use in Doubt, but AT&T Head Sees a New Step in Conquest of Nature After Years of Research."

With a screen measuring three inches by two, the likeness was excellent: "It was as if a photograph had suddenly come to life, had begun to talk, smile, nod its head and smile, look this way and that." On a larger screen, the image was not so good, the secretary's face difficult to distinguish. But his voice was clear, and he declaimed, appropriately, "Today we have, in a sense, the transmission of sight for the first time in the world's history. Human genius has now destroyed the impediment of distance in a new respect, and in a manner hitherto unknown."

As a slightly ominous hint of the likely caliber of entertainment, Hoover's formal broadcast was then followed by a link from a studio in New Jersey, featuring a vaudeville act by "A. Dolan, a comedian, as a stage Irishman with side whiskers and a broken pipe [who] did a monologue in brogue . . . then made a quick change and came back in blackface with a new line of quips in negro dialect."

The *Times* reporter did not indicate any particular hostility to the nature of this performance; he did, however, go on to write that the commercial future of television, "*if it has one*," was thought to be "largely in public entertainment—super-news reels flashed before audiences at the moment of occurrence, together with dramatic and musical acts shot on the ether waves in sound and picture at the instant they are taking place at the studio." It would actually be seven years later, in 1934, that the word *live* was first used to describe this phenomenon, which would become a high-cost hallmark of much early television.

What the *Times* had seen was a demonstration of "mechanical television," which featured rotating scanners and offered low-resolution images and in the hurly-burly of the time was invented first. In tandem, and so in competition, others were developing so-called electronic television, which was the brainchild of a remarkable inventor named Philo Farnsworth. The son of devout Mormon farmers, raised on a remote homestead in rural Utah, Farnsworth experimented for years on the use of cathode-ray tubes for making and receiving electronic images, and a TV system based on this technology eventually won the day. The tussle between the two systems was on occasion difficult and divisive, recalling the struggle between Edison and Westinghouse, between DC and AC, in the field of power generation. To this day, feelings remain strong among supporters of the two camps, one side complaining that the other stole this or that.

But the simple fact is that technicians worked tirelessly over the ensuing years, swiftly erasing any lingering public skepticism over the coming of television and confirming the general accuracy of the *Times*' forecast. In 1939 David Sarnoff, who fifteen years before had so adroitly recognized the value of radio as a point-to-mass broadcasting medium, recognized in television exactly the same thing. He had at his fingertips (thanks in part to his own laboratories at RCA, which had developed and improved cathode-ray-based receivers) a technically suitable mechanism for performing the same broadcasting task as radio, only with moving pictures, too.

His company already owned station WEAF as the New York City flagship for NBC Radio. Now he inaugurated the cumbersomely styled W2XBS, with an antenna at the top of the newly built Empire State Building, for the sole purpose of transmitting television. Sarnoff himself, never backward at coming forward, decided that he would go on air first, ostensibly to show off the new RCA pavilion at the World's Fair across the East River in Queens. "It is with a feeling of humbleness," he said, "that I come to this moment of announcing the birth of a new art so important in its implications that it is bound to affect all society."

The signal carrying his speech from the Avenue of Progress went out to only about two hundred sets and barely more than a thousand viewers. But it was quite evidently the start of something unimaginably big. World War II interrupted developments, though some kind of a network was begun during the conflict, with connecting lines opened to stations in Pennsylvania and Schenectady. Once peace had returned and the country was flush with returned men and released cash, the postwar commercial boom got into full swing, and television began its remarkable takeoff into the stratosphere.

Its staggering potential was first publicly recognized by a former newspaper reporter, Wayne Coy, whom President Truman had appointed to run the Federal Communications Commission. In 1948 he made a famously prescient declaration: "Make no mistake about it: television is here to stay. It is a new force unloosed in the land. I believe it is an irresistible force." It seemed at the time that Mr. Coy might be sticking his neck out. The television industry was looking far from robust: there were just 172,000 TV receivers in the entire country and only 28 broadcasting stations, compared with the 1,600 that transmitted radio.

But then, right on cue, the price of sets went down dramatically. In 1950 a Philco receiver with a twelve-inch screen had cost $499, but in 1955, an Admiral with a twenty-one-inch screen cost $149. It suddenly became a man's solemn duty to acquire a television and help his country's postwar economy. To do so was also an affirmation of family values, for everyone would gather in front of it and laugh and cry *together*. Suddenly, buoyed by the beginnings of well-researched cleverness of the Madison Avenue advertisers, every

American with space in a living room was demanding a set. Sales rocketed.

In 1952, eleven years before my moment with Johnny Carson, there were fifteen million televisions in America. A year later, exactly ten years before my time in Burbank, twenty-four million. By 1955, there were thirty-two million. When Carson began his career as the King of Late Night in 1962, well over 90 percent of the seventy-two million American households owned at least one television.

But while television sets in those years were cheap and became nearly ubiquitous, the making of television shows remained costly, far more costly than producing radio shows. Radio required only a quiet room, a microphone, and a transmitter, but television required studios, sets, cameras, film-processing equipment, and all manner of expensive technology and the personnel to handle and repair it. It was easy and economically feasible for many to make radio shows; but the price of producing TV allowed the industry swiftly to be dominated by a very few organizations with very deep pockets.

The consequence was profound. With so few masters and with the choices they offered their viewers at any one moment of broadcast time so necessarily circumscribed, almost everyone in America with a television was being offered exactly the same stream of warmly comforting entertainment. Television offered a sudden unification of culture such as had never been seen or imagined before in the country's history—but a kind of unification quite different from what had gone before.

Roads, railroads, telegraph systems, canals—these were creations that had allowed Americans to merge and mingle with one another on an individual basis—to become physically unified with ever-increasing ease and speed. The telegraph and telephone had similarly permitted connection via conversation. In theory the technologies behind radio and television might also allow the same thing. There is no technical reason why both kinds of devices could not be employed to let people talk to one another and see one another while doing so.

But in practice the marketplace allowed almost none of this. Both kinds of electronic media were employed to permit the dissemination of a unified mass culture. Although for about three glorious twentieth-century decades, commercial radio had played a seminally important role in doing so, everything switched across to the more exciting technology once television got properly under way. Television's role in unifying mass American culture, then in time encouraging that same culture to seep out under the doorway into the rest of the world, is now well-nigh impossible to exaggerate.

Within a decade of that first David Sarnoff broadcast, television watching had become the favorite pastime of almost half of the American people. It was a leisure activity that had ripple effects few can have imagined. Water companies had to prepare for sudden increases in consumption during commercial breaks; electrical companies, for sudden surges in demand when shows ended. Furniture companies created new kinds of chairs and sofas to help make the endless hours of watching more comfortable. The food industry came out in 1954 with premade, preplated, artificially preserved meals designed to be eaten while watching. Lexicographers noted the arrival of new-minted phrases, *boob tube* in 1963, *couch potato* in 1976, and a score of others, many migrating swiftly from slang to the mainstream. Political parties now had to groom candidates for appearances, for debates, for sudden twists and turns in policies demanded by a public that was watching all the time. Some benefited—John Kennedy, for one. Others—Richard Nixon, Joseph McCarthy—did not.

The shows that helped create this almighty mass culture have become part of the American language: *Gunsmoke, Perry Mason, The Ed Sullivan Show, Dragnet, Dallas, The George Burns and Gracie Allen*

THE MEN WHO UNITED THE STATES · 595

Show. The weekly lunacies of *I Love Lucy* transfixed the nation throughout the 1950s: forty-four million people, almost 70 percent of the TV-owning public, tuned in on January 19, 1953, to watch while Lucille Ball gave birth, in real life, to her first baby. Half as many people tuned in the following day to watch President Eisenhower's inauguration ceremony. Seventy million watched the finale of a Korean War medical comedy show, *M.A.S.H.*

In the years that followed, sporting events, political debates, moon landings, assassinations, attacks—all these were consumed by an ever-fascinated public principally via television, the griefs and triumphs shared, the conversation united. Millions got their news from men who would become trusted, avuncular figures—Edward R. Murrow, Chet Huntley, David Brinkley, Walter Cronkite—whose timbre and manner suggested an almost godlike authority. Governments feared their power; the loss of Cronkite's support for the Vietnam War was an important factor in President Nixon's decision to wind the conflict down.

But then, after just thirty years of domination by the major television networks, a domination that helped create a cultural and psychological unification of the states, there came something new: cable television. And with cable there came further changes, which began to erode TV's brief role as the electronic welder-in-chief of the country.

The technology behind cable TV is simple enough: rather than having the TV signals sent through the air by local stations affiliated with the networks that supplied most of their programs, the signals could be relayed directly to household by wire, without the need for a station in between.

In the early days, the late 1940s, these wires were generally put in for reasons of necessity. Out in the mountainous regions of the country, for example, many could simply not receive line-of-sight broadcast signals. A Pennsylvania appliance store owner and lineman named John Walson, in the Appalachian hill town of Mahanoy City, is generally credited with founding the industry in 1948. He had offered to connect those customers who had bought televisions from him but received only poor signals to an antenna he had built on a local hillside. He would connect them with a cable and through it would bring them three channels of programming for $2 a month.

Walson used a specially made high-volume coaxial cable that could carry half a dozen channels at the same time. In time the capacity of such cables increased tenfold, and by the 1970s, when cable television became a realistic commercial prospect, customers could have many scores of channels beamed directly into their televisions. Today coaxial metal cables have been replaced by fiber-optic cables. Now hundreds and potentially thousands of channels can be carried, and viewers—now usually city and suburb dwellers who buy cable by choice rather than necessity—can choose from a nearly limitless menu of entertainment, education, news, and commentary.

The result has been to dilute and disperse the unifying potential power of television even more dramatically than that of public radio. Radio had become diluted because so many geographically separated public radio stations were established, hundreds of small fiefdoms that for one reason or another proved unwilling or unable to talk to one another. Cable television, on the other hand, created no stations at all; it diluted the overall effect of the medium by virtue of the vast spectrum of choices it offered to the viewer.

The huge TV-watching population, encouraged by advertisers using clever demographically based or interest-based algorithms, swiftly began to split itself into subgroups. Some were based on demographics, with all youngsters watching this kind of television, older white people watching another kind, and young African American women yet another; others were based on interests, with those liking golf, fashion, numismatics, basketball, British comedy, and erotica tuning in to specialty channels. Cable television allowed all of these bodies of humanity to enter their own personal echo chambers, to retreat into cultural laagers and become cut off from the mainstream broadcasts for which television had been originally intended.

The consequence was immediate, dramatic, generational, and continuing, remaining a work in progress. In the 1950s and '60s, American popular culture had been unwittingly homogenized by an almost monotheistic devotion to network television. It generally ignored such criticism as that most notoriously made by Newton Minow, another chairman of the FCC, who in the 1960s derided television as a "vast wasteland." He made himself an archenemy of the networks by condemning their output as little more than "a procession of game shows, formula comedies about totally unbelievable families, blood and thunder, mayhem, violence, sadism, murder, western badmen, western good men, private eyes, gangsters, more violence, and cartoons. And endlessly commercials—many screaming, cajoling, and offending. And most of all, boredom."

Today, with cable television attracting more than half of the American population, the monoculture reared on such entertainment—for it is difficult to suppose that a modern Newton Minow would find much changed—has started to evaporate. The cultural firmament has fractured, each splinter attracting its own cohort of viewers, each wedded by fiber-optic cable or satellite transponder to the unseen broadcaster, viewers separate from one another and each group separated from the others also.

The fracturing of taste has changed the family, too. What was once an indissoluble unit gathered around the flickering blue television screen now is spoiled for choice and spoiled by choice. One parent might be watching one sport on one cable channel while another views a gentler pastime on a second set and the children are each shut away behind closed doors, watching yet other channels or, more likely, a transmission of cultural omniscience that is available by way of electronics' most recent and indisputably most profoundly important newcomer: the Internet.

The All of Some Knowledge

I was in a sheep station in a remote corner of northern Australia when I first realized the Internet's extraordinary potential. It was the mid-1990s. I was living in Hong Kong, owned a cell phone the size of my hand,

and worked on an Apple computer the size of a small filing cabinet. I had been using e-mail for the previous six years.

The Internet of the time was slow, hesitant. Enthusiasts, of which I was unashamedly one, spent hours trawling through the booths at the somewhat shady Golden Supermarket in a tenement in Wanchai to buy gadgets and listen to suggestions for making their connection marginally faster. But the basic fact was: an Internet existed, and even in those slow days, most who used it were in awe.

But on Stockholm Station, in western Queensland, people didn't have it, had never heard of it. In the farm office, there was a mechanical adding machine, handle and all, and the farmer looked mystified at the sight of my portable—or in those days, luggable—computer. The farmer's son, a seven-year-old sheep-herding Land Rover–driving boy much older than his years and named Rupert, said he had heard his by-radio School of the Air teacher, one Mrs. Bishop, describe a computer, but he had never seen one. He was in consequence more enthralled than mystified by mine, as was his pet lamb, named Gidgee, who always seemed to be gamboling cheerily along at Rupert's side. He asked me for a demonstration of what it could do. He sat down beside me, and Gidgee, obligingly charming, jumped up onto his lap and nestled there, watching, too.

First I showed him some e-mails. Though mundane now, they were quite extraordinary then, especially in the remote red heart of outback Australia. This first was a letter from Hong Kong, I explained; the next, from France; two more, from New York. Then I typed a note for a friend in San Francisco, mentioning that I had a lad named Rupert sitting beside me. "Hi, Rupert," she returned, seconds later. The boy stirred uneasily in his seat.

But now, I said: the Internet. He had no idea what it was, so I explained as best I could. Anything he wanted to know about, *anything*, I could show him with a click, in pictures, sound, text, film. So employing the Microsoft slogan of the moment, I asked him: "Where do you want to go today?"

He thought for a second. Then . . . could I perhaps show him a picture; he hesitated, presumably not wishing to sound foolish . . . a B-1 bomber? He had always been fascinated, ever since reading about them in a magazine.

I knew there had been some test aircraft at Edwards Air Force Base in California, logged on to the website, and in moments I had film clips of a B-1B soaring into the skies over the Arizona desert. Rupert's eyes were as big as dinner plates. Then he fell rather quiet, looked away for a second, then said in a nervous whisper: "Could your computer possibly . . . take me to Mars?"

It was an obsession, he confessed. He had always tried to see it in the clear nighttime skies. His dad's binoculars were helpful. But if it were possible to see a close-up . . . ?

I clicked over to the Jet Propulsion Laboratory in Pasadena, and by happy chance there was a satellite, the Sojourner, flying over the Martian surface. Every ten minutes or so, it was sending a picture, which JPL was posting on its website. Slowly, line by line, an image of the red desert landscape revealed itself. Craters, canal-like canyons, mountaintops—all were being seen for the first time here in Australia just as they were in California. Then I heard a stirring beside me.

Rupert was astonished, his eyes even wider than before. He didn't seem to know what to do—until he then placed his hands gently across Gidgee's wooly face and pointed the creature's eyes directly at the screen. He lowered his head and, desperate to share the moment with his best friend, said solemnly to the little lamb, "Look Gidgee! It's Mars!"

I left my computer behind in Australia for Rupert, who soon began sending e-mails to me, and I bought another machine the moment I arrived back home.

The new one was faster, smaller, *lighter*, and cheaper. Access to the Internet was getting easier. Converts were accumulating all the time. The addition of that one small child convinced me that the effect of connection to this parallel universe of knowledge was of seminal importance. There was no stopping it now.

And the Internet revolution has not stopped its acceleration, not for one microsecond in all the years since. Whether for ultimate good or not remains to be seen, but as a phenomenon, it seems as permanent as the pyramids, only a very great deal younger.

The dates of its early stirrings and the names of those who can lay some claim to having created it present a marvelous confusion. Almost all of its origins are American, and a great deal of them involve—as with the expeditions, the surveys, the roads, the railways, the waterways, the telegraph, and a score of other unifying events and entities—the United States government.

In this case, the government agency involved was the Defense Department. Its interest was piqued by its perceived need, following the launch of the Soviet satellites in the late 1950s, to counter a new high-technology threat. It was deemed suddenly important to the generals and admirals in Washington that their family of three huge mainframe computers at the Pentagon, at the headquarters of the Strategic Air Command in

Nebraska, and at the underground headquarters of the North American Air Defense Command in Colorado should be linked together, able to relay information about war readiness and threat assessment in real time. An agency known as the Defense Advanced Research Projects Agency was set up to investigate the possibilities of doing this. An MIT professor named Joseph Licklider, who specialized in examining how the brain converts air vibrations into the perception of sounds—psychoacoustics—was chosen to lead the team.

It can be convenient to regard the Internet that then developed as an entity that rests on three coequal pillars. There is first the physical Internet, or the hardware Internet, with its spiderwebs of fiber-optic cables and nests of routers and server farms, with its secret nodes and mirror sites and Internet exchanges. Though it is so complex that in current form it can only be a creation of many, "Lick" Licklider's contribution to its making was seminal.

For it was Licklider who in 1969 came up with core ideas behind what was to be called the ARPANET. This was the original network of defense computers, Virginia to Nebraska to Colorado and back to Virginia, a connection made back when computers were huge glass-windowed, air-cooled rooms filled with ranks of man-size towers, each one topped with whirling drums of magnetic tape that spun this way and that like the infernal clock contraptions in *Metropolis*. The ARPANET connecting these machines—devised by Larry Roberts, another half-forgotten Internet pioneer—supposedly allowed for the better defending of America and in theory permitted the nation's atomic weapons to be launched much more quickly than those owned by the Soviet Strategic Rocket Forces.

Happily that particular need never arose, and once the threat had subsided, the technology started to seep into the commercial world and inexorably and inevitably led to a civil version of ARPANET. This peaceable version eventually linked not three single room-size machines, but millions upon millions of computers and their hundreds of thousands of networks across the globe. Over the next two decades, it evolved into what was rolled out first throughout the United States in March 1990 as the modern Internet.

But the hardware is only the hardware. The computers also needed to decide, or be taught, how best to talk with one another, electronically. The way in which they do so is byzantine; the conversation that hums silently around the world today owes its existence to legions of linguists and lexicographers, if you will, who helped create it. One system eventually turned out to be particularly appropriate—a list of subatomic courtesies that let two strange machines get to know each other, along with a list of microscopic interpreters to decode the linguistic codes that are peculiar to one machine and can make them intelligible to another. The system came to be known as TCP/IP, Transmission Control Protocol and Internet Protocol. For the Internet to function properly, both are employed simultaneously, conjoined as the Internet Protocol Suite. Two Americans are most commonly associated with creating these proprieties of computer conversation, jointly becoming the Miss Manners, if you like, of the cyberworld.

Joseph Licklider, Vint Cerf, and Robert Kahn can fairly be said to have conceived and invented the basic structure of the modern Internet—with a memo from Licklider in 1963 first suggesting the need for a network of connected computers. Cerf and Kahn were awarded the Presidential Medal of Freedom in 2005; Licklider died in 1990, before the implications of the Internet were fully realized.

One was Vint Cerf, the other Robert Kahn. Working as government employees together at DARPA, the pair devised ways of slicing digitized information into tiny packets, sending these packets in cleverly arranged order down the wires connecting the computers, and then reassembling the packets in the distant computer into a precise copy of the information.

Both men eventually left government to preach the essential goodness of computer connectivity, and both have been deluged with honors, mostly from their native United States. Doubtless Licklider would have won a medley of plaudits, too, but this modest and kindly man, predictor of so much and architect of so many of the central ideas, died young, in 1990, just as the Internet was getting started. According to Larry Roberts, it was in the early 1960s that Licklider began to suggest that "everybody could use computers anywhere and get at data anywhere in the world. He didn't envision the number of computers we have today by any means, but he had the same concept—all of the stuff linked together throughout the world, that you can use a remote computer, get data from a remote computer, or use lots of computers in your job. The vision was really Lick's originally. He didn't have a clue how to build it. He didn't have any idea how to make this happen. But he knew it was important, so he sat down with me and really convinced me that it was important and convinced me into making it happen."

Both creations, the physical structure and the protocols of communication, are different in two symbolically important respects from the final supporting pylon of the Internet, the medium known as the World Wide Web. The Web is the simplest of the three baseplates of the Internet, the easiest to explain. It is a medium that offers a computer user a means* of transmitting real information—text, pictures, film, sounds—from computer to computer or device to device. It transmits this information through the physical system of the Internet that was devised at the Pentagon by Roberts and Licklider, using the protocols first made at the Pentagon by Cerf and Kahn.

But the Web does not have many creators—just one. And neither he nor its early users were Americans. Tim Berners-Lee is British, and his first customers were scientists working at the nuclear research center CERN, in Switzerland. He told them about his invention in a memo sent electronically in August 1991. The web, as he called it, "aims to allow all links to be made to any information anywhere. [It] was started to allow

---

* It is not the only means, of course; e-mail, also invented at the Pentagon, first used in 1971, is another.

high energy physicists to share data, news, and documentation. We are very interested in spreading the web to other areas, and having gateway servers for other data. Collaborators welcome!"

The Web came to California, to Stanford, a month later; it spread beyond universities and into the hands of the general public in 1993, and by the end of the year there were six hundred websites. At the end of 1994 there were nearly three thousand, including sources for information on music, cooking, and movies; an early Internet comic named Doctor Fun; a webcam pointed into a fish tank; a means of ordering pizza online; a free-speech website called Bianca's Smut Shack; and the online site of one of the world's most venerable newspapers, the *Economist*.

From the summer of 1994 on, the Internet went into accelerative overdrive, exponential and hyperkinetic. There are said to be more than six hundred million websites today, connecting the world, drilling details into and out of the most remote corners of the planet.

And the most remote corners of America, too. The physical plant of the Internet is everywhere, hidden in plain sight. Orange markers show where buried fiber-optic cables run, spearing across the remoteness but too valuable to have anyone dig anywhere close to them, so identified everywhere, even deep in forests and swamps. Google has just built an immense server farm on the site of the old drive-in movie theater in Council Bluffs, Iowa, close to the great gold-colored spike that marks the spot that Abraham Lincoln declared the starting point for the transcontinental railroad.

Lewis and Clark passed down the Columbia River in 1804, and then half a century later, the settlers and their wagons rumbled along nearby on their way to a future. Today in a flat nearby valley, there is a town called Prineville. Facebook has an enormous anonymous structure there, half dark and chill. All such centers lie behind tall razor-wire fences, are policed on the outside by guards and watchtowers and lights, have usually as the sole entrance an unmarked door with a smokers' ash receptacle beside it, and are manned on the inside by just a small corps of three or four uniformed men who pad around like keepers in some strange beastless zoo. But each one is a zoo crammed with iron mesh cages that hold hundreds and hundreds of computer servers, all of them passing data from one to another or down into the cables, up and out of them into other cables, all soundless and not a little sinister.

I was escorted around one such center in northern California. The keeper was utterly discreet about his charges, trusting that I would be sufficiently mesmerized by the millions of winking lights and the low blue light bathing the building's innards not to ask too many questions. But he did start at one point, gesturing to one especially large cage, in which the Cisco servers were all jet black and shiny, like Darth Vader's skin. "That," he said with great solemnity, "is where they store all the information on California's deadbeat dads."

Every man who has run out on child support or is late with his alimony payments or who in some other way is said to have failed his children or his former spouse—every name, every address, and every last detail of the miscreant is corralled behind the black mesh of chain mail. Every time a gathering of diodes on a server panel begins a fury of colored blinking, as happened several times while I was watching, it signified that someone, somewhere in the world, was seeking to find out something about someone whose details lie buried on slices of conducting metal within.

And while the computers are soundless in their labors, the center hums with a low-frequency rumble of motors and harmonics. Great air-conditioning systems have to be built alongside these vast new information cathedrals to keep the computers and their eternally spinning hard drives from overheating, melting, bursting into flame, and perhaps for just one critical microsecond, going disastrously offline, off the grid.

One of the newly built cathedrals of an entirely new kind of unification church—a Google server farm, housing tens of thousands of highly secure computer nodes that store and exchange at lightning speed information both for the entire American nation and, now, for the entire world.

Somewhere across the world a computer user is expecting that his click of a mouse button will yield instant access to a piece of information. If he has to wait—in a world where waiting is an intolerable new inconvenience—an analysis will show within seconds just which data center is responsible for the delay and why. To ensure that there are as few interruptions as possible, immense quantities of electricity are deployed to keep everything running with precision and perfection and permanence: 2 percent of America's electricity now goes to keeping the Internet cool, to keeping the link unbroken, for America and for the world.

Therein lies an irony, perhaps. The Internet was formed in America, based in America, a godchild of all the earlier technologies born or first used in America that helped to connect its people and landscape as one, and now its business is in connecting the world. Connecting America to the world, true, and connecting the world to America. But while many of the networks that employ the Internet—Twitter, especially—have proved themselves unimaginably adroit in linking together people and causes and helping create the new phenomenon of mass thinking that has come to be known as the *hive mind*, it is no longer a stated or perceived mission of the network to help anneal the nation that made it all into one.

If California is to feel at one, to be at one, with Maine and the Dakotas and Florida and Alabama, then it has to be hoped that the structures already settled into place by the great men and great visions of the past will continue to endure, as the republic endures. It is surely evident now that the Internet, the great new technological missionary of the age, is obliged to a future lying well beyond America. Its creators have as their unstated vision the uniting of the whole world. The *unum* that is America's proudest accomplishment today will in time become part of an even greater *pluribus*, which one day will be similarly forged by electronics into one great planetary unity.

Whether such a dream will work and will make sense remains to be seen. It may not happen, but it will surely be attempted. It may be a dream, or it may be a nightmare. In the forefront of this effort will be the century's new corps of forge masters. They will probably no longer be, as in the past, great public figures of strength and courage and determination. The days of Powell and Hayden, Lewis and Clark, Maclure and Edison and Clarence King are long over; there will be no further examples of men bent on surveying mountains or hammering railroad ties or wrenching trees from the living earth or excavating canals or listening for faint radio signals through the fog. Instead the new pioneers of unification will be technical men, hidden quietly out of sight in their blue-lit warehouses, surrounded by silent frenzies of blinking server lights.

As this dream or this nightmare unfolds, deep within these fortresses, such men of cool dispassion and quiet determination will remain fixed to their allotted tasks of a new ideal, that of making the planet one and then placing America in her proper context as a briefly glorious component of the comprehensive history of earth. The men who united the states, in their next incarnation as a part of this new ideal, will have become transmuted into the men who united the world.

Epilogue

I live in a small Northeastern American town, a lonely, lovely place so full of unusual characters and strange stories that I have often thought it amply deserving of a curious little book. But then, Sherwood Anderson wrote *Winesburg, Ohio* in 1919, in which he told tales from a similarly hidden little town in the Midwest, weaving them together into the rude fabric of a novel that has remained a classic of modern American literature ever since. In other words, the job has already been done.

Nevertheless, I fancied that it might be possible to try to write his book again. It might be done as a centennial experiment, to come out in 2019—only this time it would be set in my village of Sandisfield, Massachusetts, with a new cast and a new set of

happenings and memories. Like *Winesburg*, to protect reputations and allow for some literary license,
the place could be given a new name. I toyed with the
idea—filching from Edith Wharton's *Ethan Frome*—of
calling it *Starkfield, Massachusetts.* Under that title, a
book telling of life in a forgotten corner of America in
the twenty-first century might become a classic work of
fiction, too, to be read in the schools and living rooms
of the twenty-second. It was a fancy, little more, but
one that long lingered in my mind.

Instead, for a reason that has much to do with the
underlying theme of this book, something quite different happened. A group of us, mostly admirers of
Sherwood Anderson and all of us quite aware of the
uniquely interesting nature of our own little town and
its people, started a local monthly newspaper. The
Sandisfield Times published its first issue in April 2010,
and in the years since it has become, to the surprise
of all, an essential part of village life, required reading
for everyone—like the *Winesburg Eagle*, in fact, but a
century later.

The paper is now popular, needed, and ceaselessly
written to, and it has brought to Sandisfield something
that the village has never truly enjoyed in all of its 250
years of incorporated existence: a sense of community,
a common sense of unity.

There were reasons why it had taken so long. Geography was one: the rivers that flowed down from the Berkshire hills and through the town had long separated the tiny clusters of houses, kept people firmly apart from one another. "This is a town where you'll never be bothered if you don't want to be," someone said when I first moved here in 2001. Some days not a single car comes down our dirt road. The quiet can be deafening, though magical for being so. There is just the breeze and the birdsong and on a winter's day like this one, the cracking of the ice, and at night the screaming and yapping of the local coyotes. This is a fine place for those who value solitude.

Then again, New Englanders can be a taciturn breed, stern with newcomers. The old Puritan families, remainders of the first settlers, keep their own counsel. But this is America, and these first settlers have since been augmented by outsiders who journeyed here from just about everywhere. There are Finns and Magyars here in abundance, and Ukrainians, a handful of Scots and a Cornishman, a lady from Haiti, a Latvian, a family of newly arrived Albanians from Kosovo. The town constable's family is from Spain. Once there were enough Russian Jews here—many bent on raising chickens for the grocers of New York City—to warrant turning the old Baptist church into a synagogue. But then the Jews went down to the city, leaving the chicken farms behind. The temple that they left then became the village arts center, where a small group of enthusiasts produces plays for each other and puts on thinly attended concerts. And for years the people of Sandisfield, different, disparate, diffident, kept resolutely to themselves like this: America in ethnic microcosm, though in this corner of the country, an America somewhat disinclined to pull together as one.

But now, with the newspaper, much has changed. The forum that its pages offer is now abuzz with news, argument, and conversation. The annual town meeting, by which most New England villages govern themselves and which here used to be a sorry affair attended by almost no one, is now for the first time crowded with voters, loud with debate. The arts center staged a play about the history of the village in 2012, and for the first time in its ten-year existence, found it had to turn people away for want of room.

Neighborliness has now become commonplace, replacing distance and isolation. I write this two days after a historic snowstorm; all I can see from my study window is a blanket of pure white, the old stone walls quite submerged by drifts, the apple trees in the orchard shuddering under the blasts of wind. The early records of this town referred to it as a "howling wilderness," and on a day like this, with my windows rattling and the birds being knocked sideways in the sky, it is. Yet people still come by—some on cross-country skis, some on snowshoes—to make sure all is well; and we go out, too, knocking on doors, making certain those without power have firewood and hot soup and toddy. Everyone now has a camera, and after an event like this, the telephone rings constantly with offers of images for the coming month's newspaper. Now that there is a paper, people want to employ it to be able to remember where they were, to remind themselves what it was like, when the Great Storm of 2013 struck Sandisfield, *our town*. (And yes: Thornton Wilder's play of that name was due to be staged here later in the year—another mark of the coming change, of the growth of community.)

Whether this change is all a consequence of the new existence of the *Sandisfield Times*, I cannot entirely say, though I suspect it to be so and wish it. For if it is, then it underlines and confirms one of the themes of the previous pages, that the creation of any sense of unity among a population of potentially disharmonious settlers almost always requires *the deliberate agency of man*. Community is seldom an organic thing, especially among migrants. It needs to be nurtured, facilitated, encouraged.

In Sandisfield, a town now 250 years old, it rarely was. For most of the town's existence, such constructive agency was minimal. There was the decision to tar some of the dirt roads in the 1920s, which helped. There was the coming of the telephone in the 1930s, which only a few could initially afford. Otherwise, very little. The railroad long since passed Sandisfield by. The stagecoach was infrequent. The local inns had fallen into disrepair. The store was moribund. There was little attempt ever made to bring the townspeople together—until 2010, when the newspaper arrived. Then, almost overnight, an untapped vein of mutual feeling and goodwill was tapped. The town changed, its people becoming suddenly welded into one, turned to a single purpose with a new, united identity.

Far, far from this corner of the Berkshires, out in the great wilderness of the old American continent, there was once almost no sense of community either—until the immigrants came. There was little sense of oneness when America was peopled only by its original people. Native Americans were spread too far apart and were by geography just as isolated, though on a far larger scale, as were our villagers of today, huddled alone in their deep river valleys. And so there was little sense among the Shawnee, for instance, that they were bound in any way to the Iroquois or the Miami, little sense of brotherhood between the Comanche and the Sioux, or between the Blackfeet and the Crow. Common ancestry of the Indian people alone, the presence of common genes, was simply not sufficient to bind most of them together. Most ran their own fiefdoms, sheltered behind their palisades, warred with one another, formed uneasy alliances, never imagined the concept of continental nationhood.

But then came the migrants, then came the nation, and then with it came the gathering notion that unity was, for so complex an entity, a matter of manifest need and desire. And so the annealing began. It began even though the migrant settlers could be every bit as foreign to one another as had been the Indians, with an immigrant from Finland, say, being in genetic and cultural fact very much more different from a Sicilian, say, than ever was a Cheyenne from a Hopi or a Cree.

But as we know, this all changed. The United States was born and was slowly suffered into existence. What eventually set this new America apart from original America is that, through all of the republic's years, there existed *agencies* that were deliberately bent to the task of creating community, creating the practical means for the forging of alliances for the common good of all.

The agencies were large government bodies of power and influence that could design and build vast systems of roads, bring electricity to isolated farms, sponsor exploring expeditions involving thousands of scientists, and order into the unknown men like Lewis and Clark and demand that they ascertain the shape and nature of the nation.

Some of the agencies were individuals, men with great vision, men like George Washington, Theodore Judah, Isham Randolph, Samuel Morse, and Thomas MacDonald, whose ideas and inventions, driven by the prospect of personal fortune, in most cases, similarly helped bind ever more tightly the peoples of the country together.

Some of the agencies of man were small. Maybe they were groups or individuals who persuaded the unwilling or the recalcitrant, just as we did in our half-forgotten village in the hills, of the benefits of common purpose. Our newspaper has volunteers today whose ethnic origins are Italian, Greek, Scots, Irish, Japanese, Dutch, and Chinese. But all, in a uniquely American manner, see virtue and power in the new harmony that they have made, which manifests itself in the modest document that all can see and read on the first day of each month.

This new sense of harmony may have been a long time coming to Sandisfield, Massachusetts, and there are other communities within the country that are more isolated and forgotten than ours, where disunity is more likely to be the watchword. Yet it cannot and should not be forgotten that the story of the United States of America is still a developing one, a continuing evolution, and that the union becomes ever stronger as a result of the pressures of steady change. After all, the very notion of change informs the Preamble to the United States Constitution: "We the people . . . in order to form a more perfect union . . ."

The union, it was recognized back in the late eighteenth century, has to be made ever more perfect all the time. Our small-town newspaper is just one more step on the way. This is how it is done—our way, the American way.

Acknowledgments

As the idea for this book was born in part out of my long-held wish to become an American citizen, my first debt of gratitude must be paid to the United States government for finally approving my application to do so, and to the federal magistrate judge, the Honorable Marianne Bowler, for performing the swearing-in formalities on the deck of the great old warship the USS *Constitution.* I must also thank the captain of this venerable sailing vessel, Commander Timothy Cooper, for hosting an occasion made especially memorable by its unique maritime setting.

There were twenty-five of us, all born as new Americans on a searing-hot Boston Independence Day afternoon, in a ceremony that I feel certain not one of us will ever forget. I was the oldest of the group, by far—although my friend and countryside neighbor Sir Brian Urquhart, who kindly accompanied me to offer much-needed moral support, decided, after finding the ceremony so profoundly moving, that he should apply for citizenship too. This British Army war veteran and United Nations undersecretary general had his oath administered the following year, when he was ninety-three years old.

Shannon Such, a widely admired New York immigration lawyer who has become a good personal friend, patiently guided the reams of paperwork through the byways of the various bureaucracies. Her tireless assistant, Victoria Gelardi, managed to keep me cheerful throughout the seemingly endless and tiresome process, even when yet more papers were demanded and more fingerprints taken than I seemed to have fingers. To all, my most grateful thanks.

But all of this is prologue. The remainder of this heartfelt appreciation is for the legions who were kind enough to offer counsel, comfort, and shelter during the researching and writing of the book itself. In particular I owe much to the stimulating company of Arvid Nelson, whose knowledge not only of environmental history but also of the complexities of the American frontier is profound, and who underscored his loyalty both to me and to this book from the very outset—first by helping to refine my initial ideas, and then closely reading the completed typescript and making myriad suggestions for improvement. His unwavering support throughout has been crucial and necessary, and I thank him without reservation.

Out on the road, and back at my desk here in the Massachusetts countryside, there were many whose offers of help, both great and small, helped bring this book into being. While none can be held responsible for any failures of fact or interpretation—any mistakes are mine alone—all did their best for me as best they knew how, and I shall be eternally grateful to those who follow:

Rupert Allman, who first alerted me to the literature of the interstate highway system; Greg Ames of the St. Louis Mercantile Library, and an expert on river traffic on the Mississippi; Kurt Andersen, an enduring admirer of Omaha, Nebraska, where he was born; David Haward Bain, of Middlebury College, Vermont, who has written extensively both on railroad history and on the exploits of Clarence King; Andrew Bertalna of American University, Washington, DC, in whose custody were the papers of Thomas MacDonald; my ever-helpful and supportive friend Renee Braden of the National Geographic archives in Washington, DC; Amanda Bryden, Collections Manager of Historic New

Harmony, Indiana; James T. Campbell, a Stanford University history professor with an abiding interest in the interstate highway system, together with his delightfully hospitable parents, Ralph and Patricia Campbell, who welcomed me so warmly into their home when I met them in the tiny town of Morrison, Illinois; Kathleen Carlucci of the Thomas Edison Museum in Menlo Park, New Jersey; Jeff Carter; David Cenciotti, a writer on aviation who guided me through the complex world of in-air emergencies; Steve Colby, an expert on the Cumberland Road; Val Coleman, an elder states-man–neighbor friend in the Berkshires who read an early draft of the book and made scores of useful comments; Mark Davis of Union Pacific in Omaha; David Dolak of Columbia College in Chicago who helped my research into the Chicago ship canal; the staff of the Equinix server farm in Palo Alto, including Ally Khantzis, Melissa Neumann, and Keith Patterson; Andrea Faling, of the Nebraska State Historical Society Archives; Phillip Forbes, of the Montana-based civil engineering firm of Morrison-Maierle, Inc.; the late Philip Fradkin, biographer of Wallace Stegner and an all-round expert on the American West; Robert Germann, of the US Army Corps of Engineers; my eternally kind and enthusiastic helpmeet Leslie Gordon of the US Geological Survey in Menlo Park, California;

Tom Halstead of Gloucester, Massachusetts; map collector and atlas authority extraordinaire Derek Hayes of Vancouver, British Columbia, who kindly supplied many of the older maps used in this book; Sara Bon Harper at Monticello; Doug Hecox of the US Department of Transportation; Mary Hess, at SUNY Oswego, who is steeped in the history of the Erie Canal; my late friend Christopher Hitchens; Paul Israel, the editor of the papers of Thomas Edison; Doug Jensen, ticket agent with Amtrak at Emeryville, California, who was kindness personified; Kirk Johnson, best-known for his work at the Denver Museum of Natural History, but who now runs the Smithsonian's National Museum of Natural History in Washington, DC; Professor Markes Johnson of Williams College, an expert on early American geology; Tom and Pat Judge, farmers of Ames, Iowa; Jeffrey Key and his son Jason, of Helena, Montana, who interrupted a Sunday-morning fishing trip to show me the Gates of the Mountains; Michael Korda, who wrote so vividly about Dwight Eisenhower; Gary Mechanic, who knows much about the history of the Chicago portage; Charles Meinert, Delmar, New York, a book dealer who proved of inestimable help in locating some hard-to-find WPA guides; Arlen Miller, an Amish farmer in Nappanee, Indiana, who sports a very un-Amish interest in

Petter diesel engines; John Morrison Jr. of Morrison-Maierle, builders of Interstate 90 through Montana; Melissa Murphy, owner of Sweet Melissa's in Laramie, Wyoming, makers of the best pies in America; James O'Neal, a specialist in early radio and a critic of Aubrey Fessenden's broadcasting claims; Matthew Pearcy, historian of the US Army Corps of Engineers; Gabriele Rausse, arborist and Director of Gardens at Monticello; Marty Reuss, a specialist on the Atchafalaya River; Jack Robertson of the Monticello Library; Jeff Rosenberg, an old friend and colleague at National Public Radio; John Sciortino, who told me many interesting things at his barbershop in Council Bluffs, Iowa; Bill Siemering, a great believer in the importance of and power for good of radio, and who helped found NPR; Jennifer Sahn, editor of *Orion* magazine; Kenton Spading of the Corps of Engineers; Nelson Spencer, publisher of the redoubtable *Waterways Journal*; Ken Stewart of the South Dakota Historical Society; Earl Swift, author of *Big Roads*; Alan Thompson, of the Bureau of Land Management in Montana; Sean Visintainer of the St. Louis Mercantile Library; Monica Webb of Wired West; Stephen White, whose boundless fascination with early photography has resulted in his amassment of one of the country's more remarkable and eclectic image collections, was more than generous in allowing me to browse through and then use several of his pictures; Terry Wiltz, a captain on the Illinois waterway; Thomas Wixon, resident in Mississippi but a relative of America's first and only official geographer, Thomas Hutchins; and the ever-kindly Rex Ziak of Astoria, Oregon.

Quite overenthused by my subject and possibly carried away by the delight at my new citizenship—or, more probably, as Disraeli once said of Gladstone, *"inebriated by the exuberance of my own verbosity"*—I first delivered a typescript that was far, far too long. But by carefully employing the grace, tact, and editorial skills that I have come to know and admire over the last several years, Henry Ferris, my editor at HarperCollins, saw to it that the book was eventually distilled to a manageable size, and without losing any of the flavor and tone I first intended. I am a firm believer that all truly good books are the result of intimate and constructive collaboration between writer and editor; should this volume ever come to be kindly regarded, then it should be known that all is a consequence of the endeavors of Henry Ferris just as much as of my own.

Cole Hager, editorial assistant in New York, managed with patience and good humor the innumerable blizzard of details that always attend the creation of a book; and in London my editor Martin Redfern was supportive and enthusiastic throughout.

I am grateful, as always, to my agent, Suzanne Gluck, and to her colleague in London, Simon Trewin, as well as to Eve Atterman and Samantha Frank, peerless in their roles as legendary William Morris assistants.

And that I am finally thanking, as always, my wife, Setsuko—for her patience and forbearance, close reading, and wise advice—allows me a connection and a means of offering one final story, one that suggests yet another kind of acknowledgment:

When the young man who is now my father-in-law, Makoto Sato, left the war-ruined Japan of the early 1950s to come to America, he found himself, more by chance than intent, enrolled as a student at Kentucky State College in Frankfort, Kentucky. This was, at the time, a historically black institution—one of more than a hundred such colleges, sited predominantly in the former slave states, that offered education to African Americans who had been excluded from the federally funded Land Grant colleges. Mr. Sato was one of only three foreigners in KSC's Class of '57—a class whose commencement speaker happened to be the then little known Dr. M. L. King of Birmingham, Alabama.

On one of the research excursions for this book in 2013, I found I was due to pass through Frankfort and suggested that my eighty-four-year-old father-in-law come along. On learning that he had not been back to Kentucky for the previous fifty-six years, I called the school's head of alumni affairs—Garland Higgins—who promptly organized a red-carpet welcome that was as memorable as it was generous.

It occurred to me on leaving the campus that hot summer's day—a campus that was now teeming with white students and with non-Americans, as is presently the case with most of this country's historically black colleges—that institutions like this have also played vital roles in helping to weld the endless magical confusion of today's America into one. The kindness and vision that was expressed a half century ago by the men and women of Kentucky State College to this lone migrant from Japan, and who is now, like me, a new naturalized American too, seems to be all part of a story for which I, and countless others, must surely also feel a deep and abiding sense of gratitude.

Bibliography

While most of the books listed below provided me with specific assistance, perhaps none were of greater general value than the WPA Guides to America, of which I managed to collect more than fifty, including all of the main state guides, during the months I was writing. The quality, accuracy, and simple beauty of the writing in these small Depression-era masterpieces complement still the heroic achievement of their creation, the volumes serving as an ever-present reminder on my bookshelves that there are times in any history when government can achieve truly great and lasting things for the common good. I also made continual use of my treasured copy of the *National Atlas of the United States of America*, which I bought in Washington in 1970, during the Nixon era. The atlas, a giant of a thing, was made and published by the US Geological Survey, a government body within the Department of the Interior that for a century and a half has been intimately involved in exploring and uniting the states, thereby benefiting the entire country.

Abbott, Shirley. *The National Museum of American History.* New York: Harry N. Abrams, 1981.

Ambrose, Stephen E. *Nothing Like It in the World: The Men Who Built the Transcontinental Railroad, 1863–1869.* New York: Simon & Schuster, 2000.

Anderson, Sherwood. *Winesburg, Ohio.* New York: Huebsch, 1919; Signet Classics, 1993.

Anfinson, John O. *The River We Have Wrought: A History of the Upper Mississippi.* Minneapolis: University of Minnesota Press, 2003.

Arsenault, Raymond. *Freedom Riders: 1961 and the Struggle for Racial Justice.* New York: Oxford University Press, 2006.

Atwood, Kay. *Chaining Oregon: Surveying the Public Lands of the Pacific Northwest, 1851–1855.* Blacksburg, VA: McDonald & Woodward, 2008.

Bain, David Haward. *Empire Express: Building the First Transcontinental Railroad.* New York: Viking, 1999.

Bakeless, John. *The Eyes of Discovery: America as Seen by the First Explorers.* New York: Dover, 1961.

Barone, Michael, and Chuck McCutcheon. *The Almanac of American Politics 2012.* Chicago: University of Chicago Press, 2011.

Bartlett, Richard A. *Great Surveys of the American West.* Norman: University of Oklahoma Press, 1962.

Beebe, Lucius, and Charles Clegg. *Hear the Train Blow.* New York: Dutton, 1952.

Benson, Maxine. *From Pittsburgh to the Rocky Mountains: Major Stephen Long's Expedition 1819–1820.* Golden, CO: Fulcrum, 1988.

Bernard, Ronald M. *Sandisfield Then and Now.* Sandisfield, MA: Town of Sandisfield, 2012.

Bernstein, Peter L. *Wedding of the Waters: The Erie Canal and the Making of a Great Nation.* New York: Norton, 2005.

Berry, Trey, Pam Beasley, and Jeanne Clements, eds. *The Forgotten Expedition, 1804–1805: The Louisiana Purchase Journals of Dunbar and Hunter.* Baton Rouge: Louisiana State University Press, 2006.

Blum, Andrew. *Tubes: A Journey to the Center of the Internet.* New York: HarperCollins, 2012.

Bonnicksen, Thomas M. *America's Ancient Forests.* New York: Wiley, 2000.

Boorstin, Daniel J. *The Americans: The Democratic Experience.* New York: Random House, 1973.

Borneman, Walter R. *Rival Rails: The Race to Build America's Greatest Transcontinental Railroad.* New York: Random House, 2010.

Brands, H. W. *American Dreams: The United States since 1945*. New York: Penguin Press, 2010.

Brodie, Fawn M. *Thomas Jefferson: An Intimate History*. New York: Norton, 1974.

Burrows, Edwin G., and Mike Wallace. *Gotham: A History of New York City to 1898*. New York: Oxford University Press, 1999.

Camfield, Gregg. *The Oxford Companion to Mark Twain*. New York: Oxford University Press, 2003.

Camillo, Charles A., and Matthew T. Pearcy. *Upon Their Shoulders: A History of the Mississippi River Commission*. Vicksburg, MS: Mississippi River Commission, 2004.

Carnes, Mark, and John Garraty. *Mapping America's Past: A Historical Atlas*. New York: Henry Holt, 1996.

Cather, Willa. *O Pioneers!* Boston: Houghton Mifflin, 1913.

Chiang, Yee. *The Silent Traveller in Boston*. New York: Norton, 1959.

Clinton, Catherine. *Fanny Kemble's Civil Wars*. New York: Oxford University Press, 2000.

Coffin, Robert P. Tristram. *Kennebec: Cradle of Americans*. New York: Rinehart, 1937.

Cooley, Lyman E. *The Diversion of the Waters of the Great Lakes by Way of the Sanitary and Ship Canal of Chicago*. Chicago: Sanitary District of Illinois, 1913.

Crofutt, George A. *Crofutt's Overland Tours*. New York: Rand McNally, 1890.

Cummings, Amos Jay. *A Remarkable Curiosity: Dispatches from a New York City Journalist's 1873 Railroad Trip across the American West*. Boulder: University Press of Colorado, 2008.

Cunningham, Noble E., Jr. *In Pursuit of Reason: The Life of Thomas Jefferson*. Baton Rouge: Louisiana State University Press, 1987.

Daniels, George G. *The Spanish West*. Alexandria, VA: Time-Life Books, 1976.

Daniels, Rudolph. *Trains across the Continent*. Bloomington: Indiana University Press, 1997.

Daughan, George C. *1812: The Navy's War*. New York: Basic Books, 2011.

Davis, James E. *Frontier Illinois*. Bloomington: Indiana University Press, 1998.

Dellinger, Matt. *Interstate 69: The Unfinished History of the Last Great American Highway*. New York: Scribner, 2010.

Dolin, Eric Jay. *Fur, Fortune and Empire: The Epic History of the Fur Trade in America*. New York: Norton, 2010.

Egan, Timothy. *The Big Burn: Teddy Roosevelt and the Fire That Saved America*. Boston: Houghton Mifflin, 2009.

Ellis, Joseph J. *American Creation*. New York: Random House, 2007.

Engelman, Ralph. *Public Radio and Television in America: A Political History.* Thousand Oaks, CA: Sage Publications, 1996.

Fehrenbach, T. R. *Seven Keys to Texas.* El Paso: University of Texas Press, 1983.

Fradkin, Philip L. *Wallace Stegner and the American West.* New York: Knopf, 2008.

Frazier, Ian. *Great Plains.* New York: Picador, 1989.

Fremling, Calvin R. *Immortal River: The Upper Mississippi in Ancient and Modern Times.* Madison: University of Wisconsin Press, 2005.

Fried, Stephen. *Appetite for America: How Businessman Fred Harvey Built a Railroad Hospitality Empire That Civilized the Wild West.* New York: Bantam, 2010.

Garrett-Davis, Josh. *Ghost Dances: Proving Up on the Great Plains.* New York: Little, Brown, 2012.

Glaab, Charles N., and A. Theodore Brown. *A History of Urban America.* New York: Macmillan, 1967.

Glacken, Clarence J. *Traces on the Rhodian Shore: Nature and Culture in Western Thought from Ancient Times to the End of the Eighteenth Century.* Berkeley: University of California Press, 1967.

Glaser, Leah S. *Electrifying the Rural American West.* Lincoln: University of Nebraska Press, 2009.

Gordon, John Steele. *A Thread across the Ocean: The Heroic Story of the Transatlantic Cable.* New York: Walker, 2002.

Graham, Alan. *A Natural History of the New World: The Ecology and Evolution of Plants in the Americas*. Chicago: University of Chicago Press, 2011.

Greenblatt, Stephen. *The Swerve: How the World Became Modern*. New York: Norton, 2011.

Gunther, John. *Inside U.S.A.* New York: Curtis, 1947.

Gurasich, Marj. *Letters to Oma: A Young German Girl's Account of Her First Year in Texas, 1847*. Fort Worth: Texas Christian University Press, 1989.

Gwynne, S. C. *Empire of the Summer Moon: Quanah Parker and the Rise and Fall of the Comanches, the Most Powerful Indian Tribe in American History*. New York: Scribner, 2010.

Hadfield, Charles. *World Canals: Inland Navigation Past and Present*. New York: Facts on File, 1986.

Hahn, Thomas S., and Emory L. Kemp. *Canal Terminology in the United States*. Morgantown: West Virginia University, 1998.

Hakluyt, Richard. *Voyages to the Virginia Colonies*. London: Hutchinson, 1986.

Halberstam, David. *The Fifties*. New York: Villard Books, 1993.

Halliday, E. M. *Understanding Thomas Jefferson*. New York: HarperCollins, 2001.

Harlow, Alvin F. *Steelways of New England*. New York: Creative Age Press, 1946.

Harpster, Jack. *A Biography of William B. Ogden: The Railroad Tycoon Who Built Chicago.* Carbondale: Southern Illinois University Press, 2009.

Hauben, Michael, and Ronda Hauben. *Netizens: On the History and Impact of Usenet and the Internet.* Los Alamitos, CA: IEEE Computer Society Press, 1997.

Havighurst, Walter. *Upper Mississippi: A Wilderness Saga.* New York: Rinehart, 1937.

———. *Voices on the River: The Story of the Mississippi Waterways.* New York: Macmillan, 1964.

Haydon, Richard, ed. *Upstate Travels; British Views of 19th Century New York.* Syracuse, NY: Syracuse University Press, 1982.

Hayes, Brian. *Infrastructure: The Book of Everything for the Industrial Landscape.* New York: Norton, 2005.

Hayes, Derek. *America Discovered: A Historical Atlas of North American Exploration.* Vancouver: Douglas & McIntyre, 2004.

———. *Historical Atlas of the American West with Original Maps.* Berkeley: University of California Press, 2009.

———. *Historical Atlas of the United States with Original Maps.* Berkeley: University of California Press, 2007.

Haywood, Carl W. *Sometimes Only Horses to Eat: David Thompson, the Saleesh House Period*

1807–1812. Stevensville, MT: Rockman's Trading Post, 2008.

Heat-Moon, William Least. *PrairyErth.* Boston: Houghton Mifflin. 1991.

———. *River Horse: A Voyage across America.* New York: Houghton Mifflin, 1999.

Hegedus, Carol. *John Earl Fetzer: Stories of One Man's Search.* Kalamazoo, MI: Fetzer Institute, 2004.

Herring, George C. *From Colony to Superpower: US Foreign Relations since 1776.* New York: Oxford University Press, 2008.

Hill, Libby. *The Chicago River: A Natural and Unnatural History.* Chicago: Lake Claremont Press, 2000.

Hiltzik, Michael. *Colossus: Hoover Dam and the Making of the American Century.* New York: Free Press, 2010.

Hinckley, Helen. *Rails from the West: A Biography of Theodore Judah.* San Marino, CA: Golden West Books, 1969.

Hobson, Archie, ed. *Remembering America: A Sampler of the WPA American Guide Series.* New York: Columbia University Press, 1985.

Holbrook, Stewart H. *The Story of American Railroads.* New York: Crown, 1947.

Holloway, Marguerite. *The Measure of Manhattan.* New York: Norton, 2012.

Howe, Daniel Walker. *What Hath God Wrought: The Transformation of America, 1815–1848.* New York: Oxford University Press, 2007.

Isaacson, Walter. *Benjamin Franklin: An American Life.* New York: Simon & Schuster, 2003.

Israel, Paul. *Edison: A Life of Invention.* New York: Wiley, 1998.

Jefferson, Thomas. *Writings.* New York: Library of America, 1984.

John, Richard R. *Network Nation: Inventing American Telecommunications.* Cambridge, MA: Harvard University Press, 2010.

Kelley, Pat. *River of Lost Dreams: Navigation on the Rio Grande.* Lincoln: University of Nebraska Press, 1986.

Kennedy, David M. *Freedom from Fear: The American People in Depression and War, 1929–1945.* New York: Oxford University Press, 1999.

Kennon, Donald, ed. *The United States Capitol: Designing and Decorating a National Icon.* Athens: Ohio University Press, 2000.

King, Clarence. *Memoirs,* ed. James D. Hague. New York: Putnam, 1904.

Kittredge, William, and Annick Smith, eds. *The Last Best Place: A Montana Anthology.* Helena: Montana Historical Society, 1988.

Klein, Maury. *Union Pacific,* 3 vols. New York: Oxford University Press, 2011.

Kneiss, Gilbert. *Bonanza Railroads.* Stanford, CA: Stanford University Press, 1941.

Koeppel, Gerard. *Bond of Union: Building the Erie Canal and the American Empire.* Philadelphia: Da Capo Press, 2009.

Korda, Michael. *Ike: An American Hero.* New York: HarperCollins, 2007.

Kurlansky, Mark. *Salt: A World History.* New York: Walker, 2002.

Laborde, Adras. *Ransdell of Louisiana: A National Southerner.* New York: Benziger Brothers, 1951.

Landes, David. *The Wealth and Poverty of Nations.* New York: Norton, 1998.

Laskin, David. *The Children's Blizzard.* New York: HarperCollins, 2004.

Lavender, David. *Westward Vision: The Story of the Oregon Trail.* Lincoln: University of Nebraska Press, 1963.

LeDraoulec, Pascale. *American Pie: Slices of Life (and Pie) from America's Back Roads.* New York: HarperCollins, 2002.

Leonard, John, ed. *These United States.* New York: Nation Books, 2003.

Lewis, Meriwether, and William Clark. *The Definitive Journals of Lewis and Clark,* 8 vols, ed. Gary E. Moulton. Lincoln: University of Nebraska Press, 1991.

———. *The Journals of Lewis and Clark,* abridged, ed. Anthony Brandt. Washington DC: National Geographic Society, 2002.

Lewis, Oscar. *The Big Four: The Story of the Men Who Built the Central Pacific—Stanford, Hopkins, Huntington, Crocker.* New York: Knopf, 1938.

Lewis, Tom. *Divided Highways: Building the Interstate Highways, Transforming American Life.* New York: Penguin, 1997.

Limerick, Patricia Nelson. *The Legacy of Conquest: The Unbroken Past of the American West.* New York: Norton, 1987.

Linklater, Andro. *Measuring America: How an Untamed Wilderness Shaped the United States and Fulfilled the Promise of Democracy.* New York: Walker, 2002.

Louis-Philippe. *Diary of My Travels in America,* tr. Stephen Becker. New York: Delacorte Press, 1977.

Lyell, Charles. *A Second Visit to the United States of North America.* New York: Harper & Brothers, 1855.

———. *Travels in North America in the Years 1841–1842.* New York: Charles E. Merrill, 1909.

Mann, Charles C. *1493: How Europe's Discovery of the Americas Revolutionized Trade, Ecology and Life on Earth.* London: Granta, 2011.

Marx, Leo. *The Machine in the Garden: Technology and the Pastoral Ideal in America.* New York: Oxford University Press, 1964.

Masters, Edgar Lee. *Spoon River Anthology.* New York: Macmillan, 1916; Signet Classics, 1992.

McCague, James. *Moguls and Iron Men: The Dramatic Story of the Dreamers and Doers Who Spanned the*

American Continent with the First Transcontinental Railroad. New York: Harper & Row, 1964.

McCartney, Laton. *Across the Great Divide: Robert Stuart and the Discovery of the Oregon Trail.* New York: Free Press, 2003.

McGovern, Francis J. *This Pearl, America: An Immigrant's Memoir.* San Francisco: California Publishing Co., 2001.

McNichol, Dan. *The Roads That Built America: The Incredible Story of the U.S. Interstate System.* New York: Barnes & Noble, 2003.

McPhee, John. *The Control of Nature.* New York: Farrar, Straus & Giroux, 1989.

McPherson, James. *Battle Cry of Freedom: The Civil War Era.* New York: Oxford University Press, 1988.

Menand, Louis. *The Metaphysical Club: A Story of Ideas in America.* New York: Farrar, Straus & Giroux, 2001.

Middlekauff, Robert. *The Glorious Cause: The American Revolution, 1763–1789.* New York: Oxford University Press, 1982.

Middleton, William D., George M. Smerk, and Roberta L. Diehl, eds. *Encyclopedia of North American Railroads.* Bloomington: Indiana University Press, 2007.

Mitchell, William L. (Billy). *The Opening of Alaska.* Anchorage: Cook Inlet Historical Society, 1982.

Modelski, Andrew M. *Railroad Maps of North America: The First Hundred Years.* Washington, DC: Library of Congress, 1984.

Morgan, Ted. *A Shovel of Stars: The Making of the American West, 1800 to the Present.* New York: Simon & Schuster, 1995.

Morris, James. *Coast to Coast.* London: Faber & Faber, 1956.

Murphy, John. *The Eisenhower Interstate System.* New York: Chelsea House, 2009.

National Railway Pub. Co., *Official Railway Guide,* New York: National Railway Pub. Co., 1941.

National Ship-Canal Convention. *Proceedings.* Chicago: Tribune Company, 1863.

Neering, Rosemary. *Continental Dash: The Russian-American Telegraph.* Ganges, BC: Horsdal & Schubart, 1989.

Newton, Jim. *Eisenhower: The White House Years.* New York: Doubleday, 2011.

Nickles, John M. *Geological Literature on North America: 1785–1918.* Washington, DC: US Geological Survey, 1923.

O'Neil, Paul. *The Rivermen.* New York: Time-Life Books, 1975.

Pagnamenta, Peter. *Prairie Fever: British Aristocrats in the American West, 1830–1890.* New York: Norton, 2010.

Parkman, Francis. *The Oregon Trail: Sketches of Prairie and Rocky Mountain Life.* New York: Library of America, 1991.

Patterson, James T. *Grand Expectations: The United States, 1945–1974.* New York: Oxford University Press, 1996.

————. *Restless Giant: The United States from Watergate to Bush v. Gore.* New York: Oxford University Press, 2005.

Pauketat, Timothy R. *Cahokia: Ancient America's Great City on the Mississippi.* New York: Penguin, 2009.

Pence, Richard A., ed. *The Next Greatest Thing.* Washington, DC: National Rural Electric Cooperative Association, 1984.

Pierson, George Wilson. *American Historians and the Frontier Hypothesis.* Chicago: American Historical Association, 1941.

Powell, John Wesley. *The Exploration of the Colorado River and Its Canyons.* New York: Dover, 1961.

Raban, Jonathan. *Bad Land: An American Romance.* New York: Vintage, 1996.

————. *Driving Home: An American Journey.* New York: Pantheon, 2010.

Rabbitt, Mary C. *Minerals, Lands, and Geology for the Common Defence and General Welfare,* 3 vols. Washington, DC: US Geological Survey, 1979–1986.

Rees, Tony. *Arc of the Medicine Line: Mapping the World's Longest Undefended Border across the Western Plains.* Lincoln: University of Nebraska Press, 2007.

Reid, Robert L., ed. *Always a River: The Ohio River and the American Experience.* Bloomington: Indiana University Press, 1991.

Roberts, David. *Devil's Gate: Brigham Young and the Great Mormon Handcart Tragedy.* New York: Simon & Schuster, 2008.

Rollins, Philip Ashton, ed. *The Discovery of the Oregon Trail: Robert Stuart's Narratives of His Overland Trip Eastward from Astoria in 1812–13.* Lincoln: University of Nebraska Press, 1995.

Ronda, James P., ed. *Voyages of Discovery: Essays on the Lewis and Clark Expedition.* Helena: Montana Historical Society Press, 1998.

Rowsome, Frank, Jr. *The Verse by the Side of the Road: The Story of the Burma Shave Jingles.* Brattleboro, VT: Stephen Greene Press, 1965.

Sandweiss, Martha. *Passing Strange: A Gilded Age Tale of Love and Deception across the Color Line.* New York: Penguin, 2009.

Schecter, Barnet. *George Washington's America: A Biography through His Maps.* New York: Walker, 2010.

Shank, William H. *The Best from American Canals.* York, PA: American Canal and Transportation Center, 1991.

———. *Towpaths to Tugboats: A History of American Canal Engineering.* York, PA: American Canal and Transportation Center, 1982.

Shaw, Ronald E. *Canals for a Nation: The Canal Era in the United States, 1790–1860.* Lexington: University Press of Kentucky, 1990.

Sheriff, Carol. *The Artificial River: The Erie Canal and the Paradox of Progress, 1817–1862.* New York: Hill & Wang, 1996.

Silverman, Kenneth. *Lightning Man: The Accursed Life of Samuel F. B. Morse.* New York: Knopf, 2003.

Smiles, Samuel. *Lives of the Engineers.* London: John Murray, 1879.

Snyder, Gerald S. *In the Footsteps of Lewis and Clark.* Washington, DC: National Geographic Society, 1970.

Starr, Kevin. *Golden Gate: The Life and Times of America's Greatest Bridge.* New York: Bloomsbury, 2010.

Stavans, Ilan, ed. *Becoming Americans; Four Centuries of Immigrant Writing.* New York: Literary Classics of the United States, 2009.

Stegner, Wallace. *Beyond the Hundredth Meridian: John Wesley Powell and the Second Opening of the West.* Boston: Houghton Mifflin, 1953.

———. *Big Rock Candy Mountain.* New York: Penguin, 1991.

———. *Mormon Country.* Lincoln: University of Nebraska Press, 1981.

Stegner, Wallace, and Page Stegner. *American Places.* New York: Penguin Books, 2006.

Stein, E. P. *Flight of the Vin Fiz: Being an Account of the Wondrous Adventures of Calbraith P. Rodgers and His Flying Machine in the Grand Coast-to-*

Coast $50,000 Air Race. New York: Arbor House, 1985.

Stein, Mark. *How the States Got Their Shapes.* New York: HarperCollins, 2008.

Stone, Irving. *Men to Match My Mountains: The Opening of the Far West 1840–1900.* Garden City: Doubleday, 1956.

Stover, John F. *American Railroads.* Chicago: University of Chicago Press, 1961.

Stowe, Harriet Beecher. *Uncle Tom's Cabin; or, Life among the Lowly.* Boston: John P. Jewett, 1852; Ticknor & Fields, 1862; New York: Signet Classics, 1966.

Swift, Earl. *The Big Roads: The Untold Story of the Engineers, Visionaries, and Trailblazers Who Created the American Superhighways.* Boston: Houghton Mifflin, 2011.

Tarbell, Ida. *Owen D. Young: A New Type of Industrial Leader.* New York: Macmillan, 1932.

Taylor, Nick. *American-Made: The Enduring Legacy of the WPA, When FDR Put the Nation to Work.* New York: Bantam, 2008.

Thompson, Robert Luther. *Wiring a Continent: The History of the Telegraph Industry in the United States, 1832–1866.* Princeton, NJ: Princeton University Press, 1947.

Trollope, Anthony. *North America.* New York: Harper & Brothers, 1862; London: Penguin, 1968.

Twain, Mark. *Autobiography of Mark Twain*, ed. Robert H. Hirst. Berkeley: University of California Press, 2010.

Unruh, John D., Jr. *The Plains Across: The Overland Emigrants and the Trans-Mississippi West, 1840–1860.* Urbana: University of Illinois Press, 1979.

US Army Corps of Engineers. *Upper Mississippi Navigation Charts.* Rock Island, IL: US Army Corps of Engineers, 1989.

US Department of Commerce, *Statistical Abstract of the United States,* various volumes, 1878–2012. Washington, DC: US Department of Commerce.

Vare, Robert, ed. *The American Idea: The Best of the Atlantic Monthly.* New York: Doubleday, 2007.

Wallace, Henry A., ed. *Farmers in a Changing World: The Yearbook of Agriculture.* Washington, DC: US Department of Agriculture, 1940.

Warren, Leonard. *Maclure of New Harmony: Scientist, Progressive Educator, Radical Philanthropist.* Bloomington: Indiana University Press, 2009.

Webster, Bob, and Mike Webster. *Lewis and Clark by Air: A Pictorial Tour of the Historic Lewis and Clark Trail.* Pryor, OK: Via Planes, 2003.

Weiland, Matt, and Sean Wilsey, eds. *State by State: A Panoramic Portrait of America.* New York: HarperCollins, 2009.

Whitaker, Rogers E. M., and Anthony Hiss. *All Aboard with E. M. Frimbo: World's Greatest Railroad Buff.* New York: Kodansha, 1997.

White, Owen P. *Texas: An Informal Biography.* New York: Putnam, 1945.

White, Richard. *Railroaded: The Transcontinentals and the Making of Modern America.* New York: Norton, 2011.

Wilkins, Thurman. *Clarence King: A Biography.* New York: Macmillan, 1958.

Winchester, Simon. *American Heartbeat: Notes from a Midwestern Journey.* London: Faber & Faber, 1976.

Wood, Gordon S. *Empire of Liberty: A History of the Early Republic, 1789–1815.* New York: Oxford University Press, 2009.

Woodard, Colin. *American Nations: A History of the Eleven Rival Regional Cultures of North America.* New York: Viking, 2011.

Woodworth, Steven E. *Manifest Destinies: America's Westward Expansion and the Road to the Civil War.* New York: Knopf, 2010.

Worster, Donald. *A River Running West: The Life of John Wesley Powell.* New York: Oxford University Press, 2001.

Wright, Constance. *Fanny Kemble and the Lovely Land.* New York: Dodd, Mead, 1972.

Ziak, Rex. *In Full View: A True and Accurate Account of Lewis and Clark's Arrival at the Pacific Ocean, and Their Search for a Winter Camp along the Lower Columbia River.* Astoria, OR: Moffitt House Press, 2002.

Zittrain, Jonathan. *The Future of the Internet and How to Stop It.* New Haven: Yale University Press, 2008.

About the Author

S imon Winchester is the acclaimed author of many books, including *The Professor and the Madman*, *The Map that Changed the World*, *The Man Who Loved China*, *A Crack in the Edge of the World*, and *Krakatoa*. Those books were *New York Times* bestsellers and appeared on numerous best and notable lists. His most recent book is *Atlantic: Great Sea Battles, Heroic Discoveries, Titanic Storms, and a Vast Ocean of a Million Stories*. In 2006 Mr. Winchester was made an officer of the Order of the British Empire (OBE) by Her Majesty the Queen. He resides in western Massachusetts.